Animals and us:

50 years and more of applied ethology

Animals and us:

50 years and more of applied ethology

edited by:
Jennifer A. Brown
Yolande M. Seddon
Michael C. Appleby

Wageningen Academic
Publishers

EAN: 9789086862825
e-EAN: 9789086868285
ISBN: 978-90-8686-282-5
e-ISBN: 978-90-8686-828-5
DOI: 10.3920/978-90-8686-828-5

First published, 2016

Many thanks to all supporting sponsors, without which the publication of this book would not have been possible:

zoetis

CERTIFIED* HUMANE
RAISED & HANDLED

* Meets the Humane Farm Animal Care Program standards, which include nutritious diet without antibiotics, or hormones, animals raised with shelter, resting areas, sufficient space and the ability to engage in natural behaviors.

International Society for Applied Ethology
ΗΘΟΣ

WORLD ANIMAL PROTECTION

Noldus
Information Technology

Sir James Dunn Animal Welfare Centre
ATLANTIC VETERINARY COLLEGE • UNIVERSITY OF PRINCE EDWARD ISLAND

Livestock Behavior Research Unit
USDA ARS

Many thanks to Merial, our major sponsor:

MERIAL

A SANOFI COMPANY

Foreword: A word from our major sponsor

The topic of animal welfare has become paramount for consumers and for society as a whole. The following quote is commonly attributed to Mahatma Gandhi: 'The greatness of a nation and its moral progress can be judged by the way its animals are treated.' For the past 50 years, the study of ethology has played an important role in advancing our understanding of animal behaviour. For captive animals, including those in animal agriculture, this greater understanding has allowed us to evaluate the impact of housing and management practices on their welfare and productivity.

This book recounts the history of the ISAE, highlighting the research of applied ethologists and veterinarians who have contributed to the field. It traces the evolution of the field since the 1960s, including changes in research objectives and methodologies as a result of changing cultural views and technologies. The text explores the different roles that applied ethology has played within different cultures and economies around the globe. In developed countries, ethology research has contributed to the development of commercial production systems that better meet the needs of animals, and management practices that address welfare issues while maintaining health and efficiency. In less developed economies, ethology research focuses more on practical ways of using locally available technologies and resources to improve animal care on farms. The book concludes with a look to the future, describing new areas for research for example in lab and companion animals and how ethology can contribute to conservation and sustainability. What is certain is that animal welfare will continue to be an important issue, and the sound and credible principles of applied ethology will continue to shape and mold the way we raise and manage animals.

Merial is proud to have been a major sponsor of this important project. As a major animal health product company, one of our main objectives is to provide products that improve the lives of animals, which not only include vaccines and medications to prevent or treat disease, but also ones that help to alleviate pain. Ethology research has played an important role in the development and application of many of these products, and will continue to do so in the development of new ones, especially within the context of an ever-increasing focus on animal welfare. As a result, support of projects like this are important to Merial, as the ultimate goal is to improve the lives of animals.

Dr. Tim Nickel
Technical Service Veterinarian
Merial Canada Inc.

Table of contents

Part 2. The science of behaviour

Introduction

J.A. Brown

One of the themes for the ISAE's 50th Congress in Edinburgh is 'on the shoulders of giants'. Since applied ethology is still a young science, we are very fortunate in having many of the early members of the field still actively engaged in research and teaching. I felt particularly fortunate when several of these individuals agreed to contribute to this book, celebrating the ISAE's first 50 years, and to share their views on the history of the society, development of the science, and the challenges ahead. This is a milestone year, so what better time to take stock and get our bearings?

Part one of this book sets out to document the history of applied ethology and the ISAE, and to introduce those early giants in the field. Ian Duncan and Carol Petherick recount the beginnings of the Society for Veterinary Ethology, and trace its growth up to the present day, including some personal accounts of the major changes that took place along the way. In Chapter 2, Ruth Newberry and Vicki Sandilands take a fun look at the individuals who make up the roots, trunk and branches of applied ethology. Surprisingly, much of this history has never been written before – despite having a great story it seems that ethologists in general (too absorbed in their work, or maybe just introverted?) have rarely gone out of their way to document or promote their field.

In Part two, we explore some of the exciting ideas and central issues in applied ethology. Topics include questions such as: Why are we so interested in other animals? Do veterinarians and applied ethologists approach animal behaviour differently, and how can we best work together? How are animal welfare and applied ethology related, and what are the positive and negative implications of this relationship? How have research methods in applied ethology developed and changed over time? And how have they helped us get closer to understanding animals? While many of these ideas will be familiar to those in the field, we hope these chapters can offer some new insights and tell a story that will help to inspire new students of ethology.

While the ISAE was born in the UK, it has since become an international organisation and it aspires to greater worldwide participation. In Part three, we follow the development of the ISAE and animal behaviour research in different regions of the globe. While the authors of these sections were each given the same basic task: to document the rise of applied ethology in their region, the approaches they have followed are varied. Some authors have taken a personal and anecdotal approach to the story, while others (no doubt following the comfort of habit!) relate the story from a more technical perspective, following the progress of science through the published literature. The chapters on the development of applied ethology in Japan, by Shusuke Sato (Chapter 11), and in China by Jun Bao and Yuzhi Li (Chapter 12) are particularly refreshing as they describe an alternative perspective from the typical western world view. They give us important insights into how the culture, history and philosophical views of each nation has influenced ideas around animal behaviour and welfare, and the development of science.

J.A. Brown, Y.M. Seddon and M.C. Appleby (eds.) ***Animals and us: 50 years and more of applied ethology***
DOI 10.3920/978-90-8686-828-5_Intro, © Wageningen Academic Publishers 2016

Admittedly, there are great challenges ahead, not just for applied ethology or animal welfare science, but for all of us. The final section of this book describes the role of ethology and the ISAE in a broad context. There is a need for research in applied ethology from many perspectives. In developed countries we see a continued focus on improving the care and management of animals raised for food, to increase production and efficiency in order to feed the growing human population, while minimising waste and environmental damage, and reduce animal suffering. In developing nations, applied ethology can play a more direct role in identifying local resources or management practices to improve the care of animals, to the benefit of animals, farmers and their communities. The ISAE can be justifiably called 'farm animal centric', but we recognise that there is also a growing need for research into companion, zoo, lab and wild species. These avenues are examined by Jeremy Marchant-Forde and Bas Rodenburg in Chapter 16.

This book came about also because of concerns for the next generation. Specifically, a concern that the increasing prevalence of urban views, the growing lack of connection with agriculture, and lack of training on critical thinking, combined with a growing demand for animal welfare research could lead to poor quality science and a loss of credibility for applied ethology. It is critical that applied ethology retain a goal of objectivity in the evaluation of animal welfare. Certainly we may also play a role in the development of research questions and the interpretation of results, but we need to be clear about our perspective, approach or bias and how this colours our views. In order for our views to be considered and respected, we need sound methods and objective, critical interpretation, especially as we begin to explore complex issues such as cognitive bias and positive affect, otherwise we risk accusations of anthropomorphism, or simply becoming irrelevant.

Behaviour and welfare research have become so intertwined, it is interesting to note that the SVE began in 1966, at the same time that animal welfare science was born. Rogers Brambell was the notable chair of the Brambell Report (Command paper 2836, 1965), which so clearly laid out the initial concerns and approaches to studying animal welfare. In 1967, Professor Brambell was named an honorary member of the SVE, and addressed the annual congress with a thoughtful acceptance speech. His address is included below, and 49 years later, it is a compelling testament to his common sense, humour and humility, and foresight.

In conclusion, this book is a celebration of animal behaviour, the science of applied ethology, and the International Society for Applied Ethology. While our science may be obscure, with a title that often inspires confusion, and our society is small, with typical attendance at the international congress numbering under 500, our members share an important bond and form a strong family. At any ISAE meeting you can readily meet, share food and drink, and new ideas with the most renowned and respected 'silverbacks' of the field. I remember well my first meeting, the North American Regional meeting held in Vancouver, British Columbia, in 2005. It is always exciting to meet people with whom you share similar passions and interests, and to meet the authors of papers that you admire. I felt welcome and inspired. The sensation of having 'found my group' was tangible and confirmed by subsequent meetings. Other members seem to share this experience and once the presentations are over – safely together and with all our needs met, play behaviour erupts! The highlight of each meeting is the banquet dinner and party, where applied ethologists have been known to act like animals.

We hope you will enjoy this glimpse from the giant's shoulders: here's to a long and fruitful future for the ISAE, and to our growing understanding and appreciation of all animals.

Professor Brambell's address to the society for veterinary ethology

EDINBURGH, JUNE 1967

Mr. President, Ladies and Gentlemen,

May I tell you first how happy I am to be here today to thank you for the honour you have done me in electing me your first Honorary member and also for the opportunity of delivering this address on the occasion of your first annual meeting. I am all the more honoured by your kind invitation because, as you will know, I am not an expert in ethology and, indeed, as you will no doubt become aware in the course of this address, I am exceedingly ignorant of that subject.

I have no doubt that you invited me here because of the Report of the Technical Committee (1965) that has come to bear my name and the emphasis that it placed on the importance of behavioural studies for the welfare of animals kept under methods of intensive animal husbandry. My Committee stressed the importance of such studies for animals kept under intensive conditions, but I am well aware, of course, that they are also of great importance with traditional methods of husbandry. They are important, moreover, not only for the welfare of the animal but also, I do believe, for the welfare of animal husbandry. I have no doubt that happier and more contented animals are ones that are better doing, but well-doing in the true sense is something that is much more comprehensive than simple growth, or egg, or milk production, it relates to all aspects of the animal's well-being. Hence, I do not regard a single parameter, such as the growth rate, as an adequate measure of its well-doing in this wider sense, as it is often taken to be. It is frequently said that the growth rate of an animal for meat, or the egg production of a laying hen, or the milk production of a cow, are the only reliable objective measures of their welfare. It is argued that continued productivity should be taken as decisive evidence that the animal is not suffering; this is an over-simplified and incomplete view that should be rejected. It is an argument that would I suspect, carry little conviction with those who normally lead an active life but have been compelled to undergo a period of inactivity and restraint and have put on weight in consequence. They would hardly regard this weight increment as either a measure of their contentment or a reflection of their physical fitness. Any sufficient estimate of an animal's welfare must be based on a full and wide understanding of the ethology of the particular species in question. To make this point I want to consider first the very difficult question of what an animal feels, of its sensations, and then to go on to consider how their sufferings can be evaluated. These, as you will know, are two distinct questions of which the first is the much more answerable.

There are sound anatomical and physiological reasons for accepting that domestic animals and birds experience the same kinds of sensations as we do. The structure of the sensory organs of mammals and of their nervous systems is essentially similar to those of man and the functions of the ductless glands are known to be comparable. The sensations certainly differ in degree, for example the senses of hearing or smell may be better developed in some animals than in man or they may be capable of appreciating a greater range of stimuli. It is probable that imaginative anticipation is incomparably less well developed in most animals. They appear to live much more in the present and their feelings to be correspondingly more transitory, although many animals can both remember the past and fear the future to a recognisable extent. It is justifiable to assume that the feelings of animals are not identical with those of human beings. It is equally justifiable to assume that they feel in similar ways. The valid point where the line should be drawn between these two extremes is very difficult to determine and is, in the end, a matter of judgement. In this connection I would like to quote an opinion expressed by the late Lord Brain which we included as a footnote in our Report. He wrote: 'I personally can see no reason for conceding mind to my fellow men and denying it to animals. Mental functions, rightly viewed, are but servants of the impulses and emotions by which we live and these, the springs of life, are surely diencephalic in their neurological location. Since the diencephalon is well developed in animals and birds, I at least cannot doubt that the interests and activities of animals are correlated with awareness and feelings in the same way as my own and which may be, for all I know, just as vivid'.

Of course we need to know a good deal more about the threshold and range of each of the various sense organs with which domestic animals and birds are equipped but this is largely in the field of physiology and I don't propose to spend time on it today.

The second and much more difficult question is how the suffering of an animal is to be evaluated. In attempting to answer this question we must first consider how we judge suffering in another human being. We cannot measure the feeling of another person however well we know them. Suffering is a subjective experience and is no more measureable than loving, or hating or fearing. We judge the suffering of another person by analogy with our own, from what the person tells us and from our observations of their looks, behaviour and health. The evaluation of the suffering of an animal similarly must rest on analogy with our own and must be derived from observations of their cries, expressions, reactions, behaviour, health and productivity. Our understanding of their suffering is not different in kind but rather in degree from that which we form of the suffering of a fellow human being. Animals show unmistakable signs of suffering from pain, exhaustion, fright, frustration, rage and so forth and the better we are acquainted with them and with their behaviour, the more readily we can hope to detect these signs. We cannot determine their suffering objectively but must rely in the end on a subjective judgement. I have perhaps laboured this point because I hear so often from people who are interested in postponing action that nothing must be done to interfere with husbandry practice without scientific proof of suffering and that means, of course, objective demonstration of it. Frankly, this argument angers me because it sounds plausible but is absolute nonsense. If your child is suffering, do you want a measure of it or do you take action on your subjective judgement of its severity? The problem is, in essence, as simple as that. Nevertheless, much more extensive knowledge than we possess at present of the behaviour and ethology of the various species with which we are concerned under conditions of domestication, would enable us to judge much more sensitively and reliably the degree of suffering to which they are exposed at any time. It is in this connection that I hope the work of this Society will do much for animal welfare and for animal husbandry. Domestication of any kind necessarily involves some measure of restraint upon the animal; the kind of restraints that are imposed and their severity vary with the particular method of husbandry adopted and with the species to which this is applied but restraint undoubtedly reaches its most extreme form in modern methods of intensive husbandry, where the stress is upon limitation of space and economy of labour. In these circumstances, mechanisation can reduce the labour, provided the animals are kept under conditions of close confinement and regulated environment. Modern intensive methods may involve control of almost all the environmental conditions to which the animal is exposed from birth to slaughter. These conditions include space, diet, ventilation, illumination, bedding, companionship or isolation etc. Mutilation that handicaps the animal is claimed to be necessary in some instances to prevent vice. It is because of these various restraints that are imposed on the animal by the artificial environments of intensive husbandry that the whole question of welfare has become so crucial today. My Committee summed this up by stating 'Above and beyond all these matters stands the fact that modern intensive animal production methods, most markedly increase the responsibility of those who use them towards the animals in their charge. If any creature is wholly and continuously under control, we believe that this total human responsibility must be acknowledged; changing patterns of husbandry may mean varying degrees of frustration and discomfort to animals whose normal patterns of behaviour are still imperfectly understood. We are certain that a beginning must be made to safeguard their welfare'.

I would like now to consider some of the ways in which ethology can help with the welfare of animals. First of all, you, who are acquainted with animals on farms, will agree that it is quite astonishing how far the behaviour patterns of their remote wild ancestors are retained, even after thousands of years of domestication. It is this basic behaviour pattern of the species which is so important if we are to understand the behaviour of the modern animal. Studies, therefore, which throw light on the behaviour patterns of the wild ancestors of the modern breeds, should be given priority. Opportunities for these are increasingly limited with the disappearance of the wild descendants of the ancestors of so many domestic species but there may be some opportunity still of throwing light on such things; I am thinking, for instance, of the work of Kruijt (1962a,b; 1964) on the Red Jungle Fowl of Burma. Opportunities may offer, for instance, of studying the behaviour of feral sheep, such as the Soay sheep on St. Kilda, where Jewell (1966a,b) Boyd *et al.* (1964) Grubb and Jewell (1966) have studied their reproduction and mortality, or again, the so-called wild White Cattle of Chillingham Park, Vaynol Park and other places, may provide some opportunities of learning more about the ancestral behaviour pattern of modern cattle. Comparative studies of related species, that are still available in the wild state, may help to throw light on the probable behaviour patterns of the actual ancestors of our present breeds, where these are not available. Such studies can help us greatly in determining the probable importance in the lives of domestic animals of, for example, companionship or illumination or noise or feeding and sleeping rhythms.

It is, of course, true that genetics plays a large part in the behaviour of the animal. Too little attention may have been paid, as yet, to breeding strains the reactions of which are adapted to the particular methods of husbandry for which they are intended; for example, there is, as we all know , a great difference in the pugnacity of various strains of poultry, yet, although cannibalism is a major factor in several methods of intensive poultry husbandry and the vile practice de-beaking has had to be resorted to to prevent it, there seems little evidence of any persistent effort to breed stocks that are both docile and good egg producers, much more emphasis having been placed apparently on egg production than on docility. Again, on this question of pugnacity in its relation to cannibalism, particularly in poultry, I wonder whether enough work has been done on various methods of reducing it through studies of behaviour. (For example, I have heard it stated that the presence of a cock reduces cannibalism amongst the hens and I should like very much to know if there is any evidence that this is so. If this is true, have such things as tape-recordings of crowing cocks been tried in batteries?) Would it not be possible, with the great diversity of plastics now available, to devise some simple attachment to the tip of the beak to discourage cannibalism without seriously incommoding or hurting the bird, as does de-beaking or the spectacles that are attached by a staple driven through the nasal septum? I think much more could be done by study of the factors which underlie cannibalism and of the methods which might be adopted to reduce it. We are all aware that whereas it is a major problem in some installations, in others it is not experienced to any serious extent. How far this is due, in fact, to difference in stockmanship and how far it is due to some, as yet unidentified, factors in the environment, is still a moot point on which there are diverse opinions.

One field in which much research is needed is that of food preference. Much is known about nutritional requirements of farm animals, about the availability of food components to the animal and about digestion, but comparable attention has not been paid to the problem of preference and their effect on intake. This is a field of study which could, in my opinion, contribute greatly to the welfare of domestic animals, and I am convinced, quite as much to the welfare of farmers. The increasing use of balanced diets and relatively uniform food, raises the whole problem of how far variety in the food may contribute to appetite, and what are the qualities in the food which render it most acceptable to the animal. Even the most perfectly balanced and digestible diet cannot achieve growth rate, so important in modern intensive methods of husbandry, unless the animal chooses to eat the optimum quantity. It may well be that maximum acceptability is as important as maximum nutritional value both for the welfare and for the well-doing of the animal.

More knowledge is needed about the importance of companionship for the welfare of animals. One might expect that this would be more important in animals that came from herd-living ancestors than in those which did not and we might, therefore, imagine that this would play a larger part in the lives of cattle, for example, than say of cats. Moreover, it is likely to be more important for the young than for the adult animal, yet under methods of intensive husbandry for the production of quality white veal, it is the calf that is so notably restricted and confined and isolated in a box throughout its life. Moreover, play is obviously important to animals, especially young ones, as can be seen when they are on free range. It is frequently acknowledged to be so for pigs kept intensively and we are familiar with the hanging chains and other toys sometimes provided for their amusement. Could more be done with advantage in this direction?

What are the needs of illumination of the various species of animals, the best wavelengths, the duration of illumination and the intensity? It may be, for example, that pigs which are partly nocturnal or at least crepuscular in their natural condition, require less light than cattle, which are entirely diurnal animals. The same may be true of ducks as compared to hens, since in the wild state the mallard is active at night as well as in the day time, whereas the hen spends all the hours of darkness in roosting. It is well known that lengthening daily periods of illumination can induce laying in poultry or oestrus in ferrets and hence can have a pronounced effect on the behaviour of these animals, which suggests that conditions of illumination may well be significant for welfare.

We are even more ignorant as to the significance of the sense of smell for animals, although it is evident that this sense is much better developed in many species than in ourselves. The recent work of Bruce and Parkes (Bruce, 1959, 1960, 1961; Bruce and Parrott, 1960; Parkes and Bruce, 1961, 1962) on mice provides a striking example of the part it can play. These authors have shown that exposure of a recently mated female mouse to the smell of an alien male mouse, of a strain other than that to which its mate belonged, can block pregnancy by preventing implantation. This is a remarkable example not only of olfactory acuity but also of olfactory memory and is a clear warning that attention to the part which smells may play in the behaviour of domestic animals should not be neglected.

It is, however, the limitation of space which appears to restrict the normal behaviour pattern of the animal most seriously. This is the context in which ethology can be most rewarding. Close confinement may prevent the performance of a particular pattern of behaviour either (a) by depriving the animal of the opportunity of experiencing the necessary stimulus, so that the pattern is not initiated, or (b) the stimulus may be experienced but the performance wholly or partly frustrated. The consequence for the welfare of the animal could be very different according to which of these two alternatives was operative. It is easy to think of examples. An example of the first is provided by laying hens in batteries where the removal of the eggs prevents the urge to brood. An example of the second is provided by a cockerel in a cage in which it cannot stand upright and crow properly. The first of these two conditions might be expected to cause much less stress than the second. This points to the need to anticipate conditions that are likely to induce stress and to recognise stress when it results. I am, of course, aware that severe and prolonged stress is said to have an effect upon the adrenals, producing enlargement of these organs and derangements of the metabolisms of adrenal steroids but I am not so much concerned with recognising, after the event, that prolonged and severe stress has been experienced by the animal, as recognising the condition when it first appears and the causes of it. How can discomfort or unease or stress be recognised, that is a major problem for the ethologist? Can it be done for poultry for example, as Professor Thorpe has suggested, by careful analysis of the recorded sounds in a battery house? Can it be done by observation of the animal's posture or movements? We hold ourselves well and walk well when our morale is high and we tend to slouch with shoulders rounded when our morale is low. Can, by analogy, the stance and movements of poultry be used as a guide to their state of mind? Most of us can recognise a really happy and contented domestic animal and distinguish it from one that is very unhappy and discontented but we do so by a general and uninformed impression of its appearance and behaviour rather than by any precise characteristics. How far can the qualities on which we base our impression be used as more precise guides to the condition of the animal? We look for bright eyes, good coat, lively behaviour, lack of fear and so forth but these are not readily measurable at present and it is to find means of evaluating such features of the behaviour of an animal that attention should be directed. This is the most important problem of all for animal welfare, to recognise stress, discomfort or unease before they have become acute. It is to the solution of this problem that I believe studies on animal behaviour can contribute to welfare so much in the future.

References

Bao, J. and Li, Y., 2016. China perspective: emerging interest in animal behaviour and welfare science. In: Brown, J.A., Seddon, Y.M. and Appleby, M.C. (eds.) Animals and us – 50 years and more of applied ethology. Wageningen Academic Publishers, Wageningen, the Netherlands, pp. 241-252

Boyd, J.M., Doney, J.M., Gunn, R.G. and Jewell, P.A., 1964. The Soay sheep of the island of Hirta, St. Kilda. A study of a feral population. Proceedings of the Zoological Society of London 142: 129-164.

Bruce, H.M., 1959. An exteroceptive block to pregnancy in the mouse. Nature 184: 105.

Bruce, H.M., 1960. A block to pregnancy in the mouse by proximity to strange males. Journal of Reproduction and Fertility 1: 96-103.

Bruce, H.M., 1961. Time relations in the pregnancy-block induced in mice by strange males. Journal of Reproduction and Fertility 2: 138-142.

Bruce, H.M. and Parrott, D.M.V., 1960. The role of the olfactory sense in pregnancy block by strange males. Science 131: 1526.

Command Paper 2836, 1965. Report of the technical committee to enquire into the welfare of animals kept under intensive livestock husbandry systems. Her Majesty's Stationery Office, London, UK.

Grubb, P. and Jewell, P.A., 1966. Social grouping and home range in feral Soay sheep. Symposia of the Zoological Society of London 18: 179-210.

Jewell, P.A., 1966a. Breeding season and recruitment in some British mammals confined on small islands. Symposia of the Zoological Society of London 15: 89-116.

Jewell, P.A., 1966b. The concept of home range in mammals. Symposia of the Zoological Society of London 18: 85-109.

Kruijt, J.P., 1962a. On the evolutionary derivation of wing display in Burmese Red Junglefowl and other Gallinaceous birds. Symposia of the Zoological Society of London 8: 25-35.

Kruijt, J.P., 1962b. Imprinting in relation to drive interactions in Burmese Red Junglefowl. Symposia of the Zoological Society of London 8: 219-226.

Kruijt, J.P., 1964. Ontogeny of social behaviour in Burmese Red Junglefowl (*Gallus gallus spadiceus*) Bonnaterre. Behaviour, Suppl. XII: 1-201.

Marchant-Forde, J.N. and Rodenburg, T.B., 2016. Future directions for applied ethology. In: Brown, J.A., Seddon, Y.M. and Appleby, M.C. (eds.) Animals and us – 50 years and more of applied ethology. Wageningen Academic Publishers, Wageningen, the Netherlands, pp. 297-317

Newberry, R.C. and Sandilands, V., 2016. Pioneers of applied ethology. In: Brown, J.A., Seddon, Y.M. and Appleby, M.C. (eds.) Animals and us – 50 years and more of applied ethology. Wageningen Academic Publishers, Wageningen, the Netherlands, pp. 51-75

Parkes, A.S. and Bruce, H.M., 1961. Olfactory stimuli in mammalian reproduction. Science 134: 1049-1054.

Parkes, A.S. and Bruce, H.M., 1962. Pregnancy-block in female mice placed in boxes soiled by males. Journal of Reproduction and Fertility 4: 303-308.

Sato, S., 2016. Applied animal behaviour science in Japan and the culture of '*aigo*'. In: Brown, J.A., Seddon, Y.M. and Appleby, M.C. (eds.) Animals and us – 50 years and more of applied ethology. Wageningen Academic Publishers, Wageningen, the Netherlands, pp. 227-240

Part 1. On the shoulders of giants

1. The International Society for Applied Ethology: going strong 50 years on

J.C. Petherick[1] *and I.J.H. Duncan*[2*]
[1]*The University of Queensland, Queensland Alliance for Agriculture and Food Innovation, 25 Yeppoon Road, Rockhampton, Queensland 4702, Australia*
[2]*Department of Animal Biosciences, University of Guelph, Department of Animal Biosciences, University of Guelph, 50 Stone Road East, Guelph, ON, N1G 2W1, Canada; iduncan@uoguelph.ca*

Abstract

In 1966 in Edinburgh, Scotland, a group of far-sighted veterinarians formed the Society for Veterinary Ethology with the aim of improving knowledge about the behaviour of domesticated animals. The findings of a UK government committee, investigating intensive livestock production methods, published in 1965, had made it clear that such information was urgently needed. In this chapter we describe the development of this society and how it changed from a group of 37, mostly Scottish, founder member veterinarians to an international society with about 550 members, the majority of whom are not veterinarians. This evolution in the interests of the membership led the society in 1991 to change its name to the International Society for Applied Ethology (ISAE). The Society has had a special relationship with the journal 'Applied Animal Behaviour Science' (formerly 'Applied Animal Ethology') since 1974 and it is now the official journal of the Society. The Society has had, and continues to have representation on committees and boards relating to animal behaviour and animal welfare throughout the world, and continues to evolve and work to raise both its profile and that of applied ethology globally; a bright future for the Society and applied ethology seems assured.

Keywords: scientific society, animal behaviour, veterinary ethology, ISAE, SVE, history

1.1 Introduction

This year (2016) we celebrate the 50th anniversary of the International Society for Applied Ethology. The Society owes its existence to an animal activist, a zoologist (and a committee that he chaired) and a group of veterinarians. In 1964, Ruth Harrison published her controversial book '*Animal Machines*' in which she was very critical of intensive livestock farming methods. The public outcry was so intense that the UK Government formed a committee of enquiry under the Chairmanship of a zoologist, Professor Rogers Brambell, and it published their findings the next year (Command Paper 2836, 1965). This report was also very critical of intensive farming methods but made some claims that were not based on any hard evidence. A group of far-sighted veterinarians quickly realised that much more information on livestock behaviour was urgently required in order to address animal welfare issues associated with intensive farming.

DOI 10.3920/978-90-8686-828-5_1, © Wageningen Academic Publishers 2016

This chapter charts the history of the International Society of Applied Ethology, originally called the Society for Veterinary Ethology, explaining its formation and the key people involved. We examine how the Society has developed, listing some of the major changes that have taken place during various phases of this development. Finally, we briefly offer our perspectives for the future of the Society and suggest a few actions that may enhance the Society.

1.2 Searching

The history of the study of animal behaviour is a long one. It has been suggested that cave paintings of the Palaeolithic era in Europe reveal an early interest in animal behaviour 30,000-40,000 years ago (Curtis, 2006). Archaeological evidence from other pre-historical sites confirms a very early understanding of animal behaviour. For example, the Head-Smashed-In Buffalo Jump site in Alberta, Canada, reveals an excellent knowledge of bison behaviour by native people of the North American plains nearly 6,000 years ago. In the last thousand years there is evidence from works of art, such as medieval tapestries, depicting the hunting of animals with horses and dogs, falcons, and even ferrets, showing that human beings had a good understanding of at least some aspects of animal behaviour. A consideration of the domestication of animals over the last 15,000 years also suggests some understanding of their behaviour (Clutton-Brock, 1981; Ratner and Boice, 1975), but, of course, all these observations fall short of evidence of a science of animal behaviour.

Aristotle was probably the first person, 2,400 years ago, to observe animals objectively and record what he saw (Sparks, 1982). However, it is really only in the last 200 years that something approaching a science of animal behaviour has started to appear. Wood-Gush (1991) noted that in the late 18th century a Scottish farmer initiated rotational grazing from informal observations on his grazing cattle. It may have been anticipated that the origins of formal, scientific investigation of animal behaviour (ethology) would have been 'applied' in nature, as humans would have been highly dependent on understanding and predicting the behaviour of the animals that they farmed and hunted. An English veterinarian, William Youatt, published a book in 1839 in which he described in detail many behaviour patterns in domesticated animals. In this book, Youatt (1839, p. 55-66) writes of animals' senses, emotions, consciousness, attention, memory, association of ideas, reason, instinct, and social affections. More than 30 years before Darwin's '*The Expression of the Emotions in Man and Animals*' (1872), Youatt (1839, p. 55) wrote of the intellectual faculties, 'We are endeavouring to shew that the difference [between humans and animals] in one of the most essential of all points, is in degree and not in kind'. He also described some of the invasive research that was taking place on animals at that time, and wrote 'We are operating on animals that have, probably, as keen feelings of pleasure and of pain as ourselves' (Youatt, 1839, p. 234). So he was aware that animals (at least the vertebrates) were not just behaving in a reflexive way but had accompanying subjective feelings and that these were important for the animals' welfare.

The study of animal behaviour as a scientific discipline started to grow in the latter half of the 19th century. Darwin (1859, 1872) gave convincing accounts of the evolution of certain behaviour patterns. For example, he wrote '... I can see no good reason to doubt that female birds, by selecting, during thousands of generations, the most melodious or beautiful males, according to their standard of beauty, might produce a marked effect' (Darwin, 1859, p. 137). George John Romanes, a disciple of Darwin, elaborated and explicitly included the emotions in his account of evolution. For example, he stated '... the association of Pleasure and Pain with organic states and processes which are respectively beneficial and deleterious to the organism, is the most important function of Consciousness in the scheme of Evolution (Romanes, 1884, p. 111). Biologists like Douglas Spalding, an Englishman who worked for some years in Scotland, carried out detailed experiments on the interaction between learning and instinct and used domesticated species (chicks and pigs) in these studies (Gray, 1967; Spalding, 1872). Therefore, by the end of the 19th century, a good foundation had been laid for the building of the discipline of applied animal ethology.

1.3 Courtship

As pointed out by David Wood-Gush (1991), a pioneer of poultry reproductive behaviour and the person responsible for initiating The University of Edinburgh's MSc course in Applied Animal Behaviour and Animal Welfare, it is difficult to determine the exact beginnings of applied ethology, that is, the scientific study of the behaviour of animals used or managed by humankind that generally has the aim of solving specific problems. In fact, it seems likely that fundamental ethology and applied ethology were developing as disciplines at the same time. For example, Darwin, Spalding and Romanes did not confine themselves to fundamental studies, and many of their observations and experiments involved domesticated animals.

In the 20th century, ethology as a science was given a very solid foundation by the work of Karl von Frisch, Konrad Lorenz and Niko Tinbergen who were awarded the Nobel Prize in Physiology or Medicine in 1973 for their discoveries concerning organisation and elicitation of individual and social behaviour patterns. At the same time as these three scientists were pursuing fundamental studies, others were carrying out much more applied studies. For example, Norwegian-born Thorleif Schjelderup-Ebbe (1922) was conducting pioneering studies on social relationships in poultry, having observed the 'pecking order' behaviours in his own chickens since he was 10 years old. A little later, Alex Brownlee, a young veterinarian working on sheep diseases, at first at the Moredun Research Institute in Edinburgh and later at the Institute for Animal Health at Compton, Newbury, England, documented observations on play behaviour of guinea pigs, rabbits and calves in 1939 (Brownlee, 1939). Whilst these studies were not conducted with the aim of solving problems at the time, social hierarchy and play behaviour have since been shown to be important aspects of livestock production systems in relation to aggression and distribution of limited resources, and assessment of animal welfare, respectively. It is also noteworthy that the work by Alex Brownlee was published at a time when, according to Durant (1986, p. 1606), writing about the 'making of ethology', 'animal behaviour study barely existed as an identifiable research field in Britain'. Brownlee was later to become one of the founding members of the Society for Veterinary Ethology (SVE) which went on to become the International Society for Applied Ethology (ISAE).

The veterinary profession and domesticated animals played significant roles in the development of ethology in Britain. The work of William Youatt in the 19th century was influential, as already referenced above. Alastair Worden, a veterinarian at the Institute of Pathology in Cambridge, UK, was a driving force in the development of the Institute for the Study of Animal Behaviour (ISAB), which later became the Association for the Study of Animal Behaviour (ASAB), (Durant, 1986). In 1944, Worden recruited a large number of veterinarians to ISAB, perhaps as a consequence of a joint meeting held with the British Society of Animal Production that year. However, it appears this did not sit well with ISAB/ASAB; Durant (1986, p. 1610) recounts, 'For although Worden may have been enrolling lots of vets, and although the 'Bulletin' [ISAB's publication, which later became 'Animal Behaviour'] did as a result come to feature a considerable number of papers on the behaviour of domestic animals, Thorpe's [W.H. Thorpe, a Cambridge zoologist] growing influence was more than sufficient to ensure the Institute would remain faithful to the essentially ethological vision of its founders'. This suggests that ISAB/ASAB did not regard the study of the behaviour of domestic animals as authentic ethology.

Thus, it seems that barriers were being erected between fundamental and applied ethology very early on in the establishment of ethological study and science. Perhaps it was ISAB's focus on classical ethology (as conducted by Lorenz, von Frisch and Tinbergen) through Thorpe in the 1950s (Durant, 1986), with its emphasis on behaviour occurring in natural environments, that resulted in a dearth of publications on domestic animals in the ISAB's 'Bulletin'.

1.4 Conception and gestation

Andrew Fraser, a veterinarian working in the 1960s at the Royal (Dick) College of Veterinary Medicine in Edinburgh was responsible for founding the SVE. Fraser was keen to acquire information on the behaviour of farm animals, and in the early 1960s, his Department Head, Professor J.A.S. Spreull, encouraged him to promote domestic animal ethology. In the preface to his textbook, '*Farm Animal Behaviour*', Fraser (1974), points out that there was an obvious need for information on applied behaviour studies in farm animals by veterinary and animal science students in order to solve various animal health and behavioural problems that were appearing.

During this same period of time, Ruth Harrison (1964) published her book '*Animal Machines*' which brought to the attention of the general public the intensive ways in which animals were being farmed. The public outcry in response to the revelations in Harrison's book resulted in the formation of a Committee under the Chairmanship of Professor Rogers Brambell, a zoologist at the University College of North Wales in Bangor, to enquire into the welfare of animals kept in intensive livestock husbandry systems (Command Paper 2836, 1965). In this report, the need for a greater interest in and knowledge about the behaviour of livestock was emphasised and this was undoubtedly a major impetus for the formation of a society for veterinarians with an interest in building knowledge of domestic animal behaviour.

Thus it was that on 16th February 1966 a group of 13 veterinarians from Scotland and England met informally in Edinburgh at the invitation of Andrew Fraser (Table 1.1). This meeting followed on from a circular letter to 40 people enquiring about the veterinary interest in the study of animal behaviour and the positive responses to this letter. After much discussion it was agreed that a society should be formed to stimulate interest in the subject of animal behaviour, specifically of farmed animals. At the time it was resolved that membership should be confined to veterinarians and the name of the society was proposed to be the 'Veterinary Society for the Study of Animal Behaviour'. However, since this name was rather similar to the 'Association for the Study of Animal Behaviour' it was decided to leave the selection of a name to the first meeting.

An interim Executive Committee was elected: A.F. Fraser (Chair), W.G. Beaton (Secretary), C.T. McCrae (member) and P.G. Hignett (member) who were given the task of arranging a first meeting, sending invitations to veterinarians who might be interested and advertising the meeting in the 'Veterinary Record'.

1.5 Birth

The Society for Veterinary Ethology (SVE) was born on 4th June 1966 at the inaugural meeting which took place at Edinburgh University. We wonder if the participants at this meeting had any idea that they were acting as midwives at such a significant birth!

Table 1.1. The group of veterinarians who attended an informal meeting (February 1966) to establish the society.

Name	Affiliation or location
W.G. Beaton	University of Edinburgh, Royal (Dick) School of Veterinary Studies
M.A. Couttie	Milk Marketing Board, Stockton-on-Forest, Yorkshire
R. Ewbank	University of Liverpool, Veterinary School
A.F. Fraser	University of Edinburgh, Royal (Dick) School of Veterinary Studies
P.G. Hignett	University of Glasgow, Veterinary School
B.R. Howard	University of Edinburgh, Royal (Dick) School of Veterinary Studies
P. Imlah	University of Edinburgh, Royal (Dick) School of Veterinary Studies
C.T. McCrae	Ministry of Agriculture, Fisheries and Food, Thirsk, Yorkshire
H.S. McTaggart	University of Edinburgh, Royal (Dick) School of Veterinary Studies
W.M. Scott	Universities Federation for Animal Welfare, London
J.M. Swanney	Newlands, Perthshire
A.D. Weaver	University of Glasgow, Veterinary School
D.K.M. Young	Pitlochry, Perthshire

The meeting was chaired by Andrew Fraser and was attended by 26 veterinarians from Scotland and England who decided on the name for the society, approved a constitution and elected the following governing council for 1966/67:

Honorary President	A.F. Fraser
Honorary Senior Vice-President	R. Ewbank
Honorary Junior Vice-President	A. Brownlee
Honorary Secretary	W.G. Beaton
Honorary Treasurer	B.R. Howard
Members	J.R. Baillie
	C.T. McCrea
	A.D. McEwan
	J. Milne
	H.C.B. Reed
	J.M. Swanney

The aims of the SVE were agreed as follows:

a. the stimulation within the veterinary profession of enquiry and research into the behaviour of animals;
b. the encouragement of discussion on the veterinary aspects of animal behaviour and the publication of original work in this field;
c. the furtherance of ethological knowledge relative to the care and utilisation of animals, domesticated, captive and free-living;
d. the provision of a medium to promote exchange of information between veterinarians and between them and others concerned with behaviour and well-being of animals.

It was agreed that the annual subscription should be 10 shillings (50 pence in decimal currency) and that applicants who submitted written applications, expressing their support for and interest in the study of the behaviour of farmed animals and had paid their subscription should be elected as Founder Members (Table 1.2). It is interesting that two of these Founder Members were from the USA and this was probably a consequence of them being amongst the 40 veterinarians originally canvassed for their opinion on forming a society.

After the business meeting of the Society, a symposium was held at which four papers were presented and discussed:

- W. Ferguson Grazing behaviour of dwarf cattle in Nigeria.
- B.R. Howard The physiological basis of the hunger drive.
- D.R. Lane Some aspects of behaviour in the production and training of guide dogs.
- H.C.B. Reed The behaviour of boars in relation to their reproductive efficiency.

At a Council meeting on December 8th 1966, Professor Rogers Brambell was elected unanimously as an Honorary Member of the Society. Professor Brambell accepted the

Table 1.2. Founder members of the Society (June 1966).[1]

Name	Location	Name	Location
J.R. Baillie	Kirriemuir	B.R. Howard	Edinburgh
P.J. Barden	Edinburgh	P. Imlah	Edinburgh
W.G. Beaton	Edinburgh	R.A. Jones	Edinburgh
W.P. Blount	London	D.R. Lane	Leamington Spa
B. Boswood	Brentwood	A. Littlejohn	Edinburgh
A. Brownlee	Newbury	C.T. McCrea	Thirsk
L. Collery	Dublin	A.D. McEwan	Glasgow
M.A. Couttie	York	H.S. McTaggart	Edinburgh
M.D. Eaglesome	Lanark	J. Milne	Duns
R. Ewbank	Liverpool	D. Moodie	Ayr
L.C. Faulkner	Colorado, USA	G.S. Peyton	Canterbury
G.S. Ferguson	Edinburgh	H.C.B. Reed	Selby
W. Ferguson	Edinburgh	P.G.M. Rowntree	Liverpool
M.W. Fox	Illinois, USA	W.N. Scott	London
A.F. Fraser	Edinburgh	I.E. Selman	Glasgow
J.A. Fraser	Edinburgh	J.M. Swanney	Perth
T. Graham Marr	Edinburgh	J.C. Wilson	Edinburgh
H. Hastie	Edinburgh	D.K.M. Young	Edinburgh
T.W. Heard	Chippenham		

[1] Country shown only for non-UK members.

invitation and agreed to make a presentation at the next Scientific Meeting, to be held at Edinburgh University on 30th June 1967. Professor Brambell's talk foreshadowed many of the issues that became central to applied ethology in the following decades. For example, the relationship between productivity and animal welfare; the necessity of an understanding of the ethology of particular species for the assessment of welfare; the impacts of aspects of artificial environments, particularly restricted space, on behaviour and welfare; the role of genetics in farmed animal welfare; and the difficult matter of determining animals' feelings and whether animals are suffering. His presentation is reproduced in the introduction of this volume.

For the first few years of its existence, the SVE had two scientific meetings a year with the common pattern being for a summer meeting to be held in different locations in the UK (usually at a veterinary college), and a winter meeting which was held at the Royal Veterinary College in London. The winter meeting was arranged to take place consecutively with the winter meeting of the ASAB which was held at London Zoo. This meant that those interested in both the fundamental and applied aspects of animal behaviour could easily attend both meetings.

Andrew Fraser took on the task of producing the *Proceedings of the Society for Veterinary Ethology* from these early meetings, starting in 1966. For the first several years most of the abstracts of papers given at these meetings were published in the British Veterinary Journal, and Fraser (1974) acknowledges that these provided much of the raw material from which his textbook was constructed.

1.6 The neonatal and formative years

When first founded, the SVE had a very Scottish bias, with 23 of the 37 Founder Members based in Scotland, 11 in England, one in Eire and two in the USA. Moreover, all of the members were male, which reflected the veterinary profession in the UK at that time. What a change in 50 years, with Veterinary Colleges now reporting that 70-85% of students are female! The Society grew quite quickly and there was increasing interest in membership by non-veterinarians. In response to this, a decision was made in 1970 to extend membership to non-veterinarians and, as a consequence, the aims in the Constitution were amended to better reflect the broader range of interests of members:

a. became 'the stimulation of enquiry and research into the behaviour of animals among agricultural scientists, veterinarians, and zoologists'; and
d. became 'the provision of a medium to promote exchange of information between those concerned with behaviour and well-being of animals.'

David Wood-Gush became an 'Associate Member' of the Society in 1970 prior to it being open to non-veterinarians, and in 1972 at first a trickle and then a steady stream of non-veterinarians joined the Society. In the first decade of its existence, the society grew quite rapidly and also became more international (Figure 1.1). In order to spread the workload caused by the increasing membership, Council was expanded and by 1975/76 had the additional posts of Honorary Assistant Secretary, Honorary Membership Secretary and Honorary Librarian giving a Council of 14 compared to the original 11 members. The Presidents, Secretaries and Treasurers of the Society are shown in Table 1.3.

The position of Librarian was established in 1967, and enthusiastically filled by Alex Brownlee. The modest library consisting of a few textbooks and a collection of reprints donated by members, was housed, at first, in the library of the Royal (Dick) Veterinary School in Edinburgh and later in the library of the Poultry Research Centre in Edinburgh and then in Roslin (on the outskirts of Edinburgh). When he was in his late 70s and early 80s, Alex Brownlee walked from his home in Edinburgh to Roslin over the Pentland Hills, a distance of about 11 km each way, about once a month to deposit items in the library. He was a very keen observer of nature and, on arrival at Roslin, would tell one of us (I. Duncan) all the interesting things he had seen on his journey. This could vary from a plant species growing in an unusual location, to a bird singing out of season, to cattle behaving 'oddly' in neighbouring fields.

With technological changes, the need for a library diminished and the position of Librarian gradually evolved into the Editor (Senior and Junior) positions. The current Council comprises 13 full (voting) members and non-voting legal assessor and student

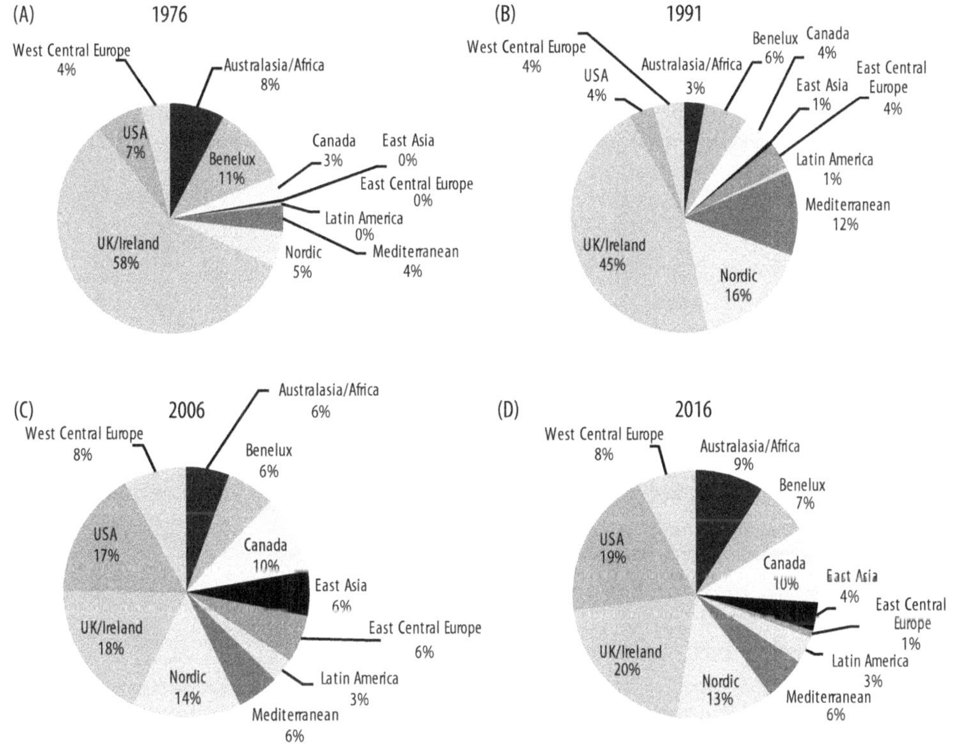

Figure 1.1. ISAE membership by region: (A) 1976 (242 members), (B) 1991 (342 members), (C) 2006 (488 members), (D) 2016 (546 members).

members. Interestingly, whilst there may no longer be a need for a physical library, the research required to prepare this history has revealed the requirement for the retention and safe storage of documents from the early period of the Society.

For the first 8 years of the Society's existence, meetings were generally held twice a year in the UK and Eire. However, in 1974 the first meeting outside of this region was held in Ghent, Belgium (Table 1.4) and this signalled the start of the full internationalisation of the Society.

1.7 The growing and exploring years

During the 1970s and 1980s, membership rapidly increased (Figure 1.1) which resulted in increased administration and, so, the position of Membership Secretary was established in 1974. Due to increasing international membership, 'National Secretaries' (now Regional Secretaries) were introduced in 1988, to aid with the development and administration of the Society at a regional level. The main responsibilities of the Regional Secretaries were, and still are, to promote and build the society in the region, to liaise and foster links with

Table 1.3. Honorary Presidents, Secretaries and Treasurers of the Society.

Year	Honorary President	Honorary Secretary	Honorary Treasurer
1966	Andrew Fraser, Scotland	Bill Beaton, Scotland	Bryan Howard, Scotland
1967	Roger Ewbank, England	'	'
1968	Jim Swanney, Scotland	'	'
1969	'	'	'
1970	Jim Baillie, Scotland	R. Jones, Scotland	'
1971	L. Hughes, England	Andrew Fraser, Scotland	'
1972	C. McCrae, Scotland	'	'
1973	Jim Milne, Scotland	'	Roger Ewbank, England
1974	George Taylor, England	Ian Duncan, Scotland	'
1975	'	'	'
1976	David Wood-Gush, Scotland	'	'
1977	Ingvar Ekesbo, Sweden	Bill Jackson, England	Barry Hughes, Scotland
1978	Laurence Collery, Eire	'	'
1979	Gerrit van Putten, the Netherlands	'	'
1980	Henrik Simonsen, Denmark	'	'
1981	Roger Ewbank, England	'	'
1982	Marie-France Boissou, France	'	'
1983	Hans Sambraus, Germany	'	Katherine Carson, Scotland
1984	Bob Baldwin, England	'	'
1985	Piet Wiepkema, the Netherlands	'	'
1986	Ian Duncan, Scotland	Alistair Lawrence, Scotland	'
1987	Don Broom, England	'	'
1988	'	'	'
1989	Bo Algers, Sweden	'	'
1990	'	'	'
1991	Herman Wierenga, the Netherlands	Mike Appleby, Scotland	Mike Mendl, England
1992	'	'	'
1993	Harold Gonyou, Canada	'	'
1994	'	'	'
1995	Jan Ladewig, Denmark	Linda Keeling, Sweden	'
1996	'	'	'
1997	Pierre Le Neindre, France	'	Jeremy Marchant-Forde, USA
1998	'	'	'
1999	Joy Mench, USA	Ute Knierim, Germany	'
2000	'	'	'
2001	Linda Keeling, Sweden	'	'
2002	'	'	'
2003	Ruth Newberry, USA	Anna Valros, Finland	'
2004	'	'	Debbie Goodwin, England
2005	Marek Špinka, Czechia	'	'
2006	'	'	'

Table 1.3. Continued.

Year	Honorary President	Honorary Secretary	Honorary Treasurer
2007	Janice Swanson, USA		'
2008	'	'	'
2009	Vicky Sandilands, Scotland	Charlotte Stewart, England	Marie Haskell, Scotland
2010	'	'	'
2011	Anna Valros, Finland	'	'
2012	'	'	'
2013	Jeremy Marchant-Forde, USA	Jean-Loup Rault, Australia	'
2014	'	'	Krista McLennan, England
2015	Birte Nielsen, France	'	'
2016	'	'	'

Table 1.4. Locations of summer meetings[1] and International Congresses and Wood-Gush Lecturers.

Year	Location of scientific meetings	David Wood-Gush Memorial Lecturers
1966	Edinburgh, Scotland	
1967	Edinburgh, Scotland	
1968	Glasgow, Scotland & Bristol, England	
1969	Edinburgh, Scotland & Cambridge, England	
1970	Edinburgh, Scotland	
1971	Cambridge, England & Bristol, England	
1972	Liverpool, England & Dublin, Eire	
1973	London, England	
1974	Ghent, Belgium	
1975	Bristol, England	
1976	Cambridge, England	
1977	Brighton, England	
1978	Skara, Sweden	
1979	Dublin, Eire	
1980	Zeist, the Netherlands	
1981	Edinburgh, Scotland	
1982	Reading, England	
1983	Tours, France	
1984	Kiel, Germany	
1985	Cambridge, England	
1986	Wageningen, the Netherlands	
1987	Tänikon, Switzerland	
1988	Skara, Sweden	
1989	Bristol, England	
1990	Montecatini, Italy	

Table 1.4. Continued.

Year	Location of scientific meetings	David Wood-Gush Memorial Lecturers
1991	Edinburgh, Scotland	
1992	Pittsburg, USA	
1993	Berlin, Germany	
1994	Foulum, Denmark	Ian Duncan, University of Guelph, Canada
1995	Exeter, England	Marian Dawkins, Oxford University, England
1996	Guelph, Canada	David Fraser, University of British Columbia, Canada
1997	Prague, Czechia	Ilan Golani, Tel Aviv University, Israel
1998	Clermont-Ferrand, France	Bernie Rollin, Colorado State University, USA
1999	Lillehammer, Norway	Paul Hemsworth, University of Melbourne, Australia
2000	Florianopolis, Brazil	Jeff Rushen, Agriculture and Agri-Food Canada, Canada
2001	Davis, USA	Temple Grandin, Colorado State University, USA
2002	Egmond aan Zee, the Netherlands	Fred Toates, The Open University, England
2003	Abano Terme, Italy	Jan Ladewig, Royal Veterinary & Agricultural University, Denmark
2004	Helsinki, Finland	Per Jensen, Linköping University, Sweden
2005	Sagamihara, Japan	Shigeru Watanabe, Keio University, Japan
2006	Bristol, England	Alistair Lawrence, Scottish Agricultural College, Scotland
2007	Merida, Mexico	Don Broom, University of Cambridge, England
2008	Dublin, Eire	Francoise Wemelsfelder, Scottish Agricultural College, Scotland
2009	Cairns, Australia	Lesley Rogers, University of New England, Australia
2010	Uppsala, Sweden	Charlotte Hemelrijk, University of Groningen, the Netherlands
2011	Indianapolis, USA	Stephen Porges, University of Illinois, USA
2012	Vienna, Austria	Norbert Sachser, University of Muenster, Germany
2013	Florianopolis, Brazil	Paulo Oliveira, Universidade Estadual de Campinas, Brazil
2014	Vitoria-Gasteiz, Spain	Volker Grimm, Helmholtz Centre for Environ. Research, Germany
2015	Sapporo Hokkaido, Japan	Shusuke Sato, Teikyo University of Science, Japan
2016	Edinburgh, Scotland	Christine Nicol, University of Bristol, England

[1] December meetings in London are not listed.

other groups and organisations in the region, to plan and coordinate regional meetings and to be the conduit between members in the regions and Council. There are currently 11 regions: Australasia/Africa; Benelux; Canada; East Asia; East Central Europe; Latin America; Mediterranean; Nordic; UK/Ireland; USA; and West Central Europe.

The schedule of a summer and winter meeting of the SVE continued, with the 'summer meeting' eventually becoming the annual International Congress. In 1988, with the establishment of the Regions, it was decided that it would be appropriate for Regions other than the UK/Ireland to hold meetings in 'winter', but it is now the situation that regional meetings are held at whatever time of year suits the organisers. The International Congress, however, generally continues to be held in July or August. The locations of the 'summer' meetings and International Congresses are given in Table 1.4.

As indicated earlier, the proceedings of the SVE meetings were initially published in the British Veterinary Journal, but in December 1974 the long relationship between the Society and the publishers Elsevier started with the launch of the journal Applied Animal Ethology under the editorship of Andrew Fraser, with David Wood-Gush as Associate Editor. In March 1984 Applied Animal Ethology was renamed Applied Animal Behaviour Science (AABS). From 1977 on, abstracts of papers presented at SVE meetings were published in Applied Animal Ethology, and later in AABS.

In 1979, the Council of Europe convened a Committee to consider the welfare of farmed animals, and the SVE was invited to send an advisor to the Committee Meetings. The official title was the Council of Europe Standing Committee of the European Convention for the Protection of Animals kept for Farming Purposes. This was a really exciting development and the Society eagerly accepted the invitation. However, there were some initial problems in selecting a representative (it turned out that Government employees were not allowed to participate). Professor Henrik Simonsen of the Royal Veterinary and Agricultural University of Copenhagen was chosen to attend. The first topic to be considered by the Committee in 1980 was the welfare of laying hens. Simonsen quickly contacted various experts in poultry welfare within the Society and they put together a comprehensive review of the topic. Simonsen presented this at a Meeting of the Committee in May 1980 and it was very well received which was a tremendous boost for the Society; suddenly, the Society was playing an important role in a supranational organisation which had powers to change how animals were kept. Also in 1980, Professor Ingvar Ekesbo took over as Chair of the Standing Committee which strengthened ties with the SVE even further. The Society has been sending an Advisor to Council of Europe meetings ever since. Different Members have acted as advisors depending on the topic under discussion but Professor Donald Broom, of the University of Cambridge, has done the lion's share of this work.

1.8 Adolescence and puberty

The 1990s was a period of substantial change for the Society which started with a change of name voted-in at the AGM of the 25th anniversary meeting (1991) in Edinburgh. A name change had been considered and discussed for many years prior to this (the earliest reference found is 1976) because many members felt that the scope of the Society was so much more than the words 'veterinary ethology' implied. Two letters published in the SVE Newsletter 39 (October 1988), both from Founder Members of the Society, reveal the nature of the argument used and depth of feeling associated with a name change. We quote those letters here:

> *At the annual general meeting of the SVE in Skara on 18th June 1988 it was decided to start moves which might eventually change the society into some sort of federal/international organisation.*
>
> *There is much to be said for this initiative. Workers in the subject are scattered across many countries and would benefit greatly from the setting up of an*

information network, from the co-ordination of their various activities and from the holding of regular international meetings.

A plea: whatever the eventual name of the organisation its title should contain the term applied ethology and not the word veterinary. The area of interest of this new organisation should be much wider than the relatively narrow discipline of veterinary ethology and full extent of this interest should be obvious in its name.

The use of the term applied ethology would have the advantage that the old SVE name could be retained for the UK society; the disadvantage is that the SVE title, as such, is recognised officially at the Council of Europe. Attempts to change here might lead to problems.

Change, however, is necessary. The study of the behaviour of animals which man uses and controls has advanced greatly in many parts of the world since SVE was founded in the UK in 1966 – there is now a real need for an international applied ethology organisation.

Yours faithfully,

Roger Ewbank,

Founder Vice President, SVE.

I respect the views provided by Roger Ewbank, my erstwhile co-founder of the SVE, regarding the naming of the society and the realities of changed circumstances. Applied ethology has indeed become an appropriate appellation. It means more than simple ethology turned practical and has its own vigour widely recognised. Much of the impetus to applied ethology's development has come from its contemporary relevance to the animal industries in need of a new objective in relation to husbandry.

The growth of this discipline, to its present state internationally, takes it well out of the limited realm of veterinary matters. Nor does veterinary teaching usually do justice to this body of knowledge; but this is unfortunately also true in many agricultural schools. So our name has remained, almost as an anachronism.

Solutions often lie in compromises. I believe that some of the newer generation of members proposed a name recently which serves this purpose, viz. 'Society of applied and Veterinary Ethology'. It has an appropriate acronym. I would wish to be counted among those who support such a change of name.

The word 'veterinary', I submit, must be retained for many reasons. With it the old and honourable society can still be recognised. In addition, if we are to give ourselves free-rein to look into such essentially veterinary matters as suffering and assessment of well-being in animals, we need it (plus the people who go with it).

Yours sincerely,

Andrew Fraser

The change of name was hotly debated at the AGM in 1991 and the members voted, with a substantial majority, to change the name from the Society for Veterinary Ethology to the International Society for Applied Ethology (ISAE).

The change of name also had consequences for parts of the Constitution which were reviewed and revised accordingly, and this procedure has continued over the last 25 years. The current aims reflect the changes that have occurred within the Society in terms of the interests and qualifications for membership, the methods used to disseminate information, and the role of the Society in education and collaborations with other scientific disciplines. The aims now read:

- to encourage and support basic and applied research into the behaviour of animals as related to the use of animals by humans. This includes domestic, laboratory, zoo, companion, pest and captive animals or managed wild animals;
- to provide an international forum in which scientists can communicate and discuss the results of the above research. This can be achieved by organising or encouraging scientific meetings and by encouraging scientific publications;
- to encourage, where appropriate, links between applied animal behaviour science and other disciplines. This can be realised by encouraging presentations, discussions and publications and by maintaining contacts with appropriate scientific societies;
- to encourage and support the teaching of animal behaviour in research and academic institutions, especially veterinary schools, departments of animal science and animal production, and departments concerned with laboratory or companion animals or captive wildlife;
- to provide a pool of expertise to national governments, international bodies, industry and to those animal welfare organisations which deal with problems involving animal behaviour. To encourage, where possible and appropriate, the assimilation of scientific knowledge so as to facilitate its use in relation to practical problems concerning the way animals are kept and cared for.

Other noticeable changes during this period were related to technological developments in electronics; the ease and speed of communication that email permitted was especially astounding! In 1991 most mass communication was still done by post which was expensive in terms of the costs of paper, envelopes, printing and postage, as well as the time spent 'stuffing envelopes'. A headline in the ISAE Newsletter 6 (March 1994) announced '*Email Network 'Applied Ethology' On Line!*' and under this heading Joe Stookey had written:

> Today a person with an email or bitnet address can gain access to information or communicate with another computer user on opposite sides

> of the globe by typing and receiving messages at a computer terminal. A personal computer attached to a phone modem is equally capable of linking you with the electronic world.

This article explained how an electronic network (bulletin board) had been established, test messages sent on 18 November 1993 and that, at the time of writing, there were 95 subscribers. Nowadays, given how rapidly electronic technology has advanced, it seems incredible that the establishment of an electronic network would be such an enormous step for the Society. Within 2 years of its establishment the bulletin board had over 300 subscribers, but as it was open to both ISAE members and non-members, it became increasingly swamped with largely subjective messages about dog behaviour and this led to the gradual withdrawal of ISAE members from it. Interestingly, this bulletin board still operates and has about 300 subscribers, although there have been no posts for some years. At its initiation, this bulletin board was an exciting way to exchange views and ideas about applied ethology, but with technological changes (e.g. email, online access to articles and search engines) it quickly became superfluous.

Another article in that same Newsletter (6; March 1994), written by Jeff Rushen, also makes fascinating reading because it now seems incredible that the benefits of electronic publishing and communication had to be strongly advocated. Rushen discussed the benefits of distributing an electronic newsletter by email, but at the time only about 25% of ISAE members had email. Indeed, one of us (J.C. Petherick) had recently moved to live and work on a remote cattle station in north Queensland, Australia and wrote a note in response to Rushen's article saying that, where I was located, I was very thankful for electricity, let alone email!

In his article, Rushen also raised the issue of the ISAE publishing its own scientific journal and recommended that the Society seriously consider publishing an electronic journal. The matter of the ISAE having its own journal was raised again at the AGM in 1997 and subsequently rejected, and once more recently, with Council establishing a Journal Assessment Working Group in January 2014. Among the aspects considered by the Group were: the society's influence on AABS; whether it covered the aims of the ISAE; the accessibility of articles, both for ISAE members and target readers; the possibility and cost of Open Access; the administrative work for the ISAE; and the review/editorial load. The outcome from the Group's investigations was a decision for the ISAE to continue the relationship with Elsevier and AABS because of the workload and financial risks associated with producing its own journal.

The publication of abstracts from the SVE meetings in AABS had continued, but the arrangement was informal and occurred in a rather piecemeal way. In 1993 discussions took place with Elsevier and a formal agreement was reached, although the first contract was not signed until 1999. It was agreed that an annual special issue of the journal would incorporate proceedings of the ISAE's International Congress and other selected material. In addition, members of the society would qualify for a personal subscription to the journal at a very favourable rate. This arrangement continues to the present. The first ISAE special issue was from the 28th International Congress held in Foulum, Denmark and was published in 1995. Unfortunately, the production of congress-derived special issues has been only sporadic since then.

In October 1991 Katherine Houpt replaced Andrew Fraser as Editor-in-Chief of AABS and links between the ISAE and the journal were strengthened when, in March 1993, Mike Appleby, who was an ISAE Council member at the time, was appointed Editor to assist Katherine Houpt by handling European submissions. In March 1997 Per Jensen, an ISAE member, took over from Mike Appleby and remains one of the Co-Editors-in-Chief.

The first International Congress to be held outside Europe was in 1992 in Pittsburgh, USA when it was held in conjunction with the American Society of Animal Science (ASAS) conference. It is noteworthy that the President of the ASAS invited an ISAE member, Ian Duncan, to give a plenary lecture to an audience of the combined societies. Since about 2004, an attempt has been made to hold congresses alternately between European and non-European countries, although there is no strict rule on this.

With the increasing number and geographical spread of members (Figure 1.1), improved communication was a focus during this period. The Council position of Communications Officer was established in 1995 to address this. The first ISAE website with its domain name, applied-ethology.org (decided by Council in December 1995), was developed and established by the then Communications Officer, John Eddison, and hosted at the University of Plymouth, UK.

Another noticeable change during this period was an increasing female membership and women taking on Council roles. In the period from 1966 to 1998 there had been just one female president, Marie-France Bouissou, in 1982 (Table 1.3). In 1995, Linda Keeling became the first female Secretary, but the society did not get its second female President until Joy Mench in 1999.

It was also in this period that the death of David Wood-Gush occurred. He died of a heart attack while checking into a hotel prior to the 1992 UK regional winter meeting in London. We recall the occasion vividly, the terrible shock that his death was to so many, and the decision to proceed with the meeting being made, as it was felt that David would have wished it. Not long after his death, the David Wood-Gush Memorial Fund was established to help support guest speakers to present at the International Congress. David's long-time friend and colleague, Ian Duncan, was the first Memorial Lecturer in 1994 (Table 1.4). Over the years, the lectures have been delivered by both ISAE and non-ISAE members, the latter being selected to provide different perspectives of ethology and stimulate ISAE members to think how they could apply different concepts and techniques to their own work. The lectures remain a highlight of the International Congress and the Society hopes that these lectures will help preserve David's memory and the work to which he dedicated so many years.

1.9 Adulthood

A major focus of the last 15 years has been improving the transparency and efficiency of operation of the Society. In 2000 Council determined that there was a need to document the operations of the Society to enable transparency, continuity, and consistency in decision-making and operations, and to obtain input from ISAE members. This was done through the co-opting of 'Parliamentarians' who developed the society's Procedural Guidelines. The inaugural Parliamentarians (later renamed Procedural Advisors), Julie Morrow, Birte Nielsen and Bill Jackson, did the hard work of drafting the first version of the Procedural Guidelines and these are now revised each September, by the Procedural Advisor, in-line with the outcomes of discussions and decisions made at Council meetings and the AGM.

Also in 2000, the duties and responsibilities of the Regional Secretaries were published in the ISAE Newsletter. Regions were encouraged to establish their own webpage with links to the ISAE site and, in 2008, there was implementation of regional webpages hosted on the ISAE home page. Benelux, Canada and UK/Ireland were the first regions to do this. Regional activity has continued to increase over the years and, in 2014, nine of the 11 regions had their own ISAE regional meetings or meetings combined with those of other societies, and the two regions that did not were those that held the annual congresses in 2013 and 2014. Joint ISAE meetings between Canada and the USA, and between East and West Central Europe are quite common.

Keeping track of members and ensuring subscription payment has been a difficult and onerous task for both the Membership Secretary and the Treasurer. Discussions about the burgeoning workload and a possible solution started in 1999 and these led to what turned out to be a major and, at times, harrowing activity that ran for about 10 years. In 2002 a decision was made to raise the membership subscription (with reduced fees for students) to cover the cost of employing a professional company to assist with the development and administration of an on-line subscription payment system. Unfortunately, a series of companies failed to fulfil contractual agreements and in December 2009, the Society was back in the situation where the work was being performed by the Membership Secretary and the Treasurer. Keelin O'Driscoll was instrumental in a major change to the ISAE which probably turned round the Society's financial situation and has subsequently allowed the Society to invest more in other operational aspects. O'Driscoll researched website design companies and what they could offer and presented the options to Council. A selection was made and O'Driscoll began working with them on designing a new website that integrated the membership database and allowed subscriptions (for membership and AABS) to be paid online, thus bringing the ISAE into the 21st century. The new website was officially launched in 2011. Inevitably there were some teething troubles, but by 2012 the major issues had been resolved, thus drastically reducing the workload of the Membership Secretary and Treasurer. Importantly too, for the first time in some years the Society showed a reasonable profit, which allowed more money to be allocated to supporting student attendance at the International Congresses.

Communication continued to be a focus in this period. The ISAE website was given a major overhaul by the then Communications Officer, Derek Haley in 2003. The website has thus become the main way that information about the Society is made available to both members and non-members, although information and Newsletters are also emailed direct to members. Communication between Council members has also become more efficient with the holding of electronic Council meetings. These allow for greater participation of Council members in a second scheduled Council meeting per annum, and also make it possible to hold discussions and make decisions on any urgent matters that arise. Initially (in the early 2000s) these electronic meetings were conducted by e-mail, which proved to be *'a pain to handle'* (as noted by a previous Secretary who shall remain anonymous!) but then Derek introduced a messaging-board system on which the various agenda items have different threads and Council members add comments pertaining to a particular agenda item within the discussion thread. This system continues to be used today.

Since 1999, of the nine Presidents, seven have been women and, since the first woman Secretary (in 1995), three of the four have been women (Table 1.3). This bias is perhaps unsurprising given that about 70% of the current ISAE membership is women, which is a dramatic contrast with the early years of the Society which were dominated by men. We do not know who was the first woman to join the Society and when this happened, but Marthe Kiley-Worthington, who published a book on farm animal behavioural problems in 1977 was one of the first to join. Moyra Williams, Deborah Howell and Marie-France Bouissou were also amongst the first women to join the Society. When one of us (J.C. Petherick) joined in the early 1980s, it was conspicuous that the membership was predominantly male and the increasing numbers of women has been particularly noticeable in just the last 15 to 20 years. It is important, however, that the Society continues to attract male members and, consequently, in recent years men have been approached to take on Council roles in order to raise the profile of men in the Society. In this same spirit of improving inclusivity, it was recognised that students, although constituting almost 20% of the membership, had no direct link to Council. In early 2013 student representatives were co-opted on Council giving students more of a voice in the operation of the Society. The establishment of the Congress Attendance Fund in 2003 was an important initiative that allows students who may not otherwise be able to afford it, to attend the annual congress, thereby further enhancing student participation in the Society.

For many years, a contentious issue has been, and continues to be, the determination of which submissions are accepted and which are presented orally and which as posters at the International Congress. It has been agreed that only abstracts demonstrating high quality science should be accepted and that Congress organisers should try to find innovative ways, such as beverages and food being served during poster-viewing sessions or some posters being also presented as a 1-minute oral presentation, to give posters a high profile, as they are often perceived as being valued less than oral presentations.

In 2002, Council decided to take action to meet the increasing number of requests for guidance on ethical matters from reviewers of abstracts for the International Congress.

There had also been the occasional appearance of an ethically dubious paper/poster at the Congress. After some discussion, it was decided to appoint an Ethics Officer who would chair an Ethics Committee to handle ethical questions. The duties of this Committee were to draw up ethical guidelines to help ISAE members make, what are sometimes, difficult decisions about the procedures involved in their studies. In addition, the Committee would examine abstracts for the International Congress that had been flagged by reviewers for ethical reasons.

The Ethics Committee published ethical guidelines (Sherwin *et al.*, 2003) and has reviewed abstracts for the annual congress since that time. Forming an Ethics Committee and drawing up ethical guidelines seem to have had the desired effect. Very few abstracts have been rejected on ethical grounds. Moreover, fewer abstracts seem to be flagged by reviewers each year. Simply getting scientists to think about the ethical implications of their research has had the effect of reducing the number of abstracts raising ethical concerns. Of the abstracts that are flagged on ethical grounds, the authors are often able to solve the problem by giving a little more explanation or justification in the abstract.

In 2007, Council made the decision to develop a web-based submission and review system for Congress abstracts, and looked for a professional provider of this service, including the production of proceedings. Negotiations were started with Wageningen Academic Publishers and the system was tested for the International Congress in Dublin in 2008. The process was refined and used for the Congress in Uppsala in 2010, and this arrangement has continued to the present time.

The ISAE continues to raise the profile of both the Society and applied ethology globally and, increasingly, the ISAE is approached to provide expertise or representatives for various forums. Currently ISAE has official representation on: the Council of Europe (CoE); the American Society for Animal Behavior (ASAB) Certification scheme for Animal Behaviorists; the Board of Trustees of the Association for the Assessment and Accreditation of Laboratory Animal Care International (AAALAC); and the Advisory Board of the Animal Welfare Judging and Assessment Contest (North America).

The activities of the CoE have reduced in recent years and it seems likely that ISAE representation on the CoE may not be required in the future (see also Chapter 14 by Donald Broom), but the Society continues to be involved in other EU-based, animal welfare-related activities. AWIN (Animal Welfare Indicators) was an EU-funded initiative that developed, integrated and disseminated information about animal welfare indicators, particularly those relating to pain. ISAE representatives attended AWIN conferences, but the project has recently come to an end. The AWARE (Animal Welfare Research in an Enlarged Europe) project, which ran from 2011 to 2014 was led by long-term ISAE member Marek Špinka and aimed to promote integration and increase the impact of farm animal welfare research. The joint ISAE East and West Central Europe regional meeting in 2013 was organised in cooperation with the AWARE project. ISAE was also represented in the EUWelnet project which aimed to develop a model for an EU welfare centre, or network of centres. The hope is that the report detailing the various scenarios will be used by the EU when, in the next few years, an animal welfare centre is established.

The society continues to strengthen its relationship with AABS. It was in 2000 that the logo of the ISAE first appeared on the cover of AABS and this really marked the recognition of AABS being the official journal of the ISAE. Also in 2000, Harold Gonyou took over from Katherine Houpt as Co-Editor-in-Chief and served 6 years (to 2005). Then Carol Petherick took over and served from 2006 to 2014. Sylvie Cloutier is the current Co-Editor-in-Chief working with Per Jensen. AABS launched its current cover on the first issue of 2008. This cover replaced the 'red graphics' one (which was on 107 volumes!) and makes it clearer that the journal is about domesticated animals.

1.10 Maturity or mortality: what does the future hold?

Whilst the future of applied ethology and the ISAE is discussed further in Part 4 of this book, we thought it appropriate to present here a few ideas of some of the opportunities and challenges facing the ISAE.

There is no doubt that the change of name of the Society from SVE to ISAE better reflected the interests of the membership at the time. However, 25 years later, there is a nagging suspicion that something very important was lost with the name change. The Society was founded by veterinarians, including practising veterinarians, who were concerned by the lack of knowledge in the field of farm animal behaviour identified by the Brambell Report (Command Paper 2836, 1965). With regard to the early veterinarian members, those involved in research tried to fill the gaps in our knowledge, those involved in teaching passed on this newly acquired knowledge to future generations of veterinarians and, most importantly, those in practice put into effect immediately the findings from the research. There is a distinct danger that, if the Society loses its connection with the veterinary profession, this chain might be broken. The Society's change of name probably had little effect on the recruitment of veterinarians who were researching and teaching in this field, and also had the very beneficial effect of attracting interested non-veterinarians. However, why would a practising veterinarian even think about joining the ISAE? It would be interesting to investigate this. How many practising veterinarians have joined the Society in the last 25 years? We suspect the number is very small. And yet practising veterinarians could add so much value to the Society. They could report on new, emerging animal behaviour problems as well as giving feedback on the efficacy of theoretical solutions to problems being suggested by the researchers. We would encourage the Society to think of ways of attracting practising veterinarians to become members (without changing the name of the Society again!).

The ISAE is a diverse society, particularly in relation to the areas of research and the species studied within our membership. This enhances the Society, but also presents challenges; in the last 50 years we have seen the emergence and/or strengthening of other organisations and groups that deal with specific aspects in which the ISAE has held interests, e.g. the human-companion animal bond; zoo animal behaviour; laboratory animal management; and behavioural problems in companion animals. Has this, or will this result in a movement of the membership to those specific organisations, resulting in a dilution or loss of ISAE expertise in some areas? Perhaps the ISAE should consider occasionally holding joint meetings with some of these organisations to forge stronger links and encourage collaboration between members. This is also applicable to fundamental ethology where, as alluded to in the Searching section above, there has been a 'disconnect' since the 1940s, at least in the UK. The organisers of the ISAE International Congress in 2006, in Bristol, UK, attempted to draw together fundamental and applied ethology in the programme, but to our knowledge, there has been no attempt by the ISAE to formally encourage closer ties between the fields, although some individual researchers do collaborate.

Developing countries contain, by far, the majority of the world's animals and there are many issues, such as those relating to animal welfare, conservation, tourism, and wild and captive animal management that could be addressed, at least partially, by applied ethology. ISAE membership in these areas is, however, minimal (Figure 1.1). Holding an International Congress in a developing country would help to raise the profile of applied ethology (and the ISAE), increase knowledge and encourage collaboration in applied ethology between those in developing and developed countries. At the AGM in 2014 a decision was made to explore the potential for holding a Congress in a developing country and it seems likely that, in the not too distant future, a Congress will be held in India, in partnership with a UK university. This would, hopefully, be just the start of a continuing process by the ISAE to help developing countries with some of their animal-related issues.

Greater involvement with the OIE is another method by which the ISAE could assist with improving animal welfare globally, but progress in achieving a formal relationship has been slow. ISAE initially made contact with the OIE in 2009 in the hope of setting-up a Memorandum of Understanding, but the OIE has continued to resist this 'formal' approach for providing experts to the ad hoc groups instead inviting individuals through channels other than ISAE. Council feels, however, that there are significant potential benefits to both organisations from a formal recognition of cooperation and plans to continue lobbying.

1.11 Conclusions

We are optimistic about the future of both applied ethology and the ISAE; attendance at ethological meetings and conferences reveal the passion that researchers have for their work and the potential that applied ethological research has for resolving problems. It is also extremely pleasing to see so many students and young scientists, who will hopefully continue a long-term career in applied ethology. Of course, there is a great deal of work to be done, for example, in education and mentoring to achieve high quality research, making the results of that research widely available and extending the message globally on the contribution that applied ethology can make to resolving animal-related issues. We are confident that the ISAE will continue making significant contributions to this work for many more decades to come.

Acknowledgments

We thank the Founder Members, and past and present members of the Society who provided assistance with documents and data used for the preparation of this history.

Disclaimer

The authors have done their best to ensure the accuracy of the information presented here by reviewing written records (Minutes of meetings and Newsletters) rather than relying on memories which, as we all know, can be notoriously unreliable! We take no responsibility for errors in those records, but apologise if we have got it wrong as a consequence of those errors.

References

Broom, D.M., 2016. World impact of ISAE: past and future. In: Brown, J.A., Seddon, Y.M. and Appleby, M.C. (eds.) Animals and us – 50 years and more of applied ethology. Wageningen Academic Publishers, Wageningen, the Netherlands, pp. 269-278.

Brownlee, A., 1939. Playfulness in animals with particular reference to cattle. Veterinary Journal 95: 286-290.

Clutton-Brock, J., 1981. Domesticated animals from early times. University of Texas Press, Austin, TX, USA.

Command Paper 2836, 1965. Report of the technical committee to enquire into the welfare of animals kept under intensive livestock husbandry systems. Her Majesty's Stationery Office, London, UK.

Curtis, G., 2006. The cave painters: probing the mystery of the world's first artists. Knopf, New York, NY, USA.

Darwin, C., 1859. The origin of species by natural selection. John Murray, London, UK.

Darwin, C., 1872. The expression of the emotions in man and animals. John Murray, London, UK.

Durant, J.R., 1986. The making of ethology: the association for the study of animal behaviour, 1936-1986. Animal Behaviour 34: 1601-1606.

Fraser, A.F., 1974. Farm animal behaviour. Baillière Tindall, London, UK.

Gray, P.H., 1967. Spalding and his influence on research in developmental behavior. Journal of the History of the Behavioral Sciences 4: 365-376.

Harrison, R., 1964. Animal machines. Ballantine Books, New York, NY, USA.

Kiley-Worthington, M., 1977. Behavioural problems of farm animals. Oriel Press, Oxford, UK.

Ratner, A.C. and Boice, R., 1975. Effects of domestication on behaviour. In: Hafez, E.S.E. (ed.) The behaviour of domestic animals. Baillière Tindall, London, UK, pp. 3-19.

Romanes, G.J., 1884. Mental evolution in animals. AMS Press, New York, NY, USA.

Schjelderup-Ebbe, T., 1922. Beiträge zur sozialpsychologie des haushuhns. Zeitschrift für Psychologie 88: 225-252.

Sherwin, C.M., Christiansen, S.B., Duncan, I.J., Erhard, H.W., Lay, D.C., Mench, J.A., O'Connor, C.E. and Petherick, J.C., 2003. Guidelines for the ethical use of animals in applied ethology studies. Applied Animal Behaviour Science 81: 291-305.

Spalding, D.A., 1872. On instinct. Nature 6: 485-486.

Sparks, J., 1982. The discovery of animal behaviour. Collins, Glasgow, UK.
Wood-Gush, D.G.M., 1991. The development of applied ethology in relation to certain agricultural animals over a century. In: Appleby, M.C., Horrell, R.I., Petherick, J.C. and Rutter, S.M. (eds.) Applied animal behaviour: past, present and future. Universities Federation for Animal Welfare, Potters Bar, UK.
Youatt, W., 1839. The obligation and extent of humanity to brutes, principally considered with reference to the domesticated animals. Republished in 2004. Edited, introduced and annotated by Rod Preece. Edwin Mellen Press, Lewiston, NY, USA.

2. Pioneers of applied ethology

R.C. Newberry[1] and V. Sandilands[2]*
[1]Department of Animal and Aquacultural Sciences, Norwegian University of Life Sciences, P.O. Box 5003, 1432 Ås, Norway; ruth.newberry@nmbu.no
[2] Scotland's Rural College, SRUC Auchincruive Estate, Ayr KA6 5HW, United Kingdom

Abstract

This chapter examines the environment in which the field of applied ethology emerged. Through brief vignettes about the lives and contributions of prominent scientists and other influential figures, we provide a historical context for today's celebration of the 50th anniversary of the International Society for Applied Ethology (ISAE, founded in Edinburgh in 1966 as the Society for Veterinary Ethology). From seeds sown by Darwin, Whitman and Heinroth in the 1800s, who set the stage for the emergence of ethology, we outline the Nobel Prize winning contributions of von Frisch, Lorenz and Tinbergen, who defined ethology as a respected discipline. We trace the separate trajectory of psychological research on learned behaviour through the works of Pavlov, Watson, Thorndike and Skinner, and observe the blending of ethology and psychology in studies by Bowlby, Ainsworth, Griffin and Panksepp. From this foundation, we then focus on pioneering research in applied ethology, as exemplified by the careers of David Wood-Gush, Donald Broom, Marian Dawkins and Temple Grandin. We briefly highlight the contributions of a selection of other prominent applied ethologists active in the early years of the Society, noting applications of their research to animal husbandry, veterinary medicine and, increasingly, animal welfare. We also discuss the roles played by non-ethologists such as Astrid Lindgren, Ruth Harrison, Peter Singer and Russell and Burch, whose works contributed to the broader cultural milieu within which pioneering applied ethology research was undertaken. We end with a summary of contributions by Hediger, J.P. Scott, the Brelands, E.O. Wilson, Goodall, and P. Bateson who, although not directly associated with the ISAE, have served as sources of inspiration for applied ethologists. As we look forward to many exciting new developments in our discipline, this chapter serves as a reminder of some of the pioneers who have paved our way.

Keywords: history of science, biography, animal behaviour, animal welfare

2.1 Introduction

It is appropriate that the International Society for Applied Ethology (ISAE) celebrates its 50th anniversary with the theme 'Standing on the Shoulders of Giants'. Who are these giants? In this chapter, we think back to the origins of applied ethology, and the pioneering scientists whose works have defined our field. From roots in the study of medicine, philosophy and natural history, applied ethology has developed as an offshoot of basic ethology on the one hand and psychology on the other hand, combined with

J.A. Brown, Y.M. Seddon and M.C. Appleby (eds.) ***Animals and us: 50 years and more of applied ethology***
DOI 10.3920/978-90-8686-828-5_2, © Wageningen Academic Publishers 2016

the practical sciences of animal husbandry and veterinary medicine. The modern field represents a mingling of these traditions, evolving as a dynamic ecosystem with a rich diversity of scientific expertise and interrelationships with other fields (Figure 2.1). The ISAE provides a focal point for fostering interactions among individuals from many different backgrounds, facilitating applications of ethology to challenges in animal production, animal welfare, biomedicine, wildlife conservation and human-animal interaction to name a few. Here, we focus on selected influential figures, recognising that our list is by no means comprehensive and that the entire ecosystem has developed through the collective contributions of a far wider network of personalities.

Figure 2.1. Conceptual applied ethology landscape. Sketch by Fritha Langford.

2.2 Seeds

We start our journey with **Charles Darwin** (1809-1882), whose theory of evolution provides the foundation for all the modern life sciences. After two years of studying medicine at the University of Edinburgh, he dropped out to pursue his fascination with natural history, leading to his participation in the voyage of 'The Beagle' and subsequent fame (and infamy) for his controversial theory. In applied ethology, his meticulous observations of domestic animals are particularly relevant to our understanding of domestication as an evolutionary process. Thanks to him, our observations of 'man-made' animals in their 'artificial' environments are on a firm biological footing! Studies of behavioural adaptations to anthropogenic environments are even becoming fashionable as awareness increases about the influence of humans in rapidly re-shaping the natural world. Let us also remember that, in true pioneering form, Darwin was writing about animal emotions long before our current preoccupation with this topic. Treat yourself to '*The Expression of the Emotions in Man and Animals*' – chances are that, in 1872, Darwin said something relevant to your latest discovery.

Although not as well-remembered as Darwin, zoologist **Charles Otis Whitman** (1842-1910) also laid seeds for the emergence of ethology as a field of study. His specialist areas included embryology, morphology and behaviour (Schuermann, 2015). He bred hundreds of types of pigeon (though failing in his attempts to breed the last remaining passenger pigeons), and examined associations between hereditary and phenotypic variation, noting examples of innate and learned behaviour (Burkhardt, 2005). He attributed his findings to orthogenetics, a now-refuted non-Darwinian hypothesis for evolution, whereby change was thought to be governed by an internal force within the organism leading towards a fixed goal. His book, '*Orthogenic Evolution in Pigeons*' (1919), published nine years after his death, is considered one of the first extensive studies in comparative ethology.

Oskar Heinroth (1871-1945) is also considered a founding pioneer in ethology (Podos, 1994). He and his wife were known for hand-rearing 286 different species of birds (about 1,000 individuals in total) in their apartment so that they could observe their development in detail (Schulze-Hagen and Birkhead, 2015). Director of the Berlin aquarium (which was destroyed in an air raid shortly before his death), he was a mentor to Konrad Lorenz.

2.3 Roots

2.3.1 Ethology

We see the full-fledged recognition of ethology as a field of study with the awarding of the 1973 Nobel Prize in Physiology or Medicine to Karl von Frisch, Konrad Lorenz and Nikolaas Tinbergen 'for their discoveries concerning organisation and elicitation of individual and social behaviour patterns' (Physiology or Medicine, 1973). These three pioneers are considered the 'fathers' of modern ethology.

Karl von Frisch (1886-1982) first studied medicine but then turned to zoology, obtaining his doctorate in 1910 (Nobel Lectures, 1992). His speciality was in assessing the acuity of animal senses, initially in fish but more famously in the honeybee. He demonstrated that bees not only distinguish between tastes, smells and colours, but also navigate using the sun and communicate with one another about food sources using dances (Von Frisch, 1973). If you are ever having a hard time convincing colleagues of the veracity of your hypothesis, take heart in the fact that von Frisch's famous waggle dance theory was considered unbelievable initially, though the evidence continues to provide corroboration (e.g. Riley *et al.*, 2005).

Konrad Lorenz (1903-1989) was first diagnosed as being a tumour in his 42-year-old mother's uterus but eventually turned out to be a healthy baby boy (Krebs and Sjölander, 1992). Like von Frisch, he turned from medicine to animals, gaining a doctorate in zoology in 1933. Lorenz inherited Heinroth's particular interest in ducks and geese, and penchant for closely observing hand-reared birds. He examined and described (often for the first time) how behaviour patterns were influenced by evolution (having apparently read Darwin's theory of evolution when he was just 10 years old). Indeed, his use of behavioural observation helped to re-establish this technique as a respectable scientific method. He is particularly remembered for his work on imprinting, a phenomenon originally described by Douglas Spalding (1841-1877) in chickens and re-discovered by Heinroth. Although many of Lorenz's ideas have not withstood the test of time, he was nevertheless a charismatic champion of ethology. His stories about animals in '*King Solomon's Ring*' (1952) and '*Man Meets Dog*' (1954) continue to delight readers of all ages.

Nikolaas (Niko) Tinbergen (1907-1988) was influenced to study animal behaviour by fellow bird lover, Jan Verwey (1899-1981), and by the works of von Frisch. In his biographical description for the Nobel Prize, he said that 'at both secondary school and at University, I only just scraped through, with as little effort as I judged possible without failing' (Tinbergen, 1973). This goes to show that not every famous ethologist is a born academic. Tinbergen, like Lorenz, had a keen fondness for nature. He first met Lorenz in 1936 and they instantly hit it off, corresponding for many years. Tinbergen took up a post at the University of Oxford, established the journal 'Behaviour' and explored links between animal behaviour and neurophysiology, psychology and human behaviour. Notable contributions include '*The Study of Instinct*' (1951), describing elegant experiments conducted on wild animals (especially gulls) under natural conditions, and his 1963 paper, '*On aims and methods of Ethology*', in which he lays out what have become known as Tinbergen's Four Questions. These are the four levels of inquiry at which behaviour can be explained for a comprehensive understanding of the biology of behaviour (i.e. causation, survival value and evolution, already discussed by Julian Huxley (1887-1975), to which he added ontogeny). Students of ethology take heed: sooner or later, you will be questioned about the 'four questions'. Tinbergen was the DPhil supervisor of distinguished ethologist Aubrey Manning, OBE (1930-, Emeritus Professor of Natural History at the University of Edinburgh).

2.3.2 Psychology

Whereas the European ethology tradition emphasised instinctual constraints on behaviour through careful observation of the behaviour of unrestrained animals, psychology contributed to our understanding of the role of environmental effects through experimental studies on mechanisms of learning.

A foundation for this work was established through an important discovery by **Ivan Pavlov** (1849-1936). After abandoning a religious career, Pavlov gained a doctorate in physiology in 1883. Although it was in physiology that he truly made his mark, he stumbled across a key theory of learning – that of classical conditioning. In studying how eating stimulated salivary and gastric secretions in dogs, he noticed that the dogs began to salivate at the mere appearance of a staff member in a lab coat. This lead to the realisation that experience had taught the dogs to expect food upon arrival of the staff member. Pavlov received the Nobel Prize for Physiology or Medicine in 1904 for his work on the physiology of digestion, just one of his many accolades, others including the Order of the Legion of Honour from France in 1915 (Nobel Lectures, 1967).

Building upon the findings of Pavlov, **John B. Watson** (1878-1958) developed the influential 'radical behaviorism' school of thinking in psychology, which emphasised the role of nurture (environment) on behaviour to the exclusion of nature (genetics). He advocated that psychologists should focus on the objective analysis of observable behavioural actions and avoid subjective introspection about the mind and consciousness, which he considered unscientific (a view also held by Tinbergen). From a troubled youth, Watson went on to attain a PhD in 1903 and was appointed chair of psychology at John Hopkins University at the age of 30, remaining until asked to step down due to an extramarital affair with his graduate student whom he later married (Watson, 1999). He saw no dividing line between 'man and brute' and readily applied his ideas to child rearing, including that of his own children who were treated as young adults and given limited affection for fear that it would spoil them. His Little Albert experiment, demonstrating classical conditioning of an 11-month-old boy to fear a white rat, is well-known though criticised (among other reasons) because he failed to subsequently desensitise the child, a process that he demonstrated instead with another boy, Peter.

Edward Thorndike (1874-1949) a contemporary of Watson, complemented understanding of classical conditioning through his investigations on instrumental learning. Gaining his PhD from Columbia University in 1898, his studies of cats in puzzle boxes are well known to those who study behaviour (Reinemeyer, 1999). He demonstrated that, whereas initially an accidental behaviour (e.g. pulling a string) resulted in a desired effect (opening a door), the response occurred more quickly in subsequent tests. This observation lead to his Law of Effect (responses producing a desirable effect are more likely, and those producing an undesirable effect less likely, to occur again) as elaborated in his book '*Animal Intelligence*' (1911).

Thorndike's puzzle box experiments intrigued **B.F. (Burrhus Frederic) Skinner** (1904-1990), who is known for his theory of operant conditioning (McLeod, 2015). Skinner used boxes in which an animal learned to change its behaviour through reinforcement provided after a desired response. Thus, his training boxes became known as Skinner boxes. In contrast to Pavlov, Skinner demonstrated that behaviour was affected by events following the behaviour. An opponent of punishment techniques, he introduced shaping to develop behaviours in animals using positive reinforcement, and determined how varying schedules of reinforcement could affect learning. In 1929, Skinner attended a lecture by Pavlov at Harvard University and obtained a signed photo of him which held pride of place in Skinner's office (Specter, 2014). Skinner's procedures are used not only in animal training but applied to humans as well.

2.3.3 Bridging ethology and psychology

John Bowlby's (1907-1990) theory of attachment is increasingly cited by applied ethologists. Trained in psychiatry, psychology and psychoanalysis, his research was influenced by his own childhood experience of having an affectionate nursemaid who left when he was 4 years old, seeing his mother for only a short period each day, prolonged absences of his surgeon father during World War I and being sent off to boarding school at the age of 7. His clinical observations on effects of separation of young children from their parents during World War II contributed to his growing realisation about the importance of a warm, nurturing relationship with a primary caregiver during early development. He rejected the predominant 'cupboard love' view of attachment in psychoanalysis, being inspired by the ethological principles set out by Tinbergen and Lorenz and, in turn, inspiring ethologists such as Robert Hinde, CBE (1923-, Emeritus Professor, University of Cambridge) to investigate attachment in animals.

Bowlby's protégé, **Mary Ainsworth** (1913-1999), obtained a PhD in psychology at the University of Toronto in 1939. After moving to London so her husband could complete his PhD, she joined Bowlby's research team, subsequently conducting studies on mother-child attachment and weaning practices in Uganda and pursuing a successful academic career in the USA. She is best known for the Strange Situation test, a standardised method for assessing the nature of the bond between mother and child that has, more recently, been adapted for use in dog research (e.g. Prato-Previde *et al.*, 2003).

Donald Griffin (1915-2003) pioneered the field of cognitive ethology. Influenced by his uncle, a Harvard professor of biology, home-schooled Griffin published his first paper (on bat migration and homing) as a Harvard freshman in 1934 (Gross, 2005). After making significant contributions on bird navigation and echolocation in bats, and repeating some of von Frisch's experiments with bees (satisfying himself as to von Frisch's conclusions and helping to disseminate von Frisch's work to the English-speaking world), Griffin turned his attention to questions about animal thinking and consciousness. The prevailing behaviourist attitude in the USA made these topics taboo, with scientists daring to mention them being labelled anthropomorphic and promptly marginalised. However, Griffin's reputation as a rigorous scientist, and his elegant arguments and examples in books such as '*The Question of Animal Awareness*' (1976) and '*Animal Thinking*' (1984), opened the door to discussion, leading to legitimisation of this topic as a serious field of study. He was elected to the American National Academy of Sciences in 1960 and established the Institute for Research in Animal Behaviour at Rockefeller University in 1965, which became a leading centre for behavioural research.

Jaak Panksepp (1943-), author of '*Affective Neuroscience: The Foundations of Human and Animal Emotions*' (1998), has become a beacon for applied ethologists interested in the emotions of animals. Jaak grew up in the USA after his family fled Estonia following Soviet occupation at the end of World War II. As an undergraduate, his night job as an orderly in a psychiatric ward left a profound impression, stimulating a lifelong passion to find cures for mental disorders. Resisting academic brainwashing to toe the prevailing behaviourist line, he was inspired by Darwin and the ethological perspective to consider feelings as an evolutionary adaptation. He discovered that rats would press a lever to self-stimulate brain areas generating behaviour related to seeking and avoid stimulating areas generating behavioural expressions of social aggression and fear, supporting his position that the former reflected good feelings and the latter, bad feelings. His discovery, with graduate student Jeffrey Burgdorf, that play-mimicking tickling of rats induced 50 kHz ultrasonic trills, has led to the use of these popularly dubbed 'laughter' vocalisations as an indicator of positive welfare. His research has contributed an understanding of the role of the sub-cortical brain in generating emotional behaviour and new approaches to the treatment of depression and attention deficit hyperactivity disorder. He was awarded The Order of the White Star from the Government of Estonia in 2011.

2.4 The emerging and evolving field of applied ethology

2.4.1 Drivers for applied ethology research

Post World War II, both the booming economy and birth rate led to concerns about the world food supply, and efforts to apply growing knowledge about genetics, physiology and animal development to animal production. Thus, early studies in applied ethology typically focussed on understanding behaviour to improve animal production. This effort stimulated studies in areas such as sexual behaviour, mother-young bonding, methods of animal handling and housing, competition and aggression, and behavioural genetics. Many of these applied studies were conducted by researchers based at agricultural and veterinary institutions, often without a traditional background in ethology or psychology. For example, reproductive biologist Elsayed S.E. Hafez (1922-) became interested in behaviour, leading him to produce three editions of the useful book, '*The Behaviour of Domestic Animals*' (1975).

As animal husbandry became more industrialised, the consequences of extreme selection for productivity and close confinement of animals in ecologically 'barren' environments became apparent. Adverse effects on animal health prompted the increased involvement of veterinarians in applied ethology research. Whereas many ethologists flocked towards studies of the behavioural ecology of wild animals, with its applications to wildlife management and conservation, some used domestic animals in their research due to the opportunities presented for more controlled experiments, becoming drawn into welfare issues of captive rearing. Thus, the earlier emphasis of applied ethologists on animal productivity shifted towards a greater focus on animal welfare. Although applied ethology remained central to the study of animal welfare, the latter was a multidisciplinary venture, involving cross-fertilisation with researchers having expertise in fields such as stress physiology, comparative psychology, neurobiology, engineering and ethics. These changes are reflected in the evolving careers of applied ethologists over the past 50 years.

2.4.2 Four tall trees in the field of applied ethology

Here we highlight four 'household names' of applied ethology, gender-balanced and in chronological order, whose careers have spanned the 'pioneering period'.

David G.M. Wood Gush (1922-1992, Figure 2.2) trained in botany at the University of the Witwatersrand before moving to Scotland to pursue studies in genetics at the University of Edinburgh (Sandilands, 2004). After completing his PhD on gene flow in wood lice in 1952, he catapulted into an appointment as the first Head of Ethology at the Poultry Research Centre outside of Edinburgh (now the Roslin Institute). Strongly influenced by both Lorenz and Tinbergen, David's early research years focussed on causation of behaviour in domestic fowl. In 1970, he became the first non-veterinarian elected to the Society for Veterinary Ethology (now ISAE, see Chapter 1; Petherick and Duncan, 2016), serving as Chairman of the Society in 1976. In 1978, David moved to the University of Edinburgh's School of Agriculture and he and his students branched out to study the behaviour and welfare of domestic pigs, sheep, cattle and horses in

Figure 2.2. David Wood-Gush, during a visit to the Swedish Pig Park. Photo by Claus Nielsen.

addition to chickens. One of his methods of studying domestic animals was to release them into wild or semi-natural environments and study how they behaved when freed from confinement. This approach was used to study hens released on Holy Island off the west coast of Scotland and pigs in the Edinburgh Pig Park. He also developed a research group on pain in domestic animals, particularly related to amputation (beaks, horns), influenced by his own wartime experience of limb amputation and its consequences. David supervised and worked alongside many PhD students and colleagues, attracting visiting scientists from all over the world (among them, Shusuke Sato from Japan, L.C. Hsia from Taiwan, and Justin Lynch from Australia) and inspiring a future generation of applied ethologists. He published in a wide range of journals (e.g. Lake and Wood-Gush, 1956) and produced three books ('*Behaviour of the Domestic Fowl*', 1971; '*Elements of Ethology*', 1983 and, with Pat Monaghan, '*Managing the Behaviour of Animals*', 1991). In 1990, David and colleagues established an MSc programme in Applied Animal Behaviour and Animal Welfare, which helps to continue the tradition of Edinburgh as a Mecca for our field. Active to the end, David died suddenly in 1992 after arriving in London to attend the ISAE Winter Meeting.

Donald Broom (1942-, Figure 2.3A) was a student of the eminent professor of ethology, W.H. Thorpe (1902-1986) at the University of Cambridge, where he was awarded a 'Blue' for his national ranking in modern pentathlon. After rising to the position of Reader at the University of Reading, in 1986 he became the first Professor of Animal Welfare in the world, in the Department of Clinical Veterinary Medicine at Cambridge. From this platform, he has played a major role in defining and cultivating animal welfare as a scientific discipline rooted in applied ethology. With initial studies on responses of domestic chicks to startling stimuli, his research diversified to topics ranging from effects

Figure 2.3. (A) Don Broom, world's first professor of animal welfare; (B) Marian Dawkins, a champion of the 'animal's point of view.'

of housing on calf and dry sow welfare to the role of behaviour in the transmission of fish parasites and bovine tuberculosis. He has held many key positions including serving as scientific advisor to the Council of Europe Standing Committee on the Welfare of Animals Kept for Farming Purposes, chairing the O.I.E. (World Organisation for Animal Health) Working Group on Land Animal Transport and chairing the European Union Scientific Veterinary Committee Animal Welfare Section. His books include '*Stress and Animal Welfare*' (1993, with K.G. Johnson), '*Domestic Animal Behaviour and Welfare*' (with Andrew F. Fraser, now in its 5th edition, 2015) and '*Sentience and Animal Welfare*' (2014). Now an Emeritus Professor, he continues to travel extensively, giving lectures and serving as a scientific advisor on numerous animal welfare issues around the world. A keen bird watcher, Don also devotes time to charitable work on conservation and silvopastoral farming in South America.

Starting as a student of Tinbergen at the University of Oxford, **Marian Stamp Dawkins**, CBE (1945-, Figure 2.3B) is now Professor of Animal Behaviour, Head of the Animal Behaviour Research Group and Director of the John Krebs Field Laboratory at her *alma mater*. A champion of the 'animal's point of view', she developed the idea of using animal preferences and consumer-demand theory to understand what resources animals want and their degree of motivation to obtain them. She used these techniques to evaluate, for example, the motivation of caged hens to access litter (Dawkins, 1983). Careful not to step beyond the scientific evidence, she reasoned that if an animal was highly motivated to perform a behaviour but could not, it was likely to suffer (see '*Animal Suffering: The Science of Animal Welfare*', 1980). Furthermore, in observing chickens on commercial farms, she has shown that improving farm animal welfare can also benefit the producer by improving the efficiency of production (Dawkins *et al.*, 2004), of note considering the importance of 'stepping lightly on the planet'. Always a source of innovative ideas, and ready to take us to task if we forget the 'critical' in 'critical anthropomorphism', Marian's current work includes automatic on-farm monitoring of welfare based on measures of optical flow.

Temple Grandin (1947-, Figure 2.4) is also known for taking the 'animal's point of view', literally, by climbing into the pen and getting down on her hands and knees to view the surroundings from the animal's perspective. She obtained her PhD from the University of Illinois in 1979, and is now Professor of Livestock Behaviour and Welfare at Colorado State University. Applying knowledge gained from close observation of behavioural responses to the design of low-stress handling systems, her solid-sided, curved chute designs are in use at ranches and packing plants throughout the world. She has also developed widely-adopted scoring systems for pre-slaughter welfare assessment, serving as a consultant for many companies and organisations in writing practical welfare guidelines and training personnel in their use. Through this work, she has documented dramatic improvements in handling techniques, lowering the risk of injury to animals and increasing meat quality. She is the author of a prodigious number of popular articles, scientific papers and books about livestock handling, as well as several books about autism. Her book '*Animals in Translation*' (2005, with C. Johnson) was a New York Times bestseller. Inducted into the National Cowgirl Hall of Fame in 2010, she was the recipient of the OIE Meritorious Award in 2015. She has been featured in 'Time Magazine's 100 Most Influential People in the World', (in 2010) and is the subject of a fascinating movie, 'Temple Grandin' (2010), a 'must see' for applied ethologists (along with presentations by the real Temple).

Figure 2.4. Temple Grandin, designer of low-stress livestock handling systems. Photo by Rosalie Winard.

2.4.3 Further prominent applied ethologists

Here we highlight the contributions of some additional pioneering applied ethologists to provide a flavour of how applied ethology developed. Our coverage is by no means inclusive or consistent, and we refer you to other chapters in this book for information about additional pioneers.

In the UK, veterinary researcher **Alex Brownlee** (1903-1999) was one of the first to devote attention to the study of applied ethology, publishing his observations on play behaviour of cattle in 1954. **Roger Ewbank** did pioneering work on the social behaviour of cattle, sheep and pigs, figuring out the role of the different senses in individual recognition by, for example, placing temporary blindfolds on pigs (Ewbank *et al.*, 1974). Recognition was also a theme of **Ian Horrell**'s research, who focussed on sow-piglet recognition and effects of cross-fostering. **John Webster**, former head of the Veterinary School at the University of Bristol, established the now-famous animal welfare and behaviour group at Bristol and served as a founding member of UK's Farm Animal Welfare Council, among many other influential roles. At the University of Glasgow, **Felicity Huntingford** (1948-) conducted seminal research on the behaviour and welfare of fish.

Early students of David Wood-Gush in Edinburgh included **Barry O. Hughes** and **Ian Duncan** (1941-). Barry devoted his career to the subject of behavioural needs of laying hens, succeeding David as Head of Ethology at the Poultry Research Centre and serving many years as editor of 'British Poultry Science'. Meanwhile, Ian initially focussed on frustration in the domestic fowl. He participated in the study of hens on Holy Island with David Wood-Gush and **John Savory** (1943-), and co-authored an influential paper with Barry on '*The notion of ethological 'need', models of motivation and animal welfare*' (1988). In 1989, Ian took up a Chair, or footstool, as he referred to it, in Animal Welfare at the University of Guelph where he tackled broader issues of sentience and added to David's lineage of academic grandchildren. Others active in Edinburgh during the 'early years' included **Mike Gentle**, specialising in the study of pain, and **Bryan Jones**, known for his studies on fear in chickens.

Identification of key environmental features organising behaviour was the approach taken by Swiss ethologist, **Alex Stolba** (1947-1987), in designing the family pen housing system for pigs. Following his PhD research on hamadryas baboons in Ethiopia, he was invited to Edinburgh by David Wood-Gush, where he conducted detailed observations on pigs in the Edinburgh Pig Park. He used this knowledge to develop housing that would accommodate the natural social, breeding, nesting, maternal, foraging and defecation behaviour of sows kept in stable groups for multiple parities (Stolba and Wood-Gush, 1984). Unfortunately, Alex died suddenly soon after returning to Switzerland following a study of Merino sheep behaviour in Australia. However, the 'Stolba family pen system' was further developed in Zurich after his death. Also in Zurich, research by **Detlef Fölsch** (1937-) and colleagues raised awareness about the sensitivity of the hen's beak and identified key features of housing conditions in which hens could live harmoniously without beak trimming.

In the Netherlands, **Piet Wiepkema** (1928-), an academic grandchild of Tinbergen, broke ground in studying livestock feeding behaviour, stereotypies and stress. **Jos Metz** (1942-) used behavioural observations when designing housing and equipment (e.g. milking robots), resulting in him being dubbed the 'father of precision livestock engineering'. **Geritt van Putten** investigated normal and abnormal behaviour of the pig, amusing audiences with titles such as '*Ever been close to a nosey pig?*' (presented at the 1978 Society for Veterinary Ethology conference). Despite receiving government gag orders due to his outspoken criticism of intensive pig production methods, he received a royal honour in 1994 (De Jong, 2015). The detailed study on ontogeny of social behaviour of the Burmese Red Junglefowl by **Jaap Kruijt** (1928-2002, another of Tinbergen's 'grandchildren') at Groningen inspired extensive research into the causation of feather pecking in chickens by Danish applied ethologist **Klaus Vestergaard** (1944-1999). Klaus theorised that feather pecking resulted from redirected dustbathing behaviour, leading to great debates with Wiepkema's student **Harry Blokhuis** (1955-) whose redirected foraging behaviour hypothesis prevails today.

In Germany, **Paul Leyhausen** (1916-1998) conducted seminal research on cat behaviour as well as writing with Lorenz about motivations driving human and animal behaviour. Known for his drawings of cat postures and expressions, his book, '*Cat Behavior: The Predatory and Social Behavior of Domestic and Wild Cats*' (1979) contributed knowledge that was rapidly applied to the design of zoo enclosures. Veterinary ethologist **Klaus Zeeb** (1930-) produced films on methods of training horses and circus animals, and reported on social spacing and dominance in cattle. He was involved in the development of German guidelines for animal welfare. Fellow veterinary ethologist, **Hans H. Sambraus** (1935-), discussed practical applications of behavioural studies in veterinary medicine. He spent many hours observing behaviour to address questions such as: 'How often can a bull mate in a day?' and 'How often does a calf suckle from its mother in a day?' (Sambraus, 1998). He described effects of behavioural deprivation on abnormal behaviour in cattle and rabbits, which he attributed to a build-up of action-specific energy based on Lorenz's now-refuted psychohydraulic model of behaviour. A skilled photographer, he also produced a colour atlas illustrating rare livestock breeds.

In Norway, psychologist, zoologist and poet, **Thorleif Schjelderup-Ebbe** (1894-1976), made fascinating observations on the individual personalities, vocalisations, dominance and despotism of chickens, with his name becoming synonymous with peck orders based on a paper published in 1922. He began this research early, at the age of 6! In fact, his articles about chicken behaviour were based on records that he kept between the age of 10 and completion of his Bachelor's degree in 1917 (Price, 1995). Much later, classically trained ethologist **Bjarne O. Braastad** (1954-) branched out to develop applied ethology in Norway joined by agricultural engineer **Knut Bøe** (1956-). **Ingvar Ekesbo** (1928-) was influential in introducing ethology to veterinary studies in Sweden and played an important role in improving conditions for farm animals in Sweden. He initiated the ethological research of **Bo Algers** (1950-) and **Per Jensen** (1956-), paving the way for landmark studies in the Swedish Pig Park. Serving as the Swedish representative to the Council of Europe's Standing Committee for Farm Animal Welfare, which he chaired from 1979-1986, Ekesbo stressed the importance of taking natural behaviour into account in developing European farm animal regulations (Algers, 2005). In Denmark, Klaus Vestergaard's colleague, veterinarian **Henrik Simonsen** (1939-) highlighted problems with dehorning, tail docking, tethering and slatted floors, and demonstrated how delighted pigs could be when given the opportunity to turn on a shower at will!

In France, key figures in applied ethology start with **Jean-Pierre Signoret** (1929-2008), best remembered for his ground-breaking studies of mating behaviour and pheromones, and for attracting large audiences to his stimulating lectures. Veterinarian and leading stress biologist **Robert Dantzer** (1944-) investigated interactions between stress, behaviour and immunity, including studies on the de-arousing effect of stereotypic behaviour and the mechanisms underlying sickness behaviour and depression. From an early focus on farm animals, he now applies his expertise to research on cancer-related symptomology and serves as Editor-in-Chief of the journal 'Psychoneuroendocrinology'. Behavioural geneticist and Wood-Gush protégé, **Andrew Mills**, selected socially-motivated quail using a quail-sized treadmill and **Jean-Michel Faure** assessed space preferences by providing laying hens with a button that they could peck to expand the size of their cage. Agricultural engineer, **Pierre le Neindre** (1945-), conducted wide-ranging studies into mother-young bonding, human-animal interactions and effects of the housing environment on the welfare of ruminants. He played an influential role on the Scientific Committee of the European Food Safety Authority. Finally, **Marie-France Boisseau** studied effects of hormones and sensory stimuli on the social structure of cattle and served as the first female President of the Society for Veterinary Ethology (see Chapter 1; Petherick and Duncan, 2016).

Pioneering studies on horse behaviour were conducted by **Frank Ödberg**, **Jan Ladewig** (1946-), and by **Katherine A. Houpt** (1939-) who, in addition to studying horses and many aspects of clinical animal behaviour, is known for her textbook, '*Domestic Animal Behavior for Veterinarians and Animal Scientists*', now in its 5th edition (2011). She was the second editor of the journal 'Applied Animal Behaviour Science', the official journal of the ISAE. The founding editor of the journal, which was launched in 1974 under the name 'Applied Animal Ethology', was **Andrew F. Fraser**. He began his career practicing large animal medicine in Scotland and Jamaica and was a founding member and first President of the Society for Veterinary Ethology. He subsequently emigrated to Canada, where he served as Director of Animal Care at Memorial University in Newfoundland. In addition to writing applied ethology textbooks with Don Broom, he has written several books about horses, the most recent being '*The Behaviour and Welfare of the Horse*' (2010). **Harold Gonyou** (1952-) was the third editor of the journal. Serving as the Research Director of the Prairie Swine Center in Saskatoon, he applied knowledge of behaviour to the design of pig feeders and group housing of sows.

Also in Canada, **J. (Jaromir) Frank Hurnik** (1932-) designed alternative housing systems, started the first formal undergraduate degree course in farm animal welfare in North America (at Guelph), wrote the '*Dictionary of Farm Animal Behavior*' (1995, with A. Bruce Webster and Paul B. Siegel) and played a leading role in developing the first editions of the Canadian Recommended Codes of Practice for poultry, beef and dairy cattle. Another pioneering applied ethologist in Canada is **David Fraser**, who, apart from his mere 300+ journal articles and eminently readable book '*Understanding Animal Welfare: The Science in its Cultural Context*' (2008), is notorious for his witty pig poetry. Not to be missed is the abstract to his 1991 paper with statistician, Brian Thompson, on armed sibling rivalry among suckling pigs. To give you a flavour, here are the first two sentences:

> A piglet's most precious possession
> Is the teat that he fattens his flesh on.
> He fights for his teat with tenacity
> Against any sibling's audacity.

David was invested into the Order of Canada in 2007 for his scientific contributions to the humane management of food animals.

In the USA, **James V. Craig** (1924-2003) applied his training in genetics to studies on the genetic variation of aggressiveness and cannibalism in chickens. Dairy scientist **Jack Albright** (1930-2014) is remembered for his enthusiastic pursuit of knowledge about cow behaviour and his lectures instilling a compassionate attitude towards animal care. He authored '*The Behaviour of Cattle*' (1997, with Clive Arave). **Stanley E. Curtis** (1942-2010) was an influential animal scientist at the University of Illinois. PhD supervisor to Temple Grandin, he combined research on environmental physiology with an interest in behaviour. He was a fierce critic of psychological interpretations of behaviour, leading to many lively exchanges. **W. Ray Stricklin** produced models on use of space and was a keen participant in ethical debate surrounding animal care and use, as was **Ted Friend**, who investigated effects of transportation on the behaviour and physiology of slaughter horses and circus animals. **Edward O. Price**, author of several books on domestic animal behaviour including '*Principles and Applications of Domestic Animal Behavior*' (2008), focussed on reproductive behaviour of cattle, sheep and goats, and effects of animal domestication (influenced by his failed attempt to tame a wild fox as a child).

Pioneering studies on domestic dog and wild canid behaviour were conducted by veterinary ethologist, **Michael W. Fox** (1937-), who later became Vice-President of the Humane Society of the United States and Vice-President of Humane Society International. A prolific writer, outspoken critic of animal exploitation and coiner of terms such as 'mechanomorphisation' (the treatment of animals as unfeeling machines), he was especially active in promoting animal welfare in India (Litke, 2004). Avid cyclist and energetic conservationist, **Marc Bekoff** (1945-) also conducted ethological research on canids, with a focus on play behaviour, while **Ben Hart** was active in applying ethological principles to the veterinary treatment of behaviour problems in companion animals. His wife, **Lynette Hart** is one of the first to apply ethology to the study of psychosocial benefits of companion animals. A student of Lorenz, German-born veterinary ethologist **Viktor Reinhardt** and his wife **Annie Reinhardt** made classic observations in Kenya on long-term friendships among cattle. After moving to the USA, Viktor pioneered the application of ethology to the improvement of housing for laboratory animals (Liss, 2012). As moderator of the Animal Welfare Institute's Laboratory Animal Refinement and Enrichment Forum, he has played a significant role in fostering the implementation of social housing and environmental enrichment. See Chapter 8 for further details on the North American perspective (Widowski *et al.*, 2016).

Ron Kilgour was a pioneering applied ethologist in New Zealand who, prior to his untimely death in a car accident in 1988, made classic observations on leadership in dairy cows and promoted humane pre-slaughter handling of livestock. In Australia, pioneers include **Glen McBride** (1925-), who has focussed on social organisation of domestic fowl and, more recently, evolution and consciousness. **G. Alexander** and **G. W. Arnold** could be considered the 'fathers of sheep behaviour'. **Judith Blackshaw** (1938-) engaged in diverse projects ranging from designing environments to reduce stress in pigs and cattle, to providing advice on dog parks and laboratory animal management, while **Rolf Beilharz**'s research covered topics such as social dominance and milk production in dairy cows and the genetics of fearfulness in guide dogs. **Lesley Rogers** (1943-) is well known for her research on brain laterality and its relation to behaviour, using domestic chicks as her model. She applies knowledge of brain function to questions about animal cognition and awareness. Chapter 9 provides more information on developments in Australia and New Zealand (Hemsworth *et al.*, 2016).

For pioneering developments in South America, Japan and China, we refer you to Chapters 10, 11 and 12, respectively (Galindo *et al.*, 2016; Sato, 2016; Bao and Li, 2016). Additionally, the paper by David Wood-Gush (1991) in the Society for Veterinary Ethology's 25th anniversary conference proceedings provides a fine account of early papers on the behaviour of agricultural species.

2.5 Winds across the field

While the above scientists have been instrumental in laying down the groundwork for current studies in applied ethology, our field has also been shaped by external movers and shakers who have influenced the broader cultural context within which our studies are rooted. Here we highlight a few of these individuals.

One memorable personality was **Astrid Lindgren** (1907-2002), a decorated writer best known for her children's books featuring colourful characters such as Pippi Longstocking and Emil. So beloved was she that she had an asteroid named after her, was featured on the Swedish 20 kroner note and was the inspiration for the world's largest prize for children's literature, the Astrid Lindgren Memorial Award (http://astridlindgren.se/en/). Frequently expressing her views in newspaper articles, she was known for her sharp mind, common sense and ability to sway public opinion using satirical children's tales. In a series of articles to the main Swedish evening newspaper, 'Expressen', she and veterinarian, Kristina Forslund, campaigned against intensive farming of animals, arguing that cows deserved the chance to graze outdoors in summer and feel the sunlight and pigs and chickens needed space to behave according to their natures. Their articles, following after journalist Barbro Soller's (1971) book '*Djurfabriken*' (Animal Factory), fuelled a lively public debate leading to passage of the far-reaching Swedish farm animal protection law in 1988. The Swedish Prime Minister presented the law as a gift to Astrid for her 80th birthday (albeit that she considered the final version of law 'farty', holding that it became too watered down by the time it was passed).

Ruth Harrison, OBE (1920-2000, Figure 2.5) had a comparable impact on farm animal welfare in Britain (Van de Weerd and Sandilands, 2008). She was motivated to examine farming practices after a leaflet exposing the plight of the veal calf was posted through her letterbox in 1961. Farmers, who had no reason to fear her examination, welcomed her requests for visits and allowed her husband to photograph their practices. As a result of her research, Harrison published the seminal book '*Animal Machines*' in 1964, exposing 'factory' farming practices in the UK. The information held there was so shocking that it led the government to establish a committee, chaired by F.W. Rogers Brambell (1901-1970) and including ethologist W.H. Thorpe among its members, to investigate the welfare of intensively farmed animals. The Brambell Report of 1965 (Command Paper 2836, 1965) highlighted that animals ought, at least, to have the freedom to stand up, lie down, turn around, groom themselves and stretch their limbs, which became known as the Brambell Five Freedoms . The Report also led to the establishment of the UK's Farm Animal Welfare Council (FAWC), which revised the Five Freedoms to comprise freedom from: (1) hunger and thirst, (2) discomfort, (3) pain, injury or disease and (4) fear and distress, as well as freedom (5) to express normal behaviour (FAWC, 2009). These have been guiding ideals for animal welfare assessment ever since. Ruth Harrison remained keenly interested in farm animal welfare, serving on the FAWC until she was 70. She supported research through the Farm Animal Care Trust, which she founded in 1967, and travelled extensively to observe progress on projects such as the Family Pen

Figure 2.5. (A) Ruth Harrison at a Danish mink farm in 1997; (B) Ruth Harrison and Klaus Vestergaard observe a sow at the Swedish Pig Park in 1988.

System for pigs at the University of Edinburgh and research on humane stunning at the University of Bristol.

Whereas Astrid Lindgren and Ruth Harrison called for reforms in farming practices, **Peter Singer** (1946-) called for abolition. Considered a father of the animal rights movement along with **Tom Regan** (1938-), he studied philosophy at Melbourne and Oxford. In '*Animal Liberation: A New Ethics for our Treatment of Animals*' (1975), he set out his position on moral and ethical responsibility towards animals, while damning our treatment of them in farming and scientific research. In contrast to the rights-based philosophy laid out by Regan in '*The Case for Animal Rights*' (1983), Singer adopted the philosophy of utilitarianism, in which our actions should benefit us and those affected by them the most, while causing the least harm. His edited volume, '*In Defense of Animals: The Second Wave*' (1985) further explores this theme, considering how and why an animal's (or person's) interests should be valued. No stranger to controversy for his views on speciesism, bestiality, abortion and euthanasia, Singer is currently Professor of Bioethics at Princeton University. He has been on Time Magazine's list of 100 most influential people (2005) and was appointed a Companion of the Order of Australia in 2012.

While Singer and Regan opposed animal use in scientific research, Oxford-trained ethologist **William (Bill) M.S. Russell** (1925-2006) and his laboratory assistant, microbiologist **Rex L. Burch** (1926-1996), proposed reforms from within the scientific community. In 1954, they were commissioned by the Universities Federation for Animal Welfare to investigate methods for making laboratory techniques more humane. Their extensive study led to the 1959 publication of '*The Principles of Humane Experimental Technique*', a classic work in which they advanced the concept of the 3Rs, Replacement, Reduction and Refinement. The book was largely ignored at the time. Russell carried on to become a psychoanalyst (after marrying his therapist), agronomist, classicist, translator of Japanese books, science fiction novelist, and professor in sociology at the University of Reading (Richmond, 2006). He was known for singing portions of his lectures in rhyming couplets, participating in radio shows and as the cake-eating professor in a children's book. Meanwhile, Burch set up a respected microbiological testing service in Sheringham, Norfolk (Balls, 1996). Their 3Rs work finally gained traction in the 1980s and the 3Rs have now been adopted by governments and scientific organisations worldwide, providing an ethical framework for avoiding unnecessary animal use and promoting the welfare of animals used in scientific research. Don't miss the 'tuneful polymath' Bill Russell singing about ethology and the 3Rs, Gilbert and Sullivan style, on YouTube (https://www.youtube.com/watch?v=i7n6_JCOWo8).

2.6 Fertiliser

Although not directly associated with the ISAE, the following individuals conducted behavioural research that has inspired researchers in applied ethology.

Heini Hediger (1908-1992) is considered the 'father of zoo biology' (Maple, 1992). He claimed that this 'calling' was due to his parents taking him frequently to his local zoo so that he became imprinted on the zoo! His formal education was in zoology and animal psychology and he developed a career directing zoos, including those in Bern, Basel and Zurich. He studied subjects such as sleep and dreaming, suffering, thinking and fear in animals, quite forward thinking for his time. He was especially interested in spacing between animals and making the enclosure a refuge and home range rather than a prison. He emphasised quality over quantity in the design of zoo enclosures and encouraged zoo animal trainers to promote positive behaviour and opportunities to examine and treat animal health. Hediger's key publications include the books '*Wild Animals in Captivity*' (1950) and '*Man and Animal in the Zoo*' (1969).

J.P. (John Paul) Scott (1909-2000) studied at Wyoming, Oxford and Chicago. He established the Division of Behavior Studies at the Jackson Laboratory and headed the Center for the Study of Social Behavior at Bowling Green University (Dewsbury, 2001). His early work included studies on embryology of guinea pigs (his first exposure to critical periods), aggression in different strains of mice, leadership and dominance in goats and social behaviour of sheep. He was influenced in this work by W.C. Allee (1885-1955, of 'Allee effect' fame – that fitness increases with stocking density up to a point), who was investigating social hierarchies in chickens at the time. However, he is best remembered for seminal studies on the behavioural development of dogs, including the timing and importance of the socialisation period. These studies were conducted in collaboration with behavioural geneticist John L. Fuller (1910-1992) at the Jackson Laboratory and their 1965 book, '*Genetics and the Social Behavior of the Dog*', has become a classic. Scott played a pivotal role in establishing the Animal Behavior Society (in 1964), receiving its highest honour, the Distinguished Animal Behaviorist Award, in 1990.

Marian Breland (1920-2001) and her husband **Keller Breland** (1915-1965) were students of B.F. Skinner at the University of Minnesota, helping him with his studies on operant conditioning. In 1947, they left their graduate studies to start a successful company called Animal Behavior Enterprise (ABE). The company, which grew to have over 40 employees, was first based in Minnesota and later in Arkansas, where they created a tourist attraction known as IQ-Zoo. They used positive reinforcement to train animals from over 140 species for animal exhibits and advertising campaigns. For example, chickens were trained to dance, pigs to put coins in piggy banks, and racoons to play basketball. More seriously, they developed teaching methods for people with developmental disabilities and for animals used by the US military. The latter included training 'reconnaissance ravens' to take surveillance photographs and 'ambush patrol pigeons' to detect snipers (Marr, 2002). Their 1961 paper '*The Misbehavior of Organisms*' (named after Skinner's '*The Behavior of Organisms*' (1938), which Marian helped to edit) is classic reading for students of applied ethology, showing that radical behaviorism is insufficient to predict how animals will respond. After Keller's death, Marian married zoologist Robert (Bob) E. Bailey (1936-), Director of Animal Training for the US Navy and later General Manager of ABE. Marian returned to complete a PhD in 1978 at the University of Arkansas (on acuity of bird vision) and subsequently became a professor at Henderson State University

(Timberlake, 2003). See the IQ-Zoo (2016) website for examples and videos of the Breland-Bailey training methods.

Edward O. Wilson (1929-), Professor Emeritus at Harvard University, is known as the 'father of sociobiology'. An eye injury at the age of 7 led him to the study of insects, which he could observe close-up (Academy of Achievement, 2013). His works include studies on pheromonal communication in ants, species equilibrium on islands and the role of kin selection in shaping the behaviour of eusocial insects. A prolific writer, his ability to popularise science has earned him two Pulitzer Prizes. However, it is his classic 1975 book addressing evolutionary mechanisms underlying cooperative behaviour, '*Sociobiology: The New Synthesis*', and writings on the biophilia hypothesis (including our desire to interact with animals), that have probably been most inspirational to budding applied ethologists.

Jane Goodall, DBE (1934-) first studied wild chimpanzees in Tanzania (then Tanganyika) in 1960, under the guidance of Louis S.B. Leakey (1903-1972) who reasoned that understanding chimpanzees could enlighten the evolutionary history of humans. Through her long-term studies following families over generations at Gombe Stream, she dispelled myths about chimpanzees, shedding light on their diet, tool use and complex social structure. Under the tutelage of Robert Hinde at Cambridge (where she was one of only a handful of students permitted to enrol in PhD studies without an undergraduate degree), she completed a PhD in ethology in 1965 based on her field research. She went on to write many books, including '*In the Shadow of Man*' (1971) and '*The Chimpanzees of Gombe: Patterns of Behavior*' (1986). Eschewing the anthropomorphism phobia of the day, she named the chimpanzees and wrote about their individual personalities, emotions and experiences, allowing readers to gain personal insights into their lives and serving as a role model for aspiring female applied ethologists. Popular articles and documentaries, starting with an article in National Geographic magazine with stunning photographs by Hugo van Lawick (whom she subsequently married and divorced), made her and the chimpanzees household names. More recently, she has devoted tireless effort towards campaigning for the welfare of chimpanzees in captivity as well as promoting their conservation in the wild. As a United Nations 'Messenger of Peace', her lectures to packed auditoriums resonate with her rendition of a chimpanzee greeting call and continue with a message of hope founded on personal action and responsibility. Through 'The Jane Goodall Institute' and 'Roots and Shoots' programme, Goodall continues her work to integrate wildlife protection with the needs of local people and encourage young people to make a difference (see http://www.janegoodall.org).

Sir Patrick Bateson (1938-) is an emeritus professor of ethology at the University of Cambridge (and small-scale breeder of Egyptian Mau cats). A student of Robert Hinde, he and his collaborators have made many important contributions to basic ethology, including mechanisms of imprinting and kin recognition, mate choice and (with PhD student, Mike Mendl) play in cats. However, he is also noted for his scholarly contributions on applied issues. He conducted an investigation into '*The behavioural and physiological effects of culling red deer*' (1997) for the National Trust, was commissioned by the British Kennel Club and Dogs Trust to conduct an '*Independent Inquiry into Dog Breeding*'

(2010) and chaired a '*Review of Research Using Non-Human Primates*' for the major British biomedical funding agencies (2011). He is known for 'Bateson's cube', a decision-making tool for assessing the value of a proposed project based on predicted level of animal suffering, likelihood of benefit and importance of the research. With Paul Martin, he co-authored '*Measuring Behaviour: An Introductory Guide*' (2007), mandatory reading for all applied ethologists. His daughter, Melissa Bateson, is following in his footsteps as a professor of ethology focussing on animal welfare issues.

2.7 Life in the canopy

All students of applied ethology have been influenced intellectually by the pioneers discussed above, and will find many personal connections as well. One of us (RN), for example, recalls growing up with a steady supply of honey from family friend Dr. Rolf Boch, a student of von Frisch; learning from Alex Stolba to pay attention to the subtle glances of pigs foraging in the Edinburgh Pig Park; sleeping on Jane Goodall's bed at Gombe Stream (who was away on a lecture tour at the time); being regaled with stories of scientific feuds over the nature of emotions by Jaak Panksepp; and being encouraged to inhale a whiff of CO_2 by Ruth Harrison (instantly getting her point!).

Likewise, VS remembers approaching Don Broom with trepidation at a conference to ask him a question. Upon introducing herself, he shocked her with 'Yes, I know who you are'. Other memories include listening to a rousing rendition of Robert Burns' poem 'Tam O'Shanter' by Ian Duncan at a conference dinner, hearing stories about David Wood-Gush when he had been course leader of the MSc programme at the University of Edinburgh and counting beards (on the men) at every ISAE conference since 2001 and deciding it is probably the most bearded meeting she attends.

Anyone in this field is fortunate that it is relatively small, making it easy for us to get to know one another. We are generally a friendly bunch – just watch the ISAE conference dinner dancing! Applied ethologists from around the world form a vibrant, interconnected network. While finding support and safe harbour within the ISAE, we are also colonising new regions and fields of study, and continuing to attract new inhabitants to our field. Female membership has increased dramatically in recent years, making it likely that many more 'mothers of applied ethology' will be featured when a comparable article is written in another 50 years. We end with a quote from the immortal Konrad Lorenz (1981, Preface) '... the development of a science resembles that of a coral colony. The more it thrives and the faster it grows ... – the vestiges of the founders and the contributions of the early discoverers – become overgrown and obscured by their own progeny'. While thriving and looking forward, we hope that this chapter will serve as a reminder from where we have come.

Disclaimer

The authors accept no responsibility for the omission of many important applied ethologists not long enough in the tooth to be included, and apologies to those included despite the fine condition of their dentition.

References

Academy of Achievement, 2013. Edward O. Wilson, Ph.D. biography. Exploring the origins of human nature. Academy of Achievement, Washington, DC, USA. Available at: http://www.achievement.org/autodoc/page/wil2bio-1.

Albright, J.L. and Arave, C.W., 1997. The behavior of cattle. CAB International, Wallingford, UK, 306 pp.

Algers, B., 2005. A short biography of Ingvar Ekesbo. International Society for Applied Ethology. Available at: http://tinyurl.com/j9ot3k5.

Balls, M., 1996. Rex Leonard Burch (1926-1996): humane scientist, prophet, dreamer, visionary. Animal Welfare Information Center Newsletter 7(2): 17-18. Available at: http://tinyurl.com/ztmroot.

Bao, J. and Li, Y., 2016. China perspective: emerging interest in animal behaviour and welfare science. In: Brown, J.A., Seddon, Y.M. and Appleby, M.C. (eds.) Animals and us – 50 years and more of applied ethology. Wageningen Academic Publishers, Wageningen, the Netherlands, pp. 241-252.

Bateson, P.P.G., 1997. The behavioural and physiological effects of culling red deer. Report to the Council of the National Trust, London, UK, 77 pp.

Bateson, P., 2010. Independent inquiry into dog breeding. Report to the British Kennel Club and Dogs Trust, 65 pp. Available at: http://tinyurl.com/humtnlv.

Bateson, P., 2011. Review of research using non-human primates. Report of a panel chaired by Professor Sir Patrick Bateson FRS, 53 pp. Available at: http://tinyurl.com/jtuewa7.

Breland, K. and Breland, M., 1961. The misbehaviour of organisms. American Psychologist 16: 681-684.

Broom, D.M., 2014. Sentience and animal welfare. CABI, Wallingford, UK.

Broom, D.M. and Fraser, A.F., 2015. Domestic animal behaviour and welfare, 5th edition. CABI, Wallingford, UK.

Broom, D.M. and Johnson, K.G., 1993. Stress and animal welfare. Kluwer Academic Publishers, Dordrecht, the Netherlands.

Brownlee, A., 1954. Play in domestic cattle in Britain: an analysis of its nature. British Veterinary Journal 110: 48-68.

Burkhardt Jr., R.W., 2005. Patterns of behavior: Konrad Lorenz, Niko Tinbergen, and the founding of ethology. The University of Chicago Press, Chicago, IL, USA.

Command Paper 2836, 1965. Report of technical committee to enquire into the welfare of animals kept under intensive livestock husbandry systems. Her Majesty's Stationary Office, London, UK. Available at: http://tinyurl.com/gmvjyb9.

Darwin, C.R., 1872. The expression of the emotions in man and animals. John Murray, London, UK, 374 pp. Available at: http://tinyurl.com/jtwdnn2.

Dawkins, M.S., 1980. Animal suffering: the science of animal welfare. Chapman and Hall Ltd., London, UK.

Dawkins, M.S., 1983. Battery hens name their price: consumer demand theory and the measurement of ethological 'needs'. Animal Behaviour 31: 1195-1205.
Dawkins, M.S., Donnelly, C.A. and Jones, T.A., 2004. Chicken welfare is influenced more by housing conditions than by stocking density. Nature 427: 342-344.
De Jong, D.M., 2015. Onderzoek naar verbeteren varkenswelzijn levert Gerrit van Putten prijs op. Dierenbescherming, Den Haag, the Netherlands. Available at: http://tinyurl.com/j8jcv8c.
Dewsbury, D., 2001. The life of J.P. Scott. Bowling Green State University, Bowling Green, OH, USA. Available at: http://tinyurl.com/jyazc4d.
Ewbank, R., Meese, G.B. and Cox, J.E., 1974. Individual recognition and the dominance hierarchy in the domesticated pig: the role of sight. Animal Behaviour 22: 473-480.
Farm Animal Welfare Council, 2009. Five freedoms. Available at: http://tinyurl.com/hqwdko9.
Fraser, A.F., 2010. The behaviour and welfare of the horse, 2nd revised edition. CABI Publishing, Wallingford, UK.
Fraser, D., 2008. Understanding animal welfare: the science in its cultural context. Wiley-Blackwell, Chichester, UK.
Fraser, D. and Thompson, B.K., 1991. Armed sibling rivalry among suckling piglets. Behavioral Ecology and Sociobiology 29: 9-15.
Galindo, F., Tadich, T., Ungerfeld, R., Hötzel, M.J. and Miguel-Pacheco, G., 2016. The development of applied ethology in Latin America. In: Brown, J.A., Seddon, Y.M. and Appleby, M.C. (eds.) Animals and us – 50 years and more of applied ethology. Wageningen Academic Publishers, Wageningen, the Netherlands, pp. 211-225.
Goodall, J., 1971. In the shadow of man. William Collins and Sons, Glasgow, UK.
Goodall, J., 1986. The chimpanzees of Gombe: patterns of behavior. Harvard University Press, Cambridge, MA, USA.
Grandin, T. and Johnson, C., 2005. Animals in translation: using the mysteries of autism to decode animal behavior. Scribner, New York, NY, USA.
Griffin, D.R., 1976. The question of animal awareness, 2nd edition. Rockefeller University Press, New York, NY, USA.
Griffin, D.R., 1984. Animal thinking. Harvard University Press, Cambridge, MA, USA.
Gross, C.G., 2005. Donald R. Griffin 1915-2003. Biographical memoirs 86. National Academy of Science, Washington, DC, USA. Available at: http://tinyurl.com/jtlwymh.
Hafez, E.S.E., 1975. The behaviour of domestic animals, 3rd edition. Baillière Tindall, London, UK.
Harrison, R., 1964. Animal machines: the new factory farming industry. Vincent Stuart Publishers, London, UK.
Hediger, H., 1950. Wild animals in captivity. Butterworth's Scientific Publications, London, UK.
Hediger, H., 1969. Man and animal in the zoo. Delacorte Press, New York, NY, USA.
Hemsworth, P.H., Mellor, D.J., Coleman, G.J., Beausoleil, N.J., Fisher, A.D. and Stafford, K.J., 2016. Australian and New Zealand perspectives. In: Brown, J.A., Seddon, Y.M. and Appleby, M.C. (eds.) Animals and us – 50 years and more of applied ethology. Wageningen Academic Publishers, Wageningen, the Netherlands, pp. 193-209.
Houpt, K.A., 2011. Domestic animal behavior for veterinarians and animal scientists, 5th edition. Wiley-Blackwell, Ames, IA, USA.
Hughes, B.O. and Duncan, I.J.H., 1988. The notion of ethological 'need', models of motivation and animal welfare. Animal Behaviour 36: 1696-1707.
Hurnik, J.F., Webster, A.B. and Siegel, P.B., 1995. Dictionary of farm animal behavior, 2nd edition. Iowa State University Press, Ames, IA, USA.

IQ-Zoo, 2016. Robert E. Bailey and the University of Central Arkansas, Conway, AR, USA. Available at: http://www3.uca.edu/iqzoo/iqzoo.htm.
Krebs, J.R. and Sjölander, S., 1992. Konrad Zacharias Lorenz. 7 November 1903-27 February 1989. Biographical Memoirs of Fellows of the Royal Society 38: 210-228.
Lake, P.E. and Wood-Gush, D.G.M., 1956. Diurnal rhythms in semen yields and mating behaviour in the domestic cock. Nature 178: 853.
Leyhausen, P., 1979. Cat behavior: the predatory and social behavior of domestic and wild cats. Taylor and Francis / Garland STPM Press, New York, NY, USA.
Liss, C., 2012. A phenomenon named Viktor Reinhardt: AWI salutes a visionary advocate for better treatment of animals in research. AWI Quarterly Summer, Animal Welfare Institute, Washington, DC, USA. Available at: http://tinyurl.com/hnlsnvu.
Litke, H., 2004. Dr. Michael W. Fox, biographical interview. Animal News Center. Available at: http://www.drfoxvet.net/.
Lorenz, K., 1952. King Solomon's ring: new light on animal ways. Methuen and Co. Ltd., London, UK.
Lorenz, K., 1954. Man meets dog. Methuen and Co. Ltd., London, UK.
Lorenz, K., 1981. The foundations of ethology. Springer-Verlag, Vienna, Austria.
Maple, T.L., 1992. Prof. Dr Heini Hediger (1908-1992). Zoo Biology 11: 369-372.
Marr, J.N., 2002. Marian Breland Bailey: the mouse who reinforced. The Arkansas Historical Quarterly 61: 59-79.
Martin, P. and Bateson, P., 2007. Measuring behaviour: an introductory guide, 3rd edition. Cambridge University Press, Cambridge, UK.
McLeod, S., 2015. Skinner – operant conditioning. Available at: http://tinyurl.com/7wxntsf.
Monaghan, P. and Wood-Gush (eds.), D.G.M., 1991. Managing the behaviour of animals. Chapman and Hall, London, UK.
Nobel Lectures, 1967. Ivan Pavlov – biographical, physiology or medicine 1901-1921. Elsevier, Amsterdam, the Netherlands. Available at: http://tinyurl.com/3q7fyta.
Nobel Lectures, 1992. Karl von Frisch – biographical, physiology or medicine 1971-1980. World Scientific Publishing Co., Singapore, Malaysia. Available at: http://tinyurl.com/jax7urg.
Panksepp, J., 1998. Affective neuroscience: the foundations of human and animal emotions. Oxford University Press, Oxford, UK.
Petherick, J.C. and Duncan, I.J.H., 2016. The International Society for Applied Ethology: going strong 50 years on. In: Brown, J.A., Seddon, Y.M. and Appleby, M.C. (eds.) Animals and us – 50 years and more of applied ethology. Wageningen Academic Publishers, Wageningen, the Netherlands, pp. 27-50.
Physiology or Medicine, 1973. Press Release. Nobelprize.org. Nobel Media AB 2014. Available at: http://www.nobelprize.org/nobel_prizes/medicine/laureates/1973/press.html. Podos, J., 1994. Early perspectives on the evolution of behavior: Charles Otis Whitman and Oskar Heinroth. Ethology, Ecology and Evolution 6: 467-480.
Prato-Previde, E., Custance, D.M., Spiezio, C. and Sabatini, F., 2003. Is the dog-human relationship an attachment bond? An observational study using Ainsworth's strange situation. Behaviour 140: 225-254.
Price, E.O., 2008. Principles and applications of domestic animal behavior: an introductory text. CABI, Wallingford, UK.
Price, J., 1995. A remembrance of Thorleif Schjelderup-Ebbe. Human Ethology Bulletin 10: 1-6.
Regan, T., 1983. The case for animal rights. University of California Press, Berkeley CA, USA.

Reinemeyer, E., 1999. Edward Lee Thorndike. History of psychology archives, department of psychology. Muskingum College, New Concord, OH, USA. Available at: http://tinyurl.com/knnxrk.
Richmond, C., 2006. Bill Russell. The Guardian, 16 August 2006. Available at: http://tinyurl.com/jecj8v6.
Riley, J., Greggers, U., Smith, A., Reynolds, D. and Menzel, R., 2005. The flight paths of honeybees recruited by the waggle dance. Nature 435: 205-207.
Russell, W.M.S. and Burch, R.L., 1959. The principles of humane experimental technique. Universities Federation for Animal Welfare, Wheathampstead, UK. Available at: http://tinyurl.com/lrdyhdm.
Sambraus, H.H., 1998. Applied ethology – its task and limits in veterinary practice. Applied Animal Behaviour Science 59: 39-48.
Sandilands, V., 2004. David Wood-Gush: the biography of an ethology mentor. Applied Animal Behaviour Science 87: 173-176.
Sato, S., 2016. Applied animal behaviour science in Japan and the culture of '*aigo*'. In: Brown, J.A., Seddon, Y.M. and Appleby, M.C. (eds.) Animals and us – 50 years and more of applied ethology. Wageningen Academic Publishers, Wageningen, the Netherlands, pp. 227-240.
Schjelderup-Ebbe, T., 1922. Beiträge zur sozialpsychologie des haushuhns. Observations on the social psychology of domestic fowl. Zeitschrift für Psychologie 88: 225-252.
Schuermann, D.B., 2015. Charles Otis Whitman (1842-1910). The Embryology Project Encyclopedia, Arizona State University, Tempe, AZ, USA. Available at: http://tinyurl.com/jz5sevq.
Schulze-Hagen, K. and Birkhead, T.R., 2015. The ethology and life history of birds: the forgotten contributions of Oskar, Magdalena and Katharina Heinroth. Journal of Ornithology 156: 9-18.
Scott, J.P. and Fuller, J.L., 1965. Genetics and the social behavior of the dog. University of Chicago, Chicago, IL, USA.
Singer, P., 1975. Animal liberation: a new ethics for our treatment of animals. The New York Review of Books, New York, NY, USA.
Singer, P. (ed.), 1985. In defense of animals: the second wave. Wiley-Blackwell, Hoboken, NJ, USA.
Skinner, B.F., 1938. The behavior of organisms: an experimental analysis. D. Appleton-Century Co., New York, NY, USA, 457 pp. Available at: http://tinyurl.com/zb76s7w.
Soller, B., 1971. Djurfabriken. Rabén and Sjögren, Stockholm, Sweden.
Specter, M., 2014. Drool Ivan Pavlov's real quest. The New Yorker. Available at: http://www.newyorker.com/magazine/2014/11/24/drool.
Stolba, A. and Wood-Gush, D.G.M., 1984. The identification of behavioural key features and their incorporation into a housing design for pigs. Annales de Recherches Veterinaires 15: 287-302.
Temple Grandin, 2010. Directed by Jackson, M., Home Box Office (HBO), A Time Warner Company, New York, NY, USA.
Thorndike, E.L., 1911. Animal intelligence. The MacMillan Company, New York, NY, USA, 197 pp. Available at: http://tinyurl.com/zatcdvs.
Timberlake, W., 2003. Marian Breland Bailey: many lives. Behavioural Processes 62: 1-4.
Tinbergen, N., 1973. Nikolaas Tinbergen – biographical. Available at: http://tinyurl.com/gmldv8u.
Tinbergen, N., 1951. The study of instinct. Oxford University Press, London, UK.
Tinbergen, N., 1963. On aims and methods of ethology. Zeitschrift für Tierpsychologie 20: 410-433.
Van de Weerd, H. and Sandilands, V., 2008. Bringing the issue of animal welfare to the public: a biography of Ruth Harrison (1920-2000). Applied Animal Behaviour Science 113: 404-410.
Von Frisch, K., 1973. Decoding the language of the bee. Nobel Lecture, December 12, 1973. The Nobel Foundation, Stockholm, Sweden. Available at: http://tinyurl.com/hs6wkt9.

Watson, E., 1999. John B. Watson (1878-1958). History of psychology archives, department of psychology. Muskingum College, New Concord, OH, USA. Available at: http://tinyurl.com/3ngnqpf.
Whitman, C.O., 1919. The orthogenic evolution in pigeons. In: Riddle, O. (ed.) Posthumous works of Charles Otis Whitman. Vol. 1. Carnegie Institution of Washington, Washington, DC, USA. Available at: http://tinyurl.com/zz6gsad.
Widowski, T.M., Mench, J.A., Swanson, J.C. and Stricklin, W.R., 2016. North American perspectives. In: Brown, J.A., Seddon, Y.M. and Appleby, M.C. (eds.) Animals and us – 50 years and more of applied ethology. Wageningen Academic Publishers, Wageningen, the Netherlands, pp. 175-192.
Wilson, E.O., 1975. Sociobiology: the new synthesis. Harvard University Press, Cambridge, MA, USA.
Wood-Gush, D.G.M., 1971. The behaviour of the domestic fowl. Heinemann Educational, London, UK.
Wood-Gush, D.G.M., 1983. Elements of ethology: a textbook for agricultural and veterinary students. Chapman and Hall, London, UK.
Wood-Gush, D.G.M., 1991. The development of applied ethology in relation to certain agricultural animals. In: Appleby, M.C., Horrell, R.I., Petherick, J.C. and Rutter, S.M. (eds.) Society for veterinary ethology 25th anniversary 1966-1991. Applied Animal Behaviour: Past, Present and Future. Proceedings of the International Congress Edinburgh. Universities Federation for Animal Welfare, Wheathampstead, UK, pp. 17-24. Available at: http://tinyurl.com/zd7njsy.

Note: We have also made liberal use of websites too numerous to mention, including Wikipedia, encyclopedia.com, britannica.com, university and institutional webpages, professional biographies, CVs posted online, and biographies of ISAE Honorary Fellows at http://www.applied-ethology.org/honorary_fellows.html.

Part 2. The science of behaviour

3. Applied ethology: what's the buzz?

J. Brown
Prairie Swine Centre, Box 21057, 2105 8th Street East, Saskatoon, Saskatchewan, S7H 5N9, Canada
Department of Animal and Poultry Science, University of Saskatchewan, 51 Campus Drive, Saskatoon, Saskatchewan, S7N 5A8, Canada; jennifer.brown@usask.ca

Abstract

This chapter is essentially an introduction to the rest of the book. In it, I examine why ethologists are so passionate about their work and highlight some recent achievements and new trends in behaviour research. This chapter also reflects on the importance of our current place in time and history. The fact that, looking back, we have made great advances in methods and understanding of animals' motivations and needs, and looking forward, we seem to be poised on the brink of a new era in areas such as our understanding of the brain and emotions, individual differences and stress physiology, evolutionary biology and the influence group and environment effects, among other areas. Human-animal interaction and the nature of the human-animal bond is one area that has received recent attention. Studies indicate clear benefits of positive human interaction in animals, but also beneficial effects on humans of interacting with animals. When evaluating how animals cause benefits to human health, it appears that these effects are produced by activating neural pathways in what Jaak Panksepp has dubbed our 'care' and 'seeking' systems. Studies of positive affect are also increasing, and with them an increasing need for alternative theoretical frameworks to explain the evolution of cooperative and altruistic behaviours. Multi-level models of selection that include the effects of environment and group interactions are gaining support. As well as highlighting advances and some exciting new developments, this chapter emphasises the importance of disseminating applied ethology through international efforts and education. If applied ethology is to play an important role in defining how we use and care for animals in the future we must focus on maintaining and improving the quality of research produced, and sharing our findings. There is much work to be done and much to be gained, and if we are successful – Who can guess what the next 50 years will bring?

Keywords: animal behaviour, animal welfare, positive affect, evolutionary biology

3.1 Introduction

As our introductory chapters have shown, applied ethology is a relatively new and holistic field of science. It grew rapidly in scope through the 20th century; most notably after the rise of animal welfare science in the mid-1960s and can be expected to grow and change significantly in the future. We are at a particularly interesting and exciting time in the history of the field as our understanding of animals is growing rapidly and

DOI 10.3920/978-90-8686-828-5_3, © Wageningen Academic Publishers 2016

societal perceptions of animals and their use are changing. Recent advances in our understanding of the neural underpinnings of animal emotion have provided greater insight into cognition and positive states, and new technologies have added significantly to multidisciplinary collaboration and tools for research. At the same time, the ethics of animal use and production practices are facing increasing societal scrutiny. Added to this we have increasing global demands for food as well as demands for more sustainable agricultural practices, all of which may increase the need for research in applied ethology.

Following on the tree analogy followed in Chapter 2 (Newberry and Sandilands, 2016), this chapter explores some of the sweet and colourful fruits of applied ethology in order to illustrate why this is such an interesting field, and what makes ethologists so passionate about their work. Themes include human-animal interaction, positive emotional states, evolutionary biology, the relationship between ethology and animal welfare and future directions related to increasing education and dissemination of the science. These topics are touched on briefly here, with the goal of increasing the reader's appetite for subsequent chapters.

3.2 Why animal behaviour?

Why are we humans so attracted to observing other animals? It is a simple question, and several theories have been proposed to describe the biological basis for this attraction: biophilia, otherness, comparative studies, attachment or social support (ASAS, 2014). While we may never fully understand or describe this connection, the scientific approach has certainly expanded our knowledge and understanding of animals, and (for those of the right persuasion) studying the behaviour of animals is one of the most pleasant pass times and satisfying occupations there is.

The relationship between domestic animals and humans is essentially symbiotic. The wild progenitors of domesticated species received fitness benefits through their association with humans, and humans have similarly benefitted from the association with animals in the form of more reliable sources of nutrition (in the form of milk, meat or eggs), clothing, fertiliser, protection, companionship, assistance in hunting and pest control, among other activities. Since their initial domestication these species have been further selected, both directly and indirectly, to tolerate greater human interaction and control. On average, they show smaller brain size and reduced fear responses compared to their wild progenitors (Kruska, 2005). At the same time, the differences are largely a matter of degree since domestic species retain similar behaviour and motivations compared to their wild progenitors, and many (e.g. horses, pigs, chickens, dogs, rats and goats) can exist in feral populations given the right conditions. As Per Jensen (2016) describes in Chapter 15, many early ethologists considered domestic species to be degenerate versions of their wild cousins, and therefore not suitable subjects for behavioural research. Luckily for us, this bias was not sufficient to deter Jensen's early interest in pig behaviour.

In the case of our domestic livestock, most of which are social prey species, understanding that they are prey animals and learning to adjust our behaviour to reduce fear and promote a trusting bond has been an important step in improving animal health and productivity. Within the field of applied ethology, Paul Hemsworth and colleagues John Barnett and Grahame Coleman in Australia are well known for their early work exploring the benefits of positive handling on livestock, and demonstrating the importance of human beliefs and attitudes in reducing handling stress. In studies dating back to the 1980s they showed that positive human interactions produced measureable benefits on production and health in livestock (reviewed in Hemsworth and Coleman, 2011). Some of this research is described by Hemsworth *et al.* (2016) in Chapter 9 of this book. More recent studies have explored more subtle effects of positive human interaction on animal affect as shown by measures such as cognitive bias (Brajon *et al.*, 2015) or increased play behaviour (reviewed by Held and Špinka, 2011, and discussed below). In the study by Brajon *et al.* (2015), piglets which received regular gentle contact from human handlers showed a more positive (optimistic) cognitive bias than piglets which received rough or minimal contact treatments.

As well as documenting the effects of positive handling experience, Australian researchers also led the way in collaborative research by exploring interactions between human psychology and animal behaviour, and how to increase positive human-animal interactions by affecting the beliefs and attitudes of caretakers. Studies in dairy cattle and pigs have shown that the attitudes of human handlers are correlated with their behaviour towards animals, with a positive attitude towards animals being negatively correlated with the number of slaps, pushes and hits directed towards the animals (Coleman and Hemsworth, 2014). Based on these associations, training programmes for stockpeople were designed to help develop and support positive beliefs related to animals using cognitive-behavioural techniques. These programmes are based on the Theory of Planned Behaviour (Ajzen and Fishbein, 1980); the concept that human attitudes and actions reflect underlying beliefs, so improving stockperson behaviour requires first, an assessment of the person's underlying beliefs, second, training to address specific issues related to their beliefs and to influence behavioural change, and third, maintaining the behaviour.

More recently the human-animal bond (the effects of animals on humans) has also received attention. There is growing evidence supporting the role of interacting with nature and natural experiences in improving human health, and in particular the role of animals in this regard. Animal-assisted therapies come in many forms and their popularity is growing rapidly. Some examples include pets visiting seniors' homes, riding programs for the physically disabled, equine assisted personal development programs, and interactive psychological interventions. While the benefits of animal contact are not understood scientifically (and have perhaps been overblown by media and popular science accounts), studies in this area have demonstrated numerous benefits from owning a pet, including fewer visits to the doctor, being more physically active, feeling less lonely, and having reduced blood pressure during stressful tasks (Thodberg *et al.*, 2014).

Why would this be so? While we don't have a clear answer yet, we are getting steadily closer to understanding the biological basis of the benefits of human-animal interaction. These effects can be described in physiological, neurophysiological or psychological terms. The physiological benefits such as reduced blood pressure can possibly be explained by polyvagal theory, a theory proposed by neuroscientist, Stephen Porges (2011), who incidentally presented the ISAE's David-Wood Gush Memorial lecture on this subject in 2011. Polyvagal theory describes how the vagal and sympathetic branches of the autonomic nervous system have evolved to interact as a primary control system, allowing animals to maintain homeostasis when responding to environmental or social challenges. In the context of polyvagal theory, the calming effects of animal interaction are achieved through increased parasympathetic activation (via the ventral vagal complex) and reduced sympathetic activation. This shift in autonomic balance has profound consequences on heart rate, respiration, digestion, immune and stress physiology.

Psychologically, animal interaction has a positive impact through what neuroscientist, Jaak Panksepp (introduced in Chapter 2; Newberry and Sandilands, 2016) has described as our 'care' and 'seeking' systems (Panksepp, 1998). Stimulation of the care system activates brain systems responsive to oxytocin and endorphins and associated with the formation of social bonds. These effects are also related to the parasympathetic response and calming effects of vagal activation described above. However, it is through activation of the seeking system that we gain a real sense of pleasure from observing animals. The seeking system acts through dopaminergic pathways to generate a prevailing sense of anticipation and excitement, previously described as the reward system (Panksepp, 1998). There is a fascination in observing animals that is akin to watching other people, but is notably more direct, honest and centred in the present, and lacks the added cultural baggage of human social norms. The brain structures which compose the seeking system contain significant populations of dopaminergic neurons. In early studies where rats were wired so they could stimulate this region by pressing a lever, the rats pressed for hours (Olds and Milner, 1954) giving rise to the idea that it produced the sensation of a reward. Panksepp has suggested that seeking may be a more appropriate term because the behaviour appears to be appetitive in nature, whereas the term reward suggests more of a consummatory act.

Animals are complex and reflect an amazing variety of forms, behaviours and abilities. Their differences pique our curiosity and make them inherently interesting, while at the same time they reflect profound similarities to ourselves. As such, animals provide a rich subject matter that excites our innate enthusiasm for exploration. If enrichments are materials and activities that generate and excite exploration, then other animals may be the perfect human enrichment. This explanation may help to explain the profound interest that ethologists have for their subjects.

3.3 Animal minds

Another broad area that has seen much activity is the study of animal minds, including the development of methods and tools for assessing the mental states of animals and a new emphasis on positive affective states. These topics are reflected in themes for this year's ISAE congress in Edinburgh, where session titles feature topics such as positive welfare, consciousness, cognition and emotion, and individual differences.

Historically, behavioural studies have emphasised negative interactions and emotions, focusing on states and traits which have clear negative consequences such as aggression, stress, fear or pain. While this approach has helped us to understand problems and improve housing and management systems, positive aspects of animal behaviour have been avoided or largely ignored. This avoidance was due to cultural and scientific biases that are gradually changing. The Behaviourist school of psychology was prevalent in the early 20th century, and viewed the study of animal minds as outside the scope of science. Fraser *et al.* (1997) further traced this view to Positivist philosophy, which contends that scientific knowledge should be based on the observable phenomena of nature. At the time, it was argued that humans could not objectively measure animal thought; animal minds were described as a black box or a blank slate, and attempts to describe the emotional states of animals were criticised as being anthropomorphic (Kolakowski, 1969; Rollin, 1990). While we cannot directly observe the mental states of animals, we have developed many excellent ways to assess this indirectly. Taken to an extreme, this view would also suggest that particle physics is not a suitable area for science, however this clearly has not hindered physicists from developing ingenious ways to indirectly examine the nature of matter. Thankfully for us the study of animal minds has now become an acceptable topic of study. As Jean-Loup Rault and Rebecca Doyle (2016) describe in Chapter 5, this acceptance and the development of new methodologies and assessment techniques has opened the way for the expansion of research on what makes animals tick.

One highlight in this area is the expansion of research on play behaviour. Because play behaviour requires significant energy inputs and does not serve an immediate, obvious function, scientists initially found it difficult to explain why animals play (Held and Špinka, 2011). However, young animals in particular can spend a large amount of time in play, and it is obviously a highly motivated and self-rewarding activity (Figure 3.1). Research on play has been hindered by the fact that it is hard to control without introducing confounding variables. However, in terms of welfare assessment the fact that most young

Figure 3.1. Play behaviour is still enigmatic; how can an activity with such limited immediate function be so engaging? Theories around multi-level selection may help to explain activities such as play and altruistic behaviour. Photo credit: J. Brown and Y. Seddon.

mammals will engage in play when their basic needs are satisfied and they are provided a suitable environment has made it a valuable indicator of positive welfare. Further studies are needed to identify more clearly what motivates animals to play and determine the long term benefits of play behaviour, but from an ethologists perspective the best part is that observing play is always *fun!*

An exciting and related new area is the study of cognitive bias. Over the past decade studies on cognitive bias have exploded as this approach allows us to test the effects of different environments or management procedures on optimistic or pessimistic responses. Animals first must learn to associate specific cues with negative and positive outcomes, and are then tested by presenting them with an ambiguous intermediate cue (Bateson and Matheson, 2007). Cognitive bias is assessed based on the animal's response to the ambiguous cue: if the cue is interpreted as positive, there is a positive bias (optimism), and if it is interpreted as negative, there is a negative bias (pessimism).

While the principles behind studies of cognitive bias are clear, the approach is complicated and fraught with difficulty in its implementation (e.g. study design and time needed to train and test animals), and interpretation. These studies are in their early days, and perhaps an analogy can be made to the early preference tests which were often too simplistic, for example not taking into account animal's previous experience, the interaction between preferences and other factors such as temperature, or presenting choices not well suited to the animal (Duncan, 1978). With preference testing, much effort was invested to determine how the studies should be designed and implemented to avoid such problems (Fraser, 2008). Simple choice tests were further refined by scientists like Ian Duncan and Marian Dawkins, for example by introducing operant schemes that allowed animals to work for access to resources, or the use of consumer demand theory to explain motivation in terms of cost, which contributed greatly to the objective assessment of animals' wants and needs (Dawkins, 1983). The evaluation of cognitive bias allows us to go further and examine states of optimism and pessimism in animals. Ongoing improvement of methods will increase the reliability and sensitivity of studies, and future improvements could also involve automation of training and testing methods to reduce the time and labour required. Another pipe-dream is the use of cognitive bias for on-farm welfare assessment. While current methods for assessing cognitive bias require prior training of animals under controlled conditions, finding simpler methods that could be applied on-farm would provide a valuable tool for welfare assessment.

Positive affect and emotional states are also being explored at the level of the central nervous system. The work of neuroscientists Jaak Panksepp and Stephen Porges has helped to elucidate the pathways and neurochemistry of different emotions and responses to stress. In his ground breaking book, 'Affective neuroscience: the foundations of human and animal emotions' (1998), Panksepp presents an integrated and holistic view of the brain and behaviour by thoroughly reviewing the neuroscience literature and building an understanding of emotional systems from the 'bottom up', instead of imposing 'top down' assumptions based on historical scientific or cultural bias. Similarly, Steven Porges has developed polyvagal theory to describe the evolution of the autonomic nervous system and interactive contributions of sympathetic and parasympathetic branches (Porges,

2011). Polyvagal theory gives us an integrated view of the autonomic nervous system and behavioural and physiological responses, helping to explain responses to social interaction or threats and has been used widely in human psychology for understanding and treating trauma. It is an important concept but is not easy to summarise, and Porges' book is tough reading. Hopefully, someone will soon write 'polyvagal theory for dummies' and in the meantime several attempts to explain the theory can be found online.

3.4 Evolutionary biology

Another field that has contributed to our increasing knowledge and added to the discussion on altruistic behaviour and positive affect is evolutionary biology. Theories of reciprocal altruism and kin selection were developed in the 1960s to provide evolutionary explanations for altruistic behaviour, and in 1975 these concepts were popularised by E.O. Wilson in his book, 'Sociobiology: the new synthesis'. As Per Jensen (2016) describes in Chapter 15, when these theories were first proposed, some scholars predicted that this new approach would mean the end of classical ethology, which would be replaced by behavioural ecology and sociobiology. While new branches such as evolutionary biology and behavioural ecology did arise at this time, ethology did not disappear. Over time our increasing understanding of the importance of social and affiliative behaviour has led to questions and a re-examination of these models: are theories of kin selection and reciprocal altruism sufficient to explain the level of affiliative behaviour seen in social species? And can we explain altruistic behaviour that occurs between species: is this a form of symbiosis (e.g. involving reciprocal acts) or is altruism towards another species abnormal (non-functional expression of an inherent behaviour)?

Darwin's phrase 'survival of the fittest' was perhaps unfortunate, as we have tended to imagine fitness in terms of a competition among rival individuals, and to use evolutionary theory in the defence of class structure (social Darwinism), arguing that social status is an expression of genetic fitness. Culturally there still prevails a general acceptance that nature is 'red in tooth and claw', and often our aggressive and selfish tendencies are blamed on 'emotions' or 'animal instincts', and kindness and a sense of justice identified as a sign of restraint and result of learning that follows from 'civilisation' and our higher intellect. But these explanations tend to fail when we actually look at other animals, and we humans are often baffled to see high levels of cooperation displayed both within and between other species.

The noted primatologist, Frans de Waal, has explored positive emotions and affiliative behaviours in primates through detailed observations on primate groups at the Yerkes Research Centre in Atlanta, Georgia. Based on these observations he has produced several books on cognition, empathy and the animal origins of morality. In his book '*Good Natured*' De Wall (1996) credits Russian scientist, Petr Kropotkin, for being an early advocate of a more positive view of animal coexistence and cooperation. Perhaps essential to the development of Kropotkin's views on mutual aid were his observations on animals in harsh environments, where 'survival of the fittest' was more dependent on group cohesion in a struggle against the elements than on competition within a social hierarchy (Kropotkin, 1902).

Models of multi-level selection, which incorporate environmental factors and social interaction in estimates of fitness have recently gained (or regained) popularity. American geneticist, Bill Muir, has contributed significantly to this discussion in his use of a 'group selection' approach for poultry and the development of a 'kinder and gentler' chicken (2009). His theories consider how positive interactions within a group can contribute to the success of individual members. The term multilevel selection is often used for these effects, indicating both social and environmental interactions, and avoiding the term 'group selection' due to past criticisms of that approach (Wade *et al.*, 2010). Even E.O. Wilson, the 'father of sociobiology' has moderated his early views and rejected kin selection as providing a sufficient explanation for altruistic behaviour (Wilson, 2005). In his most recent book, '*The meaning of human existence*' (2014), Wilson argues for the central role of multilevel selection in the emergence and dominance of human civilisations.

However, the existence of multilevel selection is still far from being widely accepted. Evolutionary biologists have spent many years focusing on individual selection, and there remains much work in order to fully explain the more complex effects of how group and environment could influence selection. Many, including noted Oxford biologist Richard Dawkins, actively dispute multilevel selection as a possible mechanism for evolution, maintaining that individual selection is key to evolutionary processes. In his famous book, The Selfish Gene (1976), Dawkins modelled the costs and benefits of alternative individual strategies, and identified individual selection as being central to evolution and selfishness (at the gene level, if not always at the individual level) as an essential trait, and he strongly maintains this view today. I find this a bit disappointing as I much admire Dawkins' writing in The Selfish Gene, and especially his concept of memes: the fact that various forms of information (e.g. songs, books, habits) can be replicated and spread, much in the way that genes are promoted. In trying to conceive how multi-level selection might work, I imagine 'cultural memes' as habits within groups that could help to promote the overall welfare and cohesion of the group. Thus, with the right memes (positive social behaviours) some groups would be more successful than others, not just in terms of competing groups, but in the relative success of different groups coping in a harsh environment. However, it seems that Dawkins would disagree, and it will be interesting to see how this discussion evolves in the coming decade.

The development of new technologies has also contributed to the re-evaluation of evolutionary concepts in another way. The field of epigenetics and recent discoveries related to the effects of environmental factors on gene expression have reinstated some of the tenants of Lamarkian evolution (the idea that genes can be modified by life experience and these acquired characteristics can be inherited). While DNA are not directly affected by life experience, we now know that environmental factors can indeed influence gene expression, giving animals the ability to adapt to specific environmental stressors even before birth. As Per Jensen (2016) describes in Chapter 15, new advances in the study of epigenetics and metabolomics may finally elucidate the physiological and neurochemical steps involved in potentiating behavioural responses, bridging gaps in our knowledge on how complex behaviours are encoded in the genome.

3.5 Applied ethology and animal welfare

Another recurring theme touched on within this book is the strong association between applied ethology and animal welfare science, and the relationship between these sciences and changing cultural views on ethics and our relationship with animals. Applied ethology is by nature a holistic field and shares this trait in common with animal welfare science. In other aspects they are very different, with applied ethology focusing more on the basic science of animal behaviour, and animal welfare delving into philosophical issues of ethics and differing cultural views. This relationship can be seen as complimentary, with animal welfare being the 'yin' to applied ethology's 'yang', and the dynamic of these two perspectives adds depth to our field and gives it greater cultural relevance.

Historically, these fields arose at essentially the same time and they have remained closely linked. As described in Chapter 1 (Petherick and Duncan, 2016), societal questions regarding animal use and care arose initially in the UK in the 1960s. These concerns have since grown and spread to other regions, and combined with significant increases in livestock production have created a strong climate for research into animal welfare. The initial growth took place largely in European countries, with first the Council of Europe and then the European Union playing a significant role in establishing regulations and standards for livestock production. Some of this history and the involvement of ISAE members are described by Bo Algers (2016) in Chapter 7, and Don Broom (2016) in Chapter 14. Within Europe, the United Kingdom, Sweden and Norway have been leaders in this regard, imposing stricter rules over and above the EU standards.

Within the EU, challenges such as the BSE crisis added impetus to the development of government regulations on animal agriculture. In other developed countries such as the USA, Canada, Australia and New Zealand, free-market systems are more prevalent and governments have been reluctant to impose strict regulations or to require mandatory audits. In these countries, livestock industries and retailers take on a greater role in self-regulation in terms of developing and monitoring standards of animal care. These can be in the form of voluntary standards, first or second party audits, with third party audits at farm level for niche or branded products. However, the past decade has seen also brought significant changes in these countries, with multiple public 'ballot initiatives' in the USA, and increasing standards in abattoirs, such as those developed by the American Meat Association in partnership with Temple Grandin. More recently, countries such as Brazil, Mexico, China and Japan have begun to adopt on-farm animal welfare standards and audits. As exporters of animal products, their goals are generally more directed towards accessing export markets rather than satisfying concerns within domestic markets. On a global level, the World Organisation for Animal Health (OIE) is developing basic standards for livestock production, with standards for beef and dairy cattle and broiler chickens already in place, and swine and equine codes under development (see Chapter 14 (Broom, 2016) for discussion of the ISAE's role within the OIE).

While applied ethology and animal welfare science are closely linked, the relationship has been largely dominated by a focus on animal welfare research. Without the societal emphasis placed on animal welfare over the past 50 years (predominately in EU countries), it seems that applied ethologists would have had much less in the way of funding or cultural relevance. While welfare research has been a great boon to ethologists, the emphasis on welfare has resulted in fundamental work on behaviour being ignored to some extent. Perhaps it is time to review our focus and let fundamental behaviour research shine?

3.6 Ethics and sustainability

What direction will future research take? Will demands for animal welfare science remain a strong driving force, or will environmental sustainability or other areas drive research? Much of this will depend on our ethical standards and the health of economies; increasing cultural concern over animal use may drive more research, but this work also requires a certain level of affluence. Historically, there has been reduced interest in animal welfare during poor economic times. As noted by Jeremy Marchant-Forde and Bas Rodenburg (2016) in Chapter 16, animal agriculture in particular and the human race in general are facing serious challenges. Along with increasing population and increasing demands for food and resources, public opinion on animal welfare issues is shifting. The continued urbanisation of human populations has increased the separation of rural and urban perspectives, distancing animal agriculture and placing it under greater scrutiny; the average citizen is increasingly connected to information and technology through the internet, and is less and less connected to agriculture

Many within the agricultural sector are working to improve communication and public education about agriculture, and increase 'transparency'. Examples include 'Farm and Food Care' programs in Canada, and the American Meat Institute's 'glass walls' project in the USA, which has posted a series of online videos on farming issues, including walk-through tours of beef and pig packing plants with Temple Grandin showing each step of the process (AMI, 2013). But animal agriculture in particular has many vocal detractors and these education efforts seem like a small drop in the bucket. In response to concerns about social licence, David Fraser has suggested that farmers may need to adopt a professional approach, adopting standards of production and certification in a systematic way to demonstrate self-regulation, continuous improvement and to increase public confidence in food production (NFACC, 2016). How professional standards might be implemented remains to be seen, but in many ways this trend is already occurring with the consolidation of production and increasing vertical integration. This can be seen in initiatives such as McDonald's audits of packing plants and sustainable beef production, or for example in the adoption of group sow housing systems by Smithfield Farms in the USA and Maple Leaf Farms in Canada, which is being carefully managed and taking place well ahead of most private producers.

But what is our role as scientists? Should we stand apart and try to maintain objectivity, or is some defence of agriculture needed to address the barrage of misinformation presented in social media? While admittedly there are problems within agriculture these are not insurmountable and I much prefer the approach of continuous improvement over abolition. I hate to imagine a world dominated by humans with few other animals remaining, domesticated or otherwise. We have lived successfully and symbiotically with animals for thousands of years, and domesticated species are especially well adapted and have shown exceptional tolerance to living in close contact with humans.

Now that human activity encompasses the planet and is affecting global climate change we are affecting all animal populations, not just those managed directly, and applied ethology can theoretically be said to involve all species. We should accept that our influence is far-reaching, and take a more active role to manage domestic and wild animal populations. As discussed by Francisco Galindo *et al.* (2016) in Chapter 10, this responsibility is especially evident in Central and South America where human activity is threatening many rare and endangered species.

The past 50 years have seen immense improvements in the efficiency of animal production through genetic selection, as well as improvements in animal nutrition, management and manure management. These have increased the sustainability of farming, but there are still many opportunities for improvement. For example, feed efficiency has been shown to be a heritable trait, and there is increasing interest in genetic selection for feed efficiency. Researchers have also begun to identify and promote beneficial gut microbes with the goal of improving health and productivity and reducing the need for antibiotics.

In terms of individual consumption of meat products, average global meat consumption is expected to rise due to growth of the middle class in China, Brazil, many other of our fastest growing countries. Changing technologies to allow 'precision farming', increasing automation and more efficient production will be needed to meet these demands with higher standards of care. As previously mentioned, better training and greater professionalism are needed in order to implement these changes.

In conclusion, we can anticipate increases in the volume and efficiency of livestock production as well as continuing pressure to improve the welfare of farmed animals. Genetic selection for increased production has made great strides but is approaching biological limits of efficiency and impacting neonatal survival and healthy growth. All of these factors suggest and increasing roles for applied ethologists in terms of identifying new selection traits and in developing alternative management systems.

3.7 Multidisciplinary science and education

To conclude this chapter, I will discuss two more themes related to the future of applied ethology and the ISAE. First, the multidisciplinary and collaborative nature of applied ethology research, and second, the importance of continuing education and outreach.

Because behaviour is integrative it has always required a more holistic approach than the classical sciences. Applied ethologists are considered generalists when compared to other biological fields that are more reductive and better suited to specialisation. Both approaches have their weakness. I am reminded of the old joke about specialists and generalists: the specialist learns more and more about less and less, until eventually he/she knows everything about nothing, while the generalist knows less and less about more and more, until eventually he/she knows nothing about everything. This line of thinking in itself is a good reason for collaboration, as it is one way to promote a balance between these two extremes. Following on this theme in Chapter 15, Per Jensen (2016) discusses the importance of cross-fertilisation between applied ethology and other fields of science, including physiology, psychology, evolutionary biology and genetics. The theme of cross-disciplinary collaboration is also suggested in Tinbergen's Four questions, which emphasise four distinct approaches to understanding behaviour (Tinbergen, 1963). I would argue that for both scientific and pragmatic reasons, collaborative and multidisciplinary approaches are needed now, more than ever.

In terms of advancing the study of animal minds, we are approaching a convergence of concepts in psychology, cognition, neuroscience, immunology, developmental biology, and physiology. On the down side, each of these areas has a distinct approach to studying and conceptualising the brain and behaviour, and collaboration across fields can make research more difficult (and frustrating) due to differing conceptual approaches and terminology. However, all of these areas share significant overlap in their focus, so collaboration can allow researchers to address a broader scope of questions and give potential for new insights that will hopefully outweigh the added inconvenience. By using new tools for neural imaging and working on shared and disparate concepts we can anticipate a deeper understanding of animal behaviour and the emotional lives of animals. Collaborative work also helps to spread information about applied ethology as scientists from other fields can gain a greater understanding of the work we do, and how it adds value to animal research.

Participating in collaborative research is also arguably more interesting as it involves looking at interactions across systems: between genes and environment, behaviour and physiology, etc. Depending on the area and measures, responses may be small or individual variation large, and higher sample sizes are needed to obtain significance. This brings up a more pragmatic reason for collaboration, which is related to cost and funding support. From a funder's perspective, collaborative research is a better way of slicing research dollars to fund more participants, reduce duplication and to hopefully answer multiple questions within a single project. This model is being actively promoted by many funding agencies, and while the logistics of collaboration may be difficult it does promote greater participation and efficiencies related to shared resources and data.

The importance of communicating our science is another striking theme within this book. In Chapter 11, Shusuke Sato (2016) notes that applied ethology only expanded in Japan once some key texts of applied ethology by Niko Tinbergen and Konrad Lorenz had been translated, allowing Japanese readers to begin to appreciate and understand the subject. As with Dawkins' memes, the seeds of applied ethology were planted in the form of a few classic books, and their ideas and concepts took root. Also in Chapter 12, Jun Bao and Yuzhi Li (2016) describe the impact that Professor Zhen-Kang Han had on the development of applied ethology in China, resulting from the translation of several journal articles into Chinese and publication of an influential article on the application of ethology. Clearly having works of ethology written in various languages will help in disseminating our science. The ISAE already provides a volunteer service to help in preparing publications for authors for whom English is not their first language, but we could definitely do more to assist researchers in other countries. Actively promoting translation and the publication of key texts would be a simple and direct way to 'spread the word'.

Another opportunity for communicating about applied ethology is in the growing field of animal trainers and specialists. In Chapter 4, Daniel Mills and Roger Ewbank (2016) discuss the importance of valuing the knowledge of experienced laypersons, people who are 'good with animals'. This is an important point, especially in this era of internet communication. There are many individuals who are making successful careers marketing their own brand of horse or dog training, pet therapy, or other animal-related services. In many cases these are very talented individuals who combine marketing and people skills along with their experience in working with animals. Behavioural science and the ISAE have an important role to play in this area, first guiding the development of a common language and framework for describing and discussing alternative training methods, and also for applying scientific methods to understand the principals involved and develop testable hypotheses for evaluation and comparison of methods. As discussed in Chapter 16 (Marchant-Forde and Rodenburg, 2016), in collaboration with the Association for the Study of Animal Behaviour (ASAB) the ISAE has worked to develop a professional certification program for animal behaviourists within the UK, known as the Certification of Clinical Animal Behaviourists (CCAB). This certification is a step forward in standardising the qualifications of individuals working with the behaviour disorders of dogs, cats and other animals, ensuring that individuals working in this area have a common understanding and encouraging interaction with academics in the field. Going forward, it would be helpful to expand this role and to promote certification in other countries.

Collaboration and education are therefore two important activities that can help to demonstrate the utility of ethological approaches for understanding management and behaviour-related problems, and help to promote our science by raising awareness of ethology in other regions and within other disciplines.

3.8 Conclusions

In conclusion, applied ethology is young, dynamic and rapidly growing field of science. This chapter has introduced some exciting recent developments, and highlights some of the main themes presented in this book. The descriptions here are brief as their purpose is whet the reader's appetite with the goal of increasing the motivation to consume the remaining chapters.

Human-animal interaction is an expanding area within applied ethology, as we attempt to understand how animals respond to humans, the human psychology of animal handling, and why we humans are so attracted to observing and working with animals. One neural explanation for the excitement ethologists experience when observing animals is through activation of the seeking system; animals present us with a fascinating opportunity for exploratory behaviour. Over the past 50 years we have gradually cast aside biases against the study of animal minds, and this has become a productive new area of research. We have also begun to focus more intensely on positive emotions and affiliative behaviour. The increasing emphasis on positive and altruistic behaviour has been accompanied by a re-evaluation of evolutionary biology, and the promotion of multi-level models of genetic selection.

Since the 1960s, applied ethology and animal welfare science have essentially grown up together. While this close association has been very successful and is likely to continue, I hope that some researchers in applied ethology will return to more basic studies. However, societal factors will largely dictate the direction of future research. Factors such as the increasing rural vs urban divide, economics, and growing environmental issues will influence the importance of ethics, animal welfare, and sustainability issues. Looking ahead, there are significant opportunities for collaborative research and new discoveries with the potential to improve understanding of animals, and to benefit the mental and physical health and welfare of humans and animals. The ISAE has an important role to play in this work and in promoting education and furthering the development of high quality science in this field.

References

Ajzen, I. and Fishbein, M., 1980. Understanding attitudes and predicting social behaviour. Prentice-Hall Inc., Eaglewood Cliffs, NJ, USA.

Algers, B., 2016. Applied ethology in the EU: development of animal welfare standards and actions. In: Brown, J.A., Seddon, Y.M. and Appleby, M.C. (eds.) Animals and us – 50 years and more of applied ethology. Wageningen Academic Publishers, Wageningen, the Netherlands, pp. 155-173.

American Meat Institute (AMI), 2013. If meat plants had glass walls. Available at: http://animalhandling.org/ht/a/GetDocumentAction/i/89462.

American Society for Animal Science (ASAS), 2014. The human animal bond and domestication: through the ages ... animals in our lives. Animal Frontiers 4: 3.

Bao, J. and Li, Y., 2016. China perspective: emerging interest in animal behaviour and welfare science. In: Brown, J.A., Seddon, Y.M. and Appleby, M.C. (eds.) Animals and us – 50 years and more of applied ethology. Wageningen Academic Publishers, Wageningen, the Netherlands, pp. 241-252.

Bateson, M. and Matheson, S.M., 2007. Performance on a categorisation task suggests that removal of environmental enrichment induces 'pessimism' in captive European starlings (*Sturnus vulgaris*). Animal Welfare 16: 33-36.

Brajon, S., Laforest, J.P., Schmitt, O. and Devillers, N., 2015. The way humans behave modulates the emotional state of piglets. PLoS ONE 10: e0133408.

Broom, D.M., 2016. World impact of ISAE: past and future. In: Brown, J.A., Seddon, Y.M. and Appleby, M.C. (eds.) Animals and us – 50 years and more of applied ethology. Wageningen Academic Publishers, Wageningen, the Netherlands, pp. 269-278.

Coleman, G.J. and Hemsworth, P.H., 2014. Training to improve stockperson beliefs and behaviour towards livestock enhances welfare and productivity. Scientific and Technical Review of the Office International des Epizooties 33: 131-137.

Dawkins, M.S., 1983. Battery hens name their price: consumer demand theory and the measurement of ethological 'needs'. Animal Behaviour 31: 1195-1205.

Dawkins, R., 1976. The selfish gene. Oxford University Press, New York, NY, USA.

De Waal, F.,1996. Good natured. Harvard University Press, Cambridge, MA, USA.

Duncan, I.J.H., 1978. The interpretation of preference tests in animal behaviour. Applied Animal Ethology 4: 317-326.

Fraser, D., 2008. Understanding animal welfare: the science in its cultural context. UFAW. Wiley-Blackwell, Oxford, UK.

Fraser, D., Weary, D.M., Pajor, E.A. and Milligan, B.N., 1997. A scientific conception of animal welfare that reflects ethical concerns. Animal Welfare 6: 187-205.

Galindo, F., Tadich, T., Ungerfeld, R., Hötzel, M.J. and Miguel-Pacheco, G., 2016. The development of Applied Ethology in Latin America. In: Brown, J.A., Seddon, Y.M. and Appleby, M.C. (eds.) Animals and us – 50 years and more of applied ethology. Wageningen Academic Publishers, Wageningen, the Netherlands, pp. 211-225.

Held, S.D. and Špinka, M., 2011. Animal play and animal welfare. Animal Behaviour 81: 891-899.

Hemsworth, P.H. and Coleman, G.J., 2011. Human-livestock interactions: the stockperson and the productivity and welfare of intensively-farmed animals, 2nd edition. CABI Publishing, Oxford, UK.

Hemsworth, P.H., Mellor, D.J., Coleman, G.J., Beausoleil, N.J., Fisher, A.D. and Stafford, K.J., 2016. Australian and New Zealand perspectives. In: Brown, J.A., Seddon, Y.M. and Appleby, M.C. (eds.) Animals and us – 50 years and more of applied ethology. Wageningen Academic Publishers, Wageningen, the Netherlands, pp. 193-209.

Jensen, P., 2016. New answers – old questions; new questions – old answers: how applied ethology is cross-fertilised by other disciplines. In: Brown, J.A., Seddon, Y.M. and Appleby, M.C. (eds.) Animals and us – 50 years and more of applied ethology. Wageningen Academic Publishers, Wageningen, the Netherlands, pp. 279-295.

Kolakowski, L., 1969. The alienation of reason. A history of positivist thought. Doubleday, New York, NY, USA.

Kropotkin, P., 1902. Mutual aid. Cosimo classics, New York, NY, USA.

Kruska, D., 2005. On the evolutionary significance of encephalization in some eutherian mammals: effects of adaptive radiation, domestication, and feralization. Brain, Behavior and Evolution 65: 73-108.

Marchant-Forde, J.N. and Rodenburg, T.B., 2016. Future directions for applied ethology. In: Brown, J.A., Seddon, Y.M. and Appleby, M.C. (eds.) Animals and us – 50 years and more of applied ethology. Wageningen Academic Publishers, Wageningen, the Netherlands, pp. 297-317

Mills, D.S. and Ewbank, R., 2016. ISAE, ethology and the veterinary profession. In: Brown, J.A., Seddon, Y.M. and Appleby, M.C. (eds.) Animals and us – 50 years and more of applied ethology. Wageningen Academic Publishers, Wageningen, the Netherlands, pp. 95-111.

Muir, W.M., 2009. Genetic selection and behaviour. Canadian Journal of Animal Science 89: 182-182.

National Farm Animal Care Council (NFACC), 2016. Could animal production become a profession? National Farm Animal Care Council webinar by D. Fraser, Available at: https://www.youtube.com/watch?v=ZoKWvPY5C1c.

Newberry, R.C. and Sandilands, V., 2016. Pioneers of applied ethology. In: Brown, J.A., Seddon, Y.M. and Appleby, M.C. (eds.) Animals and us – 50 years and more of applied ethology. Wageningen Academic Publishers, Wageningen, the Netherlands, pp. 51-75.

Olds, J. and Milner, P., 1954. Positive reinforcement produced by electrical stimulation of the septal area and other regions of rat brain. Journal of Comparative and Physiological Psychology 47: 419-427.

Panksepp, J., 1998. Affective neuroscience: the foundations of human and animal emotions. Oxford University Press, New York, NY, USA.

Petherick, J.C. and Duncan, I.J.H., 2016. The International Society for Applied Ethology: going strong 50 years on. In: Brown, J.A., Seddon, Y.M. and Appleby, M.C. (eds.) Animals and us – 50 years and more of applied ethology. Wageningen Academic Publishers, Wageningen, the Netherlands, pp. 27-50.

Porges, S., 2011. The polyvagal theory: neurophysiologial foundations of emotions, attachment, communication, and self-regulation. W.W. Norton and Company, New York, NY, USA.

Rault, J.-L. and Doyle, R., 2016. Cognitive approaches and new technologies: changing methodologies in applied ethology. In: Brown, J.A., Seddon, Y.M. and Appleby, M.C. (eds.) Animals and us – 50 years and more of applied ethology. Wageningen Academic Publishers, Wageningen, the Netherlands, pp. 113-132.

Rollin, B.E., 1990. The unheeded cry. Oxford University Press, Oxford, UK.

Sato, S., 2016. Applied animal behaviour science in Japan and the culture of '*aigo*'. In: Brown, J.A., Seddon, Y.M. and Appleby, M.C. (eds.) Animals and us – 50 years and more of applied ethology. Wageningen Academic Publishers, Wageningen, the Netherlands, pp. 227-240.

Thodberg, K., Berget, B. and Lidfors, L., 2014. Research in the use of animals as a treatment for humans. Animal Frontiers 4: 43-48.

Tinbergen, N., 1963. On aims and methods of ethology. Zeitschrift für Tierpsychologie 20: 410-433.

Wade, M.J., Bijma, P., Ellen, E.D. and Muir, W., 2010. Group selection and social evolution in domesticated animals. Evolutionary Applications 3: 453-465.

Wilson, E.O., 1975. Sociobiology: the new synthesis. Harvard University Press, Cambridge, MA, USA.

Wilson, E.O., 2005. Kin selection as the key to altruism: its rise and fall. Social Research 72: 159-166.

Wilson, E.O., 2014. The meaning of human existence. W.W. Norton and Company, New York, NY, USA.

4. ISAE, ethology and the veterinary profession

D.S. Mills[1*] *and R. Ewbank*[2]
[1]*School of Life Sciences, University of Lincoln, Brayford Pool, Lincoln, Lincolnshire, LN6 7DL, United Kingdom; dmills@lincoln.ac.uk*
[2]*11 Thornley Place, Compton St., Ashbourne, Derbyshire, DE6 1PQ, United Kingdom*

This chapter is dedicated to the late H.D. Mills FRCVS (1923-2014), practising veterinarian and academic, father of DSM

Abstract

Academic interest by veterinary surgeons in animal behaviour extends back to well before the scientific surge of the early1960s, i.e. before concerns were being raised over factory farming. Historically, the veterinary profession has always had an interest in behaviour and the rise of applied ethology and the ISAE has served to not only meet this interest but also increase it. In this chapter we explore the relationship between the veterinary profession and the study of applied animal behaviour from a largely historical perspective, beginning with why understanding behaviour always has been, and will always continue to be, integral to veterinary practice. Descriptions of animal management and behavioural problems in early veterinary texts show extensive appreciation of behaviour, nutrition and management factors, as well as concerns for animal welfare, and formed the basis for the development of research and education in veterinary ethology. Since the 1960s education and knowledge of animal behaviour has spread more formally through the veterinary profession. In the last part of the chapter we discuss the development of the growing field of veterinary behavioural medicine, and the emergence of a psychobiological (rather than medical or psychological) approach to this discipline. We conclude by speculating on some of the likely areas of applied ethology that might be of most importance to the veterinary profession. Clearly it is essential that veterinarians continue to work alongside ethologists and scientists from other disciplines if the field is to continue to flourish.

Keywords: animal psychology, behavioural medicine, education, history, veterinary ethology

4.1 Introduction

The ISAE was originally set up as the Society for Veterinary Ethology, and veterinarians have featured prominently in its history. In this chapter we explore the relationship between the veterinary profession and the study of applied animal behaviour, from a largely historical perspective, beginning with why understanding behaviour always has been, and will always be, integral to veterinary practice. This is true regardless of the amount of explicit education offered on ethology to veterinary students, however education programs and scientific research will do much to improve this understanding

DOI 10.3920/978-90-8686-828-5_4, © Wageningen Academic Publishers 2016

and so we also consider the development of formal educational opportunities. In the last part of the chapter we consider the development of the growing field and practice of veterinary behavioural medicine, which seeks to manage individual patients with behavioural problems, and the emergence of a more biological (rather than medical or psychological) approach to this discipline. We conclude by speculating on some of the likely areas of applied ethology that might be of most importance to the veterinary profession.

4.2 Ethology and veterinary science: longstanding bedfellows

Practising veterinarians have always been dependent upon knowledge of certain practical aspects of the behaviour of domesticated animals to carry out their duties to society safely and efficiently. These duties include the relief of suffering in animals that are used by and under the care of humans by diagnosing, treating and, at times, preventing injury and disease, by controlling diseases causing death and loss of production in farm livestock, and by controlling (together with the medical profession) diseases which can be transmitted from non-human animals to humans (zoonoses) and the converse (anthroponoses). The academic interest of veterinary surgeons in animal behaviour extends back to well before the early1960s, i.e. before concerns were being raised over factory farming and the surge in the study of the animal behavioural sciences in their own right that arose as a result.

Veterinarians were (and still are) interested in information on:

- behaviour relevant to the safe and efficient handling of animals;
- changes observed or reported in behaviour as symptoms of injury, disease or suffering;
- species typical behaviour which may affect disease processes, such as the influence of lying behaviour on mastitis in cattle;
- disturbances of behaviour which conflict with human interests or perceptions of good management or welfare, such as heightened arousal in cats and dogs, repetitive behaviours like tongue rolling in calves, pacing in polar bears or weaving and cribbing in horses. Cribbing and other repetitive behaviour of horses have historically been referred to as 'stable vices', putting an emphasis on the horse's constitution rather than its management, when both are clearly important.

Relevant, veterinary, behavioural information (even if just snippets) was (and still is) obtained from diverse sources, with a growing role for learned societies such as ISAE. These sources include:

- direct and indirect personal experience, often passed on by word of mouth between owners, keepers, managers, veterinarians, students and others: science in this field often has to fight hard against strongly held beliefs and traditions;
- formal programmes of education, which may focus on behaviour or merely mention it in passing, in lectures and practical sessions on animal husbandry and clinical instruction, within veterinary and animal sciences programmes;
- the exponentially increasing number of publications; these include scientific papers in formal periodicals, but also non-scientific reports, papers and letters in informal periodicals, newsletters and magazines (so-called grey literature) which appear from

time to time and which are rarely retained by libraries or covered by the indexing or abstracting services, but which can create their own 'perceived wisdom', such as breed stereotypes.

It is still difficult to access much of the occasional and scattered behavioural material in the formal veterinary literature prior to the start of the abstracting and indexing journals and services in the late 1930s. One way to address this partially is to consult the occasional published catalogues of the main veterinary libraries. For example, the UK's Royal College of Veterinary Surgeons (RCVS) produced, in 1955, a 'Catalogue of Modern Works 1900-1954'. This catalogue listed some 1,870 books and booklets and yet, under the term Behaviour in the subject index, only 14 individual books were mentioned. We could be tempted to assume (perhaps unfairly) that behaviour was not seen as an important area of veterinary interest, and therefore did not really need to be formally taught or examined as a separate subject in university veterinary education in the UK in the first half of the 20th Century. We shall examine this assumption, continuing to take the UK as an example.

As early as 1904, Williams published a paper entitled '*Some Neuroses of Animals*' (Williams, 1904). This dealt largely with domesticated animals that exhibited nervous symptoms but did not show, on post mortem examination, tissue changes regarded as pathological in the early 1900s. In 1910 an English translation of Friedberger and Frohner's two volume book on '*Veterinary Pathology*' was published. It contained an extended account of 'crib-biting' and 'jibbing' in the horse. Then in 1946 the 5th English edition of Hutyra and Marek's '*Special Pathology and Therapeutics of the Diseases of Domestic Animals*' (regarded by many as a good source of information on how mainland European veterinarians regard disease in general) was published. In the section on Functional Nervous Disease in its 3rd volume, neuroses with no recognisable anatomic base, and psychoses such as anxiety psychoses, mass panic and hysteria in the dog, were covered. Around the same time, an important, lengthy, critical paper on the serious need for the veterinary profession to take on animal behaviour was printed in The British Veterinary Journal (Edwards, 1945). This paper was closely followed in the same journal, the next year, by a 12 page contribution from Barry Cross (a final year student at the Royal Veterinary College, London) on animal behaviour in relation to veterinary science. There were also a few veterinarians in the UK studying animal behaviour and publishing scientific papers on the subject in the decade before and leading up to the infamous report of the Brambell Committee in 1965 that did so much to advance the field. Alexander Brownlee, for example, produced a number of papers on play, etc. in cattle (Brownlee, 1950, 1954) which are still regarded as unique and quotable references. This was soon followed by further publications on cattle behaviour by Andrew Fraser and Roger Ewbank (Ewbank 1961, 1964; Fraser 1957a,b). Around this time, Michael Fox (a UK veterinary graduate working in the USA) was also pioneering academic publications on companion animal species, with his work on dogs (Fox, 1964a,b), following the pioneering publication of the textbook '*The Mind of the Dog*' by veterinarian Reginald Smythe in 1958. So clearly the view of veterinary ethology as a recent phenomenon is at least partly misplaced, although we do not wish to take away anything from the pioneers of applied animal ethology who have helped to develop the science behind the discipline.

4.3 Veterinary ethology: a brief historical perspective across the species

Prior to the emergence of highly effective and safe antibiotics for treating infectious disease in veterinary practice during the mid-20th century, managemental intervention was an integral part of medical treatment. Behaviour problems were also considered important at this time: see for example James Lupton's classic essay (1884) on '*Evils of modern stables*'. Long before the discovery of dopamine, which leads animals to seek out stimulation when their essential needs are met, and long before the development of intensive farming methods, Lupton stated (p. 155) that:

> It does not require any vast expenditure of thought to discover that life is action; 'to be', is synonymous with 'to do': therefore, it is a sheer necessity of existence that an animated being must be doing something ... Yet it remained for mortal perversity to rebuke the first principle of established philosophy, when stables were built, in which a breathing animal was to be treated as if it were an inanimate chattel.

Sadly it might well be argued that the Victorian stables and stable management, criticised so heavily by him, were in many ways better than modern standards, which lead to so many psychological and behaviour problems. Indeed classic texts such as *Mayhew's Illustrated Horse Doctor* (1888) (Figure 4.1) discussed important observations of the development and management of stereotypic behaviour in horses, such as cribbing, which appear to have been forgotten until recently. Thus (p. 171, emphasis in original):

> to relieve crib-biting ... Place a lump of rock-salt in the manger ... add to it a piece of large chalk ... always damp the food, and at each time of feeding, sprinkle magnesia upon it, and mingle a large handful of ground oak bark with each feed of corn. Should none of these measures prove beneficial, treat the case as one of *chronic indigestion or gastritis.*

While increasing fibre (adding oak bark in the case of Mayhew), has been a long established intervention, the use of antacids is often seen as a much more recent innovation (Mills and Macleod, 2002; Nicol *et al.*, 2002).

While these two veterinarians, Lupton and Mayhew, were perhaps exceptional in their insight, their ideas were not considered exceptional and we would do well to read and listen carefully to the observations of those who live closely with the animals they manage. As H.D. 'Henry' Mills used to observe and teach others, when dealing with those who appear to know a lot or a little, we should never confuse a person's wisdom with their knowledge. Someone with no education may be very wise and insightful, and it is our job with the benefit of scientific knowledge to evaluate and not simply dismiss ideas, just because there is currently no evidence.

An all too common failing is to treat a lack of evidence as evidence of a lack of importance, when we should perhaps err on the side with the less severe consequences (the

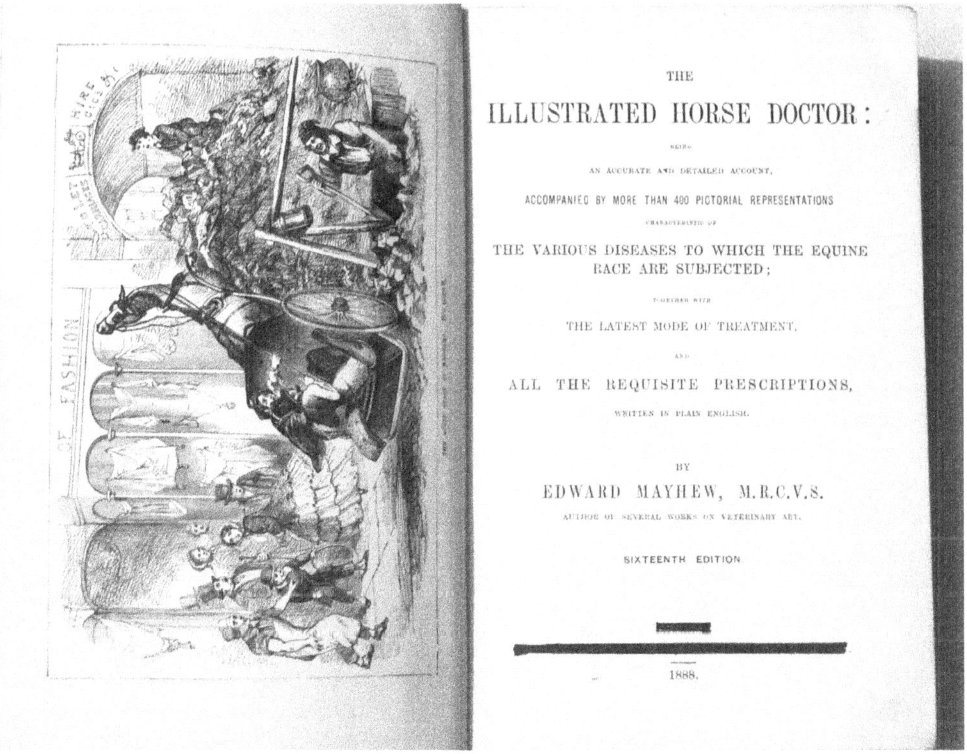
THE

ILLUSTRATED HORSE DOCTOR:

AN ACCURATE AND DETAILED ACCOUNT,

ACCOMPANIED BY MORE THAN 400 PICTORIAL REPRESENTATIONS

THE VARIOUS DISEASES TO WHICH THE EQUINE RACE ARE SUBJECTED;

THE LATEST MODE OF TREATMENT,

ALL THE REQUISITE PRESCRIPTIONS,

WRITTEN IN PLAIN ENGLISH.

BY

EDWARD MAYHEW, M.R.C.V.S.

AUTHOR OF SEVERAL WORKS ON VETERINARY ART.

SIXTEENTH EDITION

1888.

Figure 4.1. Frontispiece for Mayhew's 'Illustrated Horse Doctor'.

precautionary principle). This observation is very pertinent when considering the rise of the ethologist to work alongside the veterinary profession. This rise is often traced back to public outcry over intensive animal farming following the publication of Ruth Harrison's book '*Animal Machines*' (1964). However, it is worth noting that a couple of years earlier, there had been another direct appeal to the profession, particularly in the UK (Taylor, 1962), that animal behaviour should occupy a central place in the field of veterinary science. This was published in the Veterinary Record, a well-respected, mainstream publication read widely by many members of the profession, being the official journal of the British Veterinary Association. Also in 1962, the Association of Veterinary Students, in the UK and Eire, had shown themselves conscious of the needs of the student in relation to animal behaviour by holding a symposium on the subject (AVS Report, 1964). The first edition of Hafez's substantial (circa 600 pages), authoritative, fully documented, multi-authored book '*The Behaviour of Domestic Animals*' was also published in this year. It is perhaps fair to say that until this text was published few people had realised how much work had been done on the behaviour of the domesticated species. At the same time in France, veterinarians Brion and Ey (1964) published a similarly substantial (600 pp.) multi-author text '*Psychiatrie Animale*', which although it was never published in English was reviewed extensively in Archives of General Psychiatry (Bailey, 1964) and had a large influence on the development of the companion animal discipline in its native country (Pageat, 1996).

By contrast, Harrison made a more direct public appeal, taking a common sense look at what the veterinary profession appeared to have allowed to develop while they were supposed to be the guardians of animal welfare. There was criticism of the profession for focussing too much on the importance of physical health for animal welfare and productivity at the expense of psychological factors. A major criticism was that seemingly, there was no teaching or examination of animal behaviour within the veterinary curriculum, but of course, behavioural matters were often discussed in relation to postulated problems, such as signs and spread of disease, restraint and reaction to anaesthesia when they came up in clinical parts of the course. It seems that the RCVS (the regulatory body of the profession in the UK) had not taken action on the growing interest in animal welfare and the need for formal education on this matter within the profession. The government responded to Harrison's book by establishing a multidisciplinary committee, chaired by the eminent zoologist/immunologist Francis W. Rogers Brambell and including the ethologist William Thorpe, to examine the welfare of intensively farmed animals. The Brambell committee reported in 1965 and the Society of Veterinary Ethology (SVE, later ISAE) was formed in 1966, to add momentum to the discipline.

The backlash on the veterinary profession, which apparently was mainly voiced in committee during the 'examination' of those giving oral evidence on behalf of various veterinary organisations, could roughly be expressed in the rhetorical question: How could a profession justify its claim that it was the main guardian of animal welfare standards in the UK when many of its members, seemingly, had little knowledge of modern concepts in animal behaviour and welfare? One fairly quick response of the profession to this criticism was the British Veterinary Association's 1966 memorandum on '*The Place of Animal Behaviour in Veterinary Science*' published in the Veterinary Record. This report covered animal behaviour and research in general, and then animal behaviour in the veterinary teaching curriculum including suggestions for the coverage needed in a short behavioural course. Hafez's book was recommended and this remained the core behaviour text for veterinary students in the UK up until the late 1980s at some of the institutions where the subject was taught, e.g. Bristol. Further ideas as to the contents of short courses were published (Ewbank, 1967) in the Journal of Biological Education and also by Fox (1967).

The 1965 Brambell report established the concept of the Five Freedoms which was adapted by the Farm Animal Welfare Council in 1979 for inclusion in general animal welfare legislation globally, with the concept of 'ethological needs', later enshrined in the European Convention for the Protection of Animals Kept for Farming Purposes (1976). In that same year, Ron Kilgour published his review of the importance of psychology to the study of farm animal behaviour (Kilgour, 1976). Around the same time, reports of similar behaviour problems in zoos were beginning to be documented by zoologists such as Monika Meyer-Holzapfel (1968), interestingly in a book edited by veterinarian Michael W. Fox. However, zoos appeared to have recognised the importance of meeting ethological needs much earlier, with Young (2003) noting that the archives of Edinburgh zoo record discussion of the importance of enrichment for polar bears to express their playful and exploratory behaviour in 1911.

By contrast, interest in companion animal behaviour problems seems to have been much more patchy. As with the equine field, the work of early companion animal veterinary pioneers appears to have been largely forgotten. These included Lloyd Moss (1949) in the USA and Alastair Worden (1959) in the UK. It is interesting to note that Worden became joint editor with Barry Cross (mentioned in the introduction) of the Association for the Study of Animal Behaviour's journal from 1953 onwards. It was not perhaps until the 'modest proposal' of David Tuber, David Hothersall and Victoria Voith (1974), that any momentum was established. Tuber was an applied ethologist (often described as the first clinical animal behaviourist), Hothersall an experimental-comparative psychologist and Voith a veterinarian. Around this time in the USA, veterinarian Ben Hart and pioneer in the study of human-animal interaction Lynette Hart were also establishing a research base in this field in California. Kathe Houpt soon established a similarly influential group in the east at Cornell University, and Bonnie Beaver did likewise in Texas A&M. Figure 4.2 shows some of the early pioneers of veterinary behaviour medicine, including Ben Hart, Kathe Houpt, and Bonnie Beaver. In the 1980s, Victoria Voith joined forces with psychologist Peter Borchelt to publish a series of influential research and practical articles on companion animal behaviour problems for veterinarians explaining not only common problems but also important behavioural skills like history taking and the use of client interview forms, and the discipline of companion animal behaviour therapy was finally established. Many of the articles were subsequently collated and updated by a wider range of individuals inspired by the original work, in the text '*Readings in Companion Animal Behavior*' dedicated to David Tuber after his untimely death in 1995 (Voith and Borchelt, 1996). However, it was not until 1997 (Mills *et al.*, 1997) that the first truly global research conference was held on this subject.

Figure 4.2. Early pioneers of veterinary behavioural medicine: Ben Hart, Elizabeth Shull, RK Anderson (always known as RK), Bonnie Beaver, and Kathe Houpt.

4.4 The growth of education in veterinary ethology

Following the fall-out from Harrison's book, the SVE carried out a survey in 1968/1969 on the teaching of animal behaviour in the veterinary schools of the UK and Eire. It showed that five of the seven schools had started some sort of formal animal behaviour courses with an average duration of around eight teaching hours. The initiative to start the teaching had, in many cases, seemingly come from individual members of staff (Ewbank and Howard, 1969). In the late 1970s the UK's State Veterinary Service arranged that, in future, all its junior and as many as possible of its Divisional Veterinary Officers should attend one of a series of three day behaviour courses that it ran, and a senior veterinary officer from each of its Regions should attend a week's animal behaviour course at the University of Edinburgh (Moss, 1988). Knowledge of animal behaviour was beginning to spread through the veterinary profession.

Perhaps the main academic benefits from the Brambell Report were, and remain, first the recognition that the study of applied animal behaviour is a multidisciplinary task, and second the massive amount of relevant research funded as a result of the Report's recommendations, which helped to establish applied ethology as a respectable scientific discipline. However it was not until 1990 that the first MSc in applied animal behaviour and welfare was established, in Edinburgh, with the British Veterinary Association making available a specific veterinary bursary to encourage uptake by the profession. The need and value of this course is well illustrated by the prestigious list of alumni that have since gone on to establish their own research and educational programmes internationally.

In response to growing recognition of the need for a specific applied behaviour management programme focussed on companion animals, the Anthrozoology Institute was founded by Dr John Bradshaw at the University of Southampton in 1992, with the support of an endowment from Waltham®, and the first post graduate qualifications in companion animal behaviour counselling were made available a few years later. This Institute too has an impressive list of graduates in the field.

4.5 Caring for versus caring about animals

With the rise of clinical animal behaviour as a discipline and practice, there appears to be growing concern over the welfare of companion animals in the home and rescue environment, a topic which was previously largely ignored by those keen to make improvements in animal welfare. The possible reasons for this deserve some consideration. Animal companions, unlike other animals that humans come into contact with, are closely integrated into the family. As a result close social and emotional bonds develop, and as a result we may fail to differentiate their needs from our own. There is a useful distinction to be made here between '*caring for*' an individual and '*caring about*' them. The former is a largely emotional response driven by what we feel others need, whereas the latter is a more cognitive one which rationalises what is needed, often with a positivist focus. Unfortunately, unless we are careful our emotional responses can cloud our cognitive judgments and affect our behaviour towards others as a result. This can lead to unintentional suffering, especially when the other individual is less able to communicate their needs, as is the case with non-human animals. This may be the problem facing many companion animals: their owners *care for* them too much and perhaps need to *care about* them a bit more.

This does not mean we should not have emotional relationships with companion animals, only that we need to be aware of how these relationships might affect our responses, and recognise the importance of more objective perspectives if we really wish to provide what is best. In the case of the horse, both equitation and the management of problem behaviour have historically been largely performance-driven activities, i.e. the view of what constitutes success has focused almost exclusively on behavioural outcomes and the physical demands for this. Likewise in farm animals an excessive focus on production performance and caring about the physical needs of the animals can lead to a neglect of their psychological needs. In this regard, the pioneering work of researchers like Paul Hemsworth (e.g. Hemsworth *et al.*, 2009) has been instrumental in showing us the economic as well as ethical importance of caring for farm-stock. This represents a shift in emphasis in study from the population to the individual. Indeed, as argued later, the science of veterinary behavioural medicine is very much a science of the individual and not the mean.

The emergence of companion animal science, equitation science (particularly through the coming together of veterinarian Paul McGreevy and equitation expert Andrew Maclean, as illustrated by their 2011 text) and veterinary behavioural medicine as disciplines in their own right, means we are now in a position to elucidate the underlying processes of behavioural change within the individual, to apply them in novel ways, and more objectively to evaluate their impact on the well-being of the subject. The emergence of these scientific disciplines has been important for those who *care about* individuals and go beyond a traditional physical focus to be more broadly interested in animal welfare science. This means we can move from well-meaning intention to a more objective welfare oriented approach. This has led many veterinarians and others concerned about animal welfare to now challenge many traditional management practices for companion animals on the grounds of the harm they do to the interests of the animals concerned. This complements the similar concerns that have been expressed about farm animals over a much longer timeframe. However, companion animals are much more readily identified as individuals living in a unique environment with their keeper and so there is a different emphasis in the study of the behaviour problems that arise in this context, with much more emphasis given to individuality. Different perspectives have been taken on this as has been the case with the study of human behavioural problems (e.g. psychological versus psychiatric approaches) and these are discussed further in the next section. While the emphasis may be on companion animals, it is worth noting that the lessons learned can be applied to a wider range of species studied within applied ethology.

4.6 Traditional approaches within clinical animal behaviour

It is fair to say that the history and practice of the science of clinical animal behaviour is fragmented into different schools of thought, which have sometimes been at odds with one another. However, as illustrated in the previous section, it is clear that most progress is made in applied ethology when scientists from a range of disciplines come together. Different approaches have emerged in different regions of the world, each emphasising its own theoretical traditions. Local validation has been achieved through development of regional specialist qualifications by relevant professional bodies, such as the Animal Behavior Society in the US (Certified Applied Animal Behaviorist), the Association for the Study of Animal Behaviour (with the support of ISAE) in the UK (Certified Clinical Animal Behaviourist), the European Board of Veterinary Specialisation (Diplomat of the European College of Animal Welfare and Behavioural Medicine), and the American College of Veterinary Behavior (Diplomat of ACVB).

Until recently there have been two dominant approaches to clinical animal behaviour: the behavioural and the medical model of problem behaviour. The behavioural approach is grounded in the behaviorism of experimental psychology pioneered in the US, and has a focus on observable events. It emphasises the importance of the environment in development, with treatment focusing on reconditioning of behavioural signs. It is exemplified by the increasing number of animal trainers who use behavioural principles of reinforcement in their training, but especially by psychologists such as Susan Friedman, who have translated their practice of applied behavioural analysis from humans to non-human animals. By contrast the medical approach is grounded in the positivist reductionism typical of medicine, focusing on the identification of underlying pathologies to what is considered abnormal behaviour, which is often categorised into diagnoses. Although treatment includes forms of behaviour therapy, emphasis is also given to the need for the physical treatment of conditions, including by drugs. This approach is perhaps best illustrated by the texts of Karen Overall (1997) and Patrick Pageat (1996). However, problems arise from both of these approaches.

In behaviorism, the emphasis on the observable means that there is rarely reference to the importance of processes within the central nervous system, and so these tend to be ignored. The significance of this may be less noticeable in the experimental environment, but it may be important when dealing with animals in complex multisensory environments. The sterile, standardised experimental environments often used to study behaviour do not reflect the world companion animals tend to live in. As a result, these experiments tend to show what *can* bring about a specific behaviour, but not necessarily what commonly *does* affect behaviour in the real world. The aim of behaviorism was largely to elucidate the general laws of learning, and individual variation is often ignored, but as already indicated individuality is at the heart of clinical behaviour management: Why is this animal responding in a way that was not expected? Behaviour and stimuli are the foci of interest in behaviorism, but somewhat ironically for a field that focused on the importance of behaviour, superficially similar behaviours are often ascribed a single motivation, and details that might reflect internal processes like emotional influences are often ignored. Perhaps less obviously, but no less importantly, the focus on behaviour and stimuli has also led to biases in the way we consider the underlying mechanistic processes. There is much written about behavioural control mechanisms as if there were a direct link between stimuli and responses, when in fact behaviours result from strategic decisions made by the animal; i.e. the first decision made is to escape and only then is the behavioural method for achieving this goal determined. The behaviour is simply *one way* in which a wider strategic objective can be addressed (there are different behaviours that might allow escape, such as running away or repelling the threat with a growl), and a focus on modification of the behaviour is not necessarily an efficient way of managing the wider problem which relates to higher level cognitive and emotional influences on strategic decision making.

The medicalisation of problem behaviour in humans which has informed the development of a similar approach in non-human animals, can trace its origins to the work of Emil Kraepelin (1896) and the discovery of the infectious cause of syphilis and its relationship with dementia praecox, a schizophrenia-like condition. Around this time there was great interest in the use of new statistical methods to identify clusters of co-occurring signs and symptoms to define the characteristics of different diseases. Kraepelin applied these principles of medical diagnosis to mental problems, with a diagnosis being made on the basis of co-occurring symptoms. This still provides the dominant framework for psychiatric diagnosis in most of Europe and North America, and is encapsulated in the Diagnostic and Statistical Manual of Mental Disorders, published by the American Psychiatric Association, which is now into its 5th edition (DSM-5). While the mathematical relationship between signs and diagnosis may have considerable appeal, the question arises of whether this reflects reality. Indeed in the case of DSM-5, concerns over the reliability and validity of categories described have led the largest funder of mental health research, the National Institute for Mental Health, to state that it will no longer fund research projects that rely exclusively on DSM criteria for their definition. Implicit with the definition of distinct conditions is the notion that individuals with such conditions are qualitatively different from the 'normal' population (abnormal), when in reality they may simply differ quantitatively, e.g. have more extreme versions of adaptive traits. If this is the case then overlap between categories is likely, if gaps are to be avoided. This results in poor specificity and multiple diagnoses.

Interestingly, this seems to be exactly what is happening in some of the veterinary behaviour literature. For example, the Journal of the American Veterinary Medical Association publishes an occasional feature called 'animal behaviour case of the month,' and from 2000 to 2014 the average case of aggressive behaviour in the dog was described as having between two and three forms of aggression (average 2.4 diagnoses/case), i.e. it is implied that the case has between two and three conditions, because several terms explain why the aggressive behaviour occurs. Sometimes this is due to inconsistency in the type of classification used to describe the problem, e.g. a dog who is food guarding might be described as having both 'fear aggression' (emotional classification) and 'food related aggression' (contextual description). A consistent way of explaining the behaviour based on a description of the contexts in which a certain emotion is aroused (fear around food) would provide a clearer and more reliable classification of the problem than thinking of the animal as having two 'disorders'. The lack of a biologically meaningful qualitative difference between different diagnostic categories, such as in the previous example or when the animal's response is simply an extreme version of the normal response (i.e. differs quantitatively from the norm but not qualitatively), also explains why distinctive biological markers of different behavioural problems can be so elusive. Nonetheless, this approach remains popular with some, probably because the use of distinct categories reflects the ways that veterinarians are taught to think about medical problems. This is epitomised in Overall's more recent text (2013), in which between only 10 and 50 pages are dedicated to normal behaviour but 140 pages relate to abnormal (i.e. qualitatively different to normal) behaviour in dogs. An alternative perspective is to see these 'disorders' as potentially having fitness benefits in other contexts, such as alternative genetic combinations, alternative environments, alternative stages of development, or in the opposite sex. Just because an animal's behaviour is maladaptive, this does not mean it is malfunctional (Mills, 2003). The animal may simply be using evolutionary rules of thumb in an environment for which it does not have adaptive strategies.

In summary, with regard to clinical animal behaviour, the problem with a strict medical approach is that it focuses on the specific characteristics of abnormality, when there is often no discrete boundary. There is also a danger of letting the diagnosis of the problem result in the labelling of the individual, creating biases and self-fulfilling prophecies on how the animal's behaviour is viewed. Thus, once a dog is labelled as having some form of aggression disorder, people may behave differently around it, creating confusion and potentially exacerbating the risk of an aggressive display. Also, as the dog ages, it may become more aggressive and this may be seen as a deterioration of the disorder rather than, say, the effects of painful arthritis on irritability. The problem with a strict behavioural approach is that it focuses just on the behaviour, and there is more than the animal's behaviour to the problem, there is a dynamic with the owner. It also emphasises the importance of managing and structuring the environment, potentially underestimating the value of interventions focused on the individual. Both approaches can compromise the treatment options available and lead to certain cases receiving inefficient or inappropriate treatment.

4.7 A psychobiological approach to clinical animal behaviour

More recently, a number of individuals including the first author (Mills *et al.*, 2012) have advocated a more biological approach. This places the animal's behaviour within a broader evolutionary framework that is grounded in affective neuroscience (Panksepp, 1998), but also recognises the wider social context to the problem. As a starting point, the term 'behaviour problem' reflects the problem the behaviour is causing to others and so is a social construct based on an individual's perception of a situation, and so is not consistently definable by just objective criteria. The problem consists of clusters of behaviours, emotional states and cognitions, which are often not specific to a given complaint but which are *problematic to the owner.* So the first step to the management of the problem is the translation of the owner's perceptions into data that we can analyse in a rigorous scientific way.

A notable feature of the approach developed at Lincoln (hereafter referred to as the psychobiological approach) is the adoption of a consistent framework for applying the hypothesis testing method that is central to scientific philosophy to the assessment of emotional differentials. This replaces the more traditional verification of diagnoses through the accumulation of positive evidence in their support. To gather rigorous evidence to disprove the competing hypotheses (scientific falsification), we must be systematic in the way we gather this evidence, assess it and analyse it against the competing hypotheses that potentially explain the animal's behaviour. Central to the psychobiological approach, which we outline below, is the tenet that explanation requires us to understand and distinguish the context, motivation and emotion of the behaviour, which has not always been the case. Context defines the circumstances in which the behaviour arises; motivation describes the biological function of the behaviour; emotion relates to the significance of the event to the individual, including the associated behaviour. 'Context' is described by reference to the observable environmental circumstances surrounding the behaviour. This may relate to the triggering stimuli (immediate antecedents to the behaviour) or general circumstances preceding the incident (Setting). 'Motivation' refers to the biological function of the behaviour, i.e. it relates to the goal of the behaviour, and can only be inferred in the field setting (i.e. it cannot be measured directly). However by using the technique borrowed from Applied Behavior Analysis, of examining evidence relating to the trigger of the behaviour (antecedents), the form of the behaviour (what the behaviour looks like) and what reinforces or inhibits the behaviour (its consequences), a reliable inference can often be made. 'Emotion' refers to processes that result in an individualised response to a given situation, and has two important dimensions: emotional quality (i.e. the type of emotional response) and emotional intensity (level of emotional arousal). Like motivation, emotion cannot be measured directly in a field setting but can be inferred. This also requires a careful analysis of the behaviour, across a range of circumstances or over an extended time frame in order to elucidate information relating to the four components intrinsic to emotional reactions (Table 4.1 and discussion below).

For an emotional response to occur, an event must be appraised as having personal significance, i.e. there must be identifiable, emotionally-competent stimuli that are consistent across settings. At its simplest the event may be considered either attractive or aversive, but there are different ways in which it can be attractive or aversive, depending on the specific relationship between the emotionally competent stimulus and the subject. On the basis of affective neuroscience the following classes of emotionally competent stimulus appear to be common triggers of emotional responses, and have been proposed

Table 4.1. The four psychological components influenced by emotional arousal and their biological function.

Component	Function
Appraisal	Evaluation of events and objects
Bodily signs/arousal	System regulation and preparation
Action tendency	Direction of action
General and Specific (e.g. facial) behaviour	Communication of reaction and intention

(Mills *et al.*, 2014) as the key emotional differentials within the psychobiological approach, alongside unemotional, habitual responding:

- Desirables: things the animal wants at a given time lead to seeking out opportunities to have them.
- Frustrations: things which the animal wants but cannot access or are less than expected lead to increased focused effort aimed at achieving the goal.
- Threats: things that might harm the animal lead to the desire to avoid interaction.
- Hurts: things causing actual bodily damage lead to withdrawal and protection.
- Affiliates: individuals within the animal's social group that provide assistance and shared learning opportunities or lead to social play.
- Attachment figures and objects: those that provide safety and protection lead to a strong dependence upon them at times of uncertainty and insecurity.
- Dependents: those that are dependent on an individual animal, i.e. offspring, may invoke a range of caring behaviour directed towards them.
- Potential sexual partners: those with whom there may be breeding opportunities lead to the expression of courtship and reproductive activity.
- Undesirables: those that are thought to be a threat to the benefits of the current social group may elicit responses associated with their expulsion and exclusion.

Arousal occurs as the animal prepares to take action in response to the event. In general arousal is either increased or decreased depending on the emotion, and can be measured physiologically or behaviourally, e.g. change in heart rate, hair standing on end/lying flat, pupil dilation/constriction, trembling or relaxed muscles. The arousal seen must be consistent with the putative emotion being aroused.

The emotional response also involves a change in action tendencies, i.e. behaviours directed towards a common functional goal, e.g. there are various ways of avoiding interaction with another, such as keeping focused on the individual to ensure it comes no closer and moving away from the object to avoid contact if it approaches. These responses are consistent with the same goal and may vary according to circumstances, but indicate a single emotional response despite the behavioural diversity. Previous experience may affect an animal's emotional action tendencies e.g. if a dog has been physically punished by its owner in a certain situation, it may not show the punished behaviour (growling) but an alternative one in future (bite without a growl warning).

Finally, the event which induces an emotional response influences the signals sent by the animal. Communication signals come from a wide range of body regions and include: facial expressions, head, body, limb and tail postures, head, body, limb and tail movements as well as a range of olfactory, auditory and touch-related signals.

The anticipated or actual occurrence or termination of the stimulus is of importance to the individual's judgement of the situation and should result in a reliable response. Once as much information as possible has been gathered about these factors, the evidence is examined in relation to the features of the nine emotionally competent stimuli discussed, to see which can be excluded. Those that remain form the basis of the initial tentative diagnosis, however like any scientific hypothesis the validity of this tentative diagnosis can be questioned in light of new evidence, e.g. response to specific treatment measures.

4.8 Future priorities and conclusion

Clearly it is essential that veterinarians continue to work alongside ethologists and scientists from other disciplines if the field is to continue to flourish. The approach outlined above for evaluating the behaviour of animals in the field, provides a systematic framework for the generation of testable hypotheses, in both clinical and research settings. There has been growing interest in expressions of pain like the grimace face of rodents and more recently in horses (Dalla Costa *et al.*, 2014) and these fit well within the framework described, since they reflect the communicative aspect of the emotional quality of pain. Current work at Lincoln is looking at the uniqueness of emotional facial expressions in a wider range of emotions in both cats and dogs, with encouraging results. There is clearly a need to support this with the development of non-invasive methods of measuring physiological arousal, and while the technology already exists to do this, the funding is currently lacking. Undoubtedly the continued rise of technology will also see greater automation in the evaluation of animal behaviour, but the precise methods to be used will depend on the function. For example, at Lincoln in partnership with Nottingham, we are currently working on an app for owners which will give them details about the likelihood that their cat is in pain, which will hopefully enable more animals to get prompt veterinary care. Veterinary ethology remains a multidisciplinary field, but the breadth and importance of other academic disciplines contributing to the field is perhaps greater now than it has ever been.

Historically, the veterinary profession has always had an interest in behaviour and the rise of applied ethology and the ISAE has served to not only meet this interest but also increase it, as new relevancies are discovered. No single discipline has the right to monopolise the generation of new knowledge in this field, even if it is seen to be of direct relevance to the veterinary profession. Indeed the contribution of an ever expanding list of disciplines to the field has given, and will continue to give, new and important insights that could not be achieved by veterinarians alone. Veterinarians not only create a need, but they have also historically made important contributions to the development of applied ethology for more than a hundred years and there are no signs of this diminishing.

References

Association of Veterinary Students, 1964. Report on animal behaviour meeting. Journal of the Association of Veterinary Students 23: 73.

Bailey, P., 1964. Animal psychiatry (a descriptive review). Archives of General Psychiatry 13: 382-386.

Brion, A. and Ey, H., 1964. Psychiatrie Animale. Desclée de Brouwer, Paris, France.

British Veterinary Association, 1966. The place of animal behaviour in veterinary science. Veterinary Record 84: 458-459.

Brownlee, A., 1950. Studies on the behaviour of domestic cattle in Britain. Bulletin of Animal Behaviour 1: 11-20.
Brownlee, A., 1954. Play in domestic cattle in Britain: an analysis of its nature. British Veterinary Journal 110: 48-68.
Cross, B.A., 1946. Animal behaviour in relation to veterinary science. British Veterinary Journal 102: 397-409.
Dalla Costa, E., Minero, M., Lebelt, D., Stucke, D., Canali, E. and Leach, M.C., 2014. Development of the horse grimace scale (HGS) as a pain assessment tool in horses undergoing routine castration. PloS one 9: e92281.
Edwards, J.T., 1945. The study of animal behaviour from the veterinary aspect: or, the importance of prosecuting the study of animal behaviour, and suggestions for its development as a field of veterinary endeavour. British Veterinary Journal 101: 10-18, 33-39.
European Convention for the Protection of Animals Kept for Farming Purposes, 1976. Available at: http://tinyurl.com/z3ygsya.
Ewbank, R., 1961. The behaviour of cattle in crushes. Veterinary Record 73: 853-856.
Ewbank, R., 1967. Behavioural studies in the veterinary curriculum. Journal of Biological Education 1: 251-254.
Ewbank, R., 1964. The 'dog-sitting' posture in cattle and sheep. Veterinary Record 76: 388-393.
Ewbank, R. and Howard, B.R., 1969. The teaching of animal behaviour in the veterinary schools of the UK and Eire. Veterinary Record 84: 458-459.
Fox, M.W., 1964a. The ontogeny of behavioural and neurological responses in the dog. Animal Behaviour 12: 301-310.
Fox, M.W., 1964b. The postnatal growth of the canine brain and correlated anatomical and behavioural changes during neurogenesis. Growth 28: 135-141.
Fox, M.W., 1967. The place and future of animal behavior studies in veterinary medicine. Journal of the American Veterinary Medical Association 151: 609-615.
Fox, M.W., 1968. Abnormal behavior in animals. WB Saunders, Philadelphia, PA, USA.
Fraser, A.F., 1957a. The state of submission in cattle. British Veterinary Journal 113: 167-168.
Fraser, A.F., 1957b. The state of fight or flight in the bull. British Journal of Animal Behaviour 5: 48-49.
Friedberger, F. and Frohner, E., 1910. Veterinary pathology, 6th ed. Hurst and Blackett, London, UK.
Hafez, E.S.E., 1962. The behaviour of domestic animals. Bailliere, Tindall and Cox, London, UK.
Harrison, R., 1964. Animal machines: the new factory farming industry. Vincent Stewart Publishers Ltd., UK.
Hemsworth, P.H., Barnett, J.L. and Coleman, G.J., 2009. The integration of human-animal relations into animal welfare monitoring schemes. Animal Welfare 18: 335-345.
Hutyra, F., Marek, J. and Manninger, R., 1946. Special pathology and therapeutics of the diseases of domestic animals. Bailliere, Tindall and Cox, London, UK, pp. 431-477.
Kilgour, R., 1976. The contributions of psychology to a knowledge of farm animal behaviour. Applied Animal Ethology 2: 197-205.
Kraepelin, E., 1896. Zur Hygiene der Arbeit. Gustav Fischer, Verlag, Germany.
Lupton, J.I., 1884. Evils of modern stables. In: Fox, M.W. and Mickley, L.D. (eds.) Advances in animal welfare science 1984 (1985). Martinus Nijhoff Publishers, Dordecht, the Netherlands, pp. 155-161.
McGreevy, P.D. and McLean, A.N., 2011. Equitation science. John Wiley and Sons, New York, NY, USA.

Mayhew, E., 1888. The illustrated horse doctor. J.A. Allen and Co, London, UK, p. 171.
Meyer-Holzapfel, M., 1968. Abnormal behaviour in zoo animals. In: Fox, M.W. (ed.) Abnormal behavior in animals. W.B. Saunders, London, UK, pp. 476-503.
Mills, D.S., 2003. Medical paradigms for the study of problem behaviour: a critical review. Applied Animal Behaviour Science 81: 265-277.
Mills, D. S., Dube, M. B., and Zulch, H., 2012. Stress and pheromonatherapy in small animal clinical behaviour. John Wiley & Sons, New York, NY, USA.
Mills, D., Heath, S.E. and Harrington, L.J., 1997. Proceedings of the first International Conference on Veterinary Behavioural Medicine. Universities Federation for Animal Welfare (UFAW), Wheathampstead, UK.
Mills, D., Karagiannis, C. and Zulch, H., 2014. Stress – its effects on health and behavior: a guide for practitioners. Veterinary Clinics of North America: Small Animal Practice 44: 525-541.
Mills, D.S. and Macleod, C.A., 2002. The response of crib-biting and windsucking in horses to dietary supplementation with an antacid mixture. Ippologia 13: 33-41.
Moss, L.C., 1949. Psychoneurosis – a veterinary problem. Journal of the American Veterinary Medical Association 114: 1-2.
Moss, R., 1988. A personal view of progress in animal welfare in the past 20 years. State Veterinary Journal 42: 5-10.
Nicol, C.J., Davidson, H.P., Harris, P.A., Waters, A.J. and Wilson, A.D., 2002. Study of crib-biting and gastric inflammation and ulceration in young horses. The Veterinary Record 151: 658-662.
Overall, K.L., 1997. Clinical behavioral medicine for small animals. Mosby-Year Book, Inc, Maryland Heights, MO, USA.
Overall, K., 2013. Manual of clinical behavioral medicine for dogs and cats. Elsevier Health Sciences, Maryland Heights, MO, USA.
Pageat, P., 1996. Pathologie du comportement du chien. Editions du Point Vétérinaire, Paris, France.
Panksepp, J., 1998. Affective neuroscience: the foundations of human and animal emotions. Oxford University Press, Oxford, UK.
Royal College of Veterinary Surgeons, 1955. Catalogue of modern works 1900-1954. RCVS Library, London, UK.
Smythe, R.H., 1958. The mind of the dog. Country Life Ltd., London, UK.
Taylor, E.L., 1962. The place of animal behaviour studies in veterinary science. Veterinary Record 74: 521-524.
Tuber, D.S., Hothersall, D. and Voith, V.L., 1974. Animal clinical psychology: a modest proposal. American Psychologist 29: 762-766.
Voith, V.L. and Borchelt, P.L., 1996. Readings in companion animal behaviour. Veterinary Learning Systems, Trenton, NJ, USA.
Williams, W.O., 1904. Some neuroses of animals. Veterinary Record 9: 80-82.
Worden, A.N., 1959. Abnormal behaviour in the dog and cat. Veterinary Record 71: 966-981.
Young, R.J., 2003. Environmental enrichment for captive animals. John Wiley and Sons, New York, NY, USA.

5. Cognitive approaches and new technologies: changing methodologies in applied ethology

J.-L. Rault and R. Doyle*
Animal Welfare Science Centre, Faculty of Veterinary and Agricultural Sciences, University of Melbourne, Parkville, 3010 VIC, Australia; jean-loup.rault@unimelb.edu.au

Abstract

Having a more comprehensive understanding of animal behaviour can help us manage animals more effectively and improve their welfare. Applied ethology has adapted techniques from a wide variety of disciplines, including economics, human psychology, fundamental ethology and evolution, industrial design, engineering and computational modelling. These techniques extend across a wide variety of taxa and include all aspects of applied ethology, from farm and lab animals, to companion animals, captive vertebrates, and wild animals. Some of technological advances discussed here (although now considered basic), have greatly facilitated the study of applied behaviour. Nevertheless, most of these have not really changed the basic principles of ethological study and observation; instead they have enabled us to perform more detailed observations, to gather more data and to perform more thorough and complex analyses. In parallel, we have seen the development of less invasive behavioural approaches and physiological and neurobiological measures that provide insight into the mental states of animals. We can now gain insight into their affective states – how they feel about the situations they're exposed to. For example, knowing when animals are bored, distressed, fearful or in pain can help us to manage and reduce these negative states; and similarly, by understanding when they are in positive states, we can increase these events and thereby enhance their welfare. In time, with the expansion of behavioural observation techniques and the development of novel technologies, we will move closer to understanding how the animals we engage with feel, and how humans can positively influence these states.

Keywords: ethological methods, play, anticipatory behaviour, cognition, automation

5.1 Introduction

Having a more comprehensive understanding of animal behaviour can help us manage animals more effectively and improve their welfare. By understanding behaviour in detail we can understand better what their needs are. This includes what resources are important to them, how they interact with each other, and what implications this has for their management. We can also gain insight into their affective states – how they feel about the situations they're exposed to – which can be used to assess their welfare. For example, knowing when animals are bored, distressed, fearful or in pain can help us to manage and reduce these negative states; similarly, by understanding when they are in positive states, we can increase these events and thereby enhance their welfare. To gain

DOI 10.3920/978-90-8686-828-5_5, © Wageningen Academic Publishers 2016

a deeper understanding on why animals do what they do, applied ethology has adapted techniques from a wide variety of disciplines, including economics, human psychology, fundamental ethology and evolution, industrial design, engineering and computational modelling. These techniques extend across a wide variety of taxa and include all aspects of applied ethology, from farm and lab animals, to companion animals, captive vertebrates, and wild animals.

This section does not pretend to be an exhaustive description of all novel methods and technologies available, but rather it provides examples of how some novel techniques and technologies have been used in the field of applied ethology. Hopefully these examples will unleash the reader's creativity and encourage you to explore these technologies in new ways.

5.2 Novel methods in applied ethology

As with all scientific disciplines, the questions we ask in applied ethology and the methods used to answer them have expanded and evolved over time. While applied ethologists clearly have an interest in deepening the understanding of the animals they study, the significant growth of research in applied ethology in recent years has been largely driven by society's increasing concern for animal welfare. As we work to get closer to 'the truth' behind what drives an animal's behaviour, and how this relates to their welfare, our research has deepened. For example, the past decade has seen a clear movement away from identifying and addressing negative impacts and negative behaviours, towards a greater emphasis on positive states and behavioural changes (Boissy *et al.*, 2007). Box 5.1 contains a quote from David Fraser, a leading figure in the development of animal welfare science, illustrating the change in scientific perspective which has occurred over the past

Box 5.1. David Fraser.

Perhaps the most fundamental development I have seen in applied ethology is the acceptance that science can help us understand emotional states in animals. Like others of my generation, I had been taught that processes we cannot observe – including emotions and cognition – have no place in science. In 1971, however, I was studying the intense activity and vocalisations of piglets when they were separated from their mothers in the then-new early weaning systems. Encouraged by the work of Marian Dawkins and Ian Duncan, who were using 'suffering' and 'frustration' in the titles of scientific papers, I summoned the courage to propose that the behaviour I observed was 'an expression of distress' which could be of value for assessing animal welfare and to improve the management of weaning. If studies of pain, fear and other affective states are commonplace today, it shows how much the field has changed in 50 years.

50 years. This more comprehensive view of animals, and its impact on how we perceive behaviour, has meant that new methods and technologies needed to be developed.

5.2.1 Positive event behaviours and subtle behaviour changes

One major area of expansion in applied ethology has been around the incidence of positive behaviours. While historically much work has been done on aggressive interactions, more recent studies have begun looking at positive event behaviours. Play is the classic example of a positive behaviour. It likely results in pleasurable feelings, has long-term benefits and can be reduced or absent when fitness is compromised (Burghardt, 2005; Held and Špinka, 2011). Held and Špinka (2011) provide a comprehensive review of play behaviour, evaluating a variety of situations where the presence, absence, increase or decrease of play can reflect the impact we have on animals, and highlights the value of measuring play as well as issues in this research. A broad body of work studying play behaviour in calves demonstrates the complexities of play. For example, following painful husbandry practices and weaning, the play behaviour of calves was decreased, suggesting that these practices have a negative impact (Mintline *et al.*, 2013; Rushen and De Passillé, 2012); whereas play was increased by provision of greater milk allowance and social contact, so indicating that these are likely to be positive practices (Duve *et al.*, 2012). Play in dairy calves is also affected by age, rearing conditions, including flooring substrate and space allowances, and habituation to the test environment (Mintline *et al.*, 2012; Rushen and De Passillé, 2012; Sutherland *et al.*, 2014), and so study design and the performance of play need to be interpreted carefully.

Anticipatory behaviour reflects another positive state that is reflective of animals' needs. Anticipation is characterised by a general increase in activity, with orienting and searching behaviour patterns (Spruijt *et al.*, 2001). It can give us an insight into the value of the situation (either positive or negative), but it is also modulated by the animals underlying affective state. For example, negative states can be accompanied by a dampening of the ability to experience positive situations (anhedonia), and is reflected in reduced anticipation. Complexities exist when interpreting anticipatory behaviours as well. The degree to which they are expressed can vary depending on the internal physiological state of the animal; for instance, anticipation of feed will vary significantly depending on appetite. The increased motor activity and repeated behaviours associated with anticipation are also observed in animals performing stereotypical behaviours, so appreciating the context of the behaviour is also important.

Some recent studies have used play as the positive situation from which anticipatory behaviour was assessed (Anderson *et al.*, 2015; Chapagain *et al.*, 2014), demonstrating a useful combination of these two behavioural measures. In the study by Anderson *et al.* (2015), the behaviour of lambs prior to either receiving play or food as a reward was compared to a control group. Compared to the controls, lambs displayed more frequent short bouts of behaviours and behavioural transitions, suggesting anticipation. This anticipation decreased when the outcome differed to the expected reward as well; when play (considered to be less rewarding) was the reward instead of food (more rewarding), anticipation declined over time, which suggests that anticipation changes in response to a discrepancy of expectation.

Whole animal ethograms traditionally dominate our research, whereas the measurement of subtle behaviours focuses on the brief changes and movements that occur in the facial and sensory features of the animal. Subtle changes in ear postures and eye openings have been linked to affective state in sheep; horizontal ear postures correspond to a neutral state, and ears backward is likely associated with unpleasant uncontrollable situations, whereas ears upwards may more reflect anger (Boissy *et al.*, 2011; Reefmann *et al.*, 2009). These changes in ear posture likely reflect heightened or dampened states of alertness in the animal, and will be affected by the surrounding environment, but differences between individual animals based on their current affective state are apparent.

The interpretation of subtle changes in facial features is a common way that we assess how other humans are feeling. There is now a developing body of work measuring facial expressions in animals. Changes in facial expression in response to pain have been measured in horses, rabbits, rats, and cats (summarised in Chambers and Mogil, 2015), and there have been recent papers on pain related facial expressions in sheep (McLennan *et al.*, 2014), pigs (Lonardi *et al.*, 2015), horses (Dalla Costa *et al.*, 2014) and primates (Vick *et al.*, 2015). Most facial changes are species-specific, but there do seem to be some commonalities, including orbital tightening in response to pain. Use of these 'grimace scores' for pain assessment is especially useful in prey species, which have evolved to hide obvious signs of pain that could make them more vulnerable to predators. The investigation of these subtle changes in response to other negative affective states, and positive ones too, are likely to follow.

The interesting discovery of 'laughter' in rats by Panksepp and Burgdorf (2003) is an example of a subtle vocal signal. Panksepp and Burgdorf identified that rats had a preference for interacting with the handler when they were being 'tickled' (gentle handling that simulated play behaviour between rats), with the animals seeking out this sort of interaction. Further intrigued, they studied this behaviour in a myriad of different ways and discovered that rats made an abundance of 50-kHz 'chirps' during this type of social interaction, which are found in other social situations (e.g. sex and aggression), but are performed much more frequently during and in anticipation of play. Exploring other non-visual signals (auditory and olfactory), as well as visual, may help us to identify other subtle expressions of emotion in animals. These subjects are likely to receive more attention from applied ethologists in the future.

5.2.2 Economics of behaviour – what, when and how much?

An animal's preferences are influenced by their affective states, which are prime determinants of its welfare (Duncan and Petherick, 1991). Understanding what an animal wants, when they want it, and how much they want it can help us to understand what is valuable to them, and by inference, if we provide this to them, their welfare is likely to be enhanced. Such questions can be answered by investigating an animal's preferences and its willingness to work, also known as behavioural demand. Preference tests measure which resource an animal favours, and thus values more; whereas behavioural demand testing was developed from economic theory, and assesses how much an animal is willing to 'pay' for a given resource (Dawkins, 1983). Behavioural demand tests can therefore give us an understanding of how valuable a certain food item or access to a specific environment is, or how 'elastic' the demand is.

As reviewed by Kirkden and Pajor (2006), a variety of factors affect an animal's preferences and willingness to work for a resource. Much like anticipatory behaviours, understanding the context is again critical. For example, the desire to be with a conspecific will depend on the social relationship between the test animal and the individual that is presented, initial desires may wane as the novelty of the resource declines, and willingness to work for food will be dictated by appetite. Studies in hens have clearly shown the influence of internal state on their willingness to work to access a nest box. Studies by Cooper and Appleby (2003) found that hens' willingness to work to gain access to a nest box was low throughout most of the day, but increased significantly as time for oviposition nears. In essence, as the biological need for the bird to lay an egg increases, her internal motivation to access a comfortable place to do so does as well. Experience is also an important factor, and should be considered in preference testing as animals often inherently prefer a familiar environment over unfamiliar conditions.

Such research techniques allow us to ask animals what they want, and while these studies can be slow and laborious to administer, they provide us with important information that is otherwise difficult to obtain. Providing they are carefully conducted and interpreted, these tests can give us important insight into the resources that animals find most valuable. Through the provision of desirable resources we can either improve the way we manage animals, or potentially offset negative events by providing positive interactions before or after negative ones.

5.2.3 Understanding cognitive processing and behaviour

Our reasons for wanting to understand how animals process information are wide ranging. We find animals fascinating, so in a basic sense, understanding why they do what they do can satisfy our curiosity. From a practical standpoint, it is important to understand the capacities of domestic or captive species in order to manage them effectively. We know they aren't 'just like us', but neither are they the 'automated machines' that science once attributed to them. As we try to evaluate animal intelligence, we are starting to move away from the comparative approach – comparing what they can do to what we can do, and are becoming more interested in their capacities. We are broadening our understanding of intelligence and neural diversity. No longer are primate-specific behaviours, and their association with intelligence, the golden standard of animal intelligence. We are now much more aware and appreciative of the complex capacities of avian intelligence, for example.

Cognition and behaviour are intertwined. The study of cognition is incomplete without the consideration of behaviour; as it is an internal, unobservable process, we assess behaviour in a way that allows cognitive explanations to be drawn (Shettleworth, 2010).

Similarly, as applied ethologists, we can assess cognitively-driven behaviours to help us answer questions about our impact on animals. We know that stress can impact cognitive performance, and so by measuring performance during problem solving tasks we can assess this in real time (Mendl, 1999). For example, it has been shown that unpredictable stimuli reduce how well a sheep performs a simple task (Greiveldinger *et al.*, 2009). We also know that the spatial cognition of sheep is inhibited by novel auditory stimuli (Doyle *et al.*, 2014). These findings have practical implications, for example, when moving them through paddocks and sheep yards, past cars, motorbikes or dogs, all of which can be novel and unpredictable.

We can also assess how longer-term management practices affect animal behaviour and cognition. It is well known that the early life social conditions of laboratory species can impact their cognitive performance later on. The same effect has been demonstrated in calves; where individually housed calves were more reactive to novelty (Gaillard *et al.*, 2014) and had poorer performance in a reversal learning task when housed individually, than when in pairs or groups (Gaillard *et al.*, 2014; Meagher *et al.*, 2015). This body of work reflects the fact that socially barren environments may impact the cognitive performance, and subsequent behaviour, of animals later in life. This can be a particular problem if we then manage these animals in complex environments that demand significant habituation, learning and recall. Evidence of this is also often seen when older sows previously kept in stall housing are moved into group housing. Many older animals fail to learn how to access the electronic feeder system, whereas younger animals learn the system comparatively easily. A large body of work from Harry Harlow's laboratory in the 1960s and 1970s show similar and more extreme behavioural issues when Rhesus macaques were socially deprived.

How we manage animals can also influence their emotional or affective state, and how they process information. Human studies have consistently shown that how an individual feels influences their cognitive processes, including attention, memory and judgement (Paul *et al.*, 2005). In particular, the judgement of ambiguous information can be biased if the individual is in a positive affective state, resulting in a more optimistic judgement and inflated expectation of a positive outcome. In contrast, individuals in a negative affective state display a more pessimistic judgement and show greater expectation of a negative outcome. Over the last 10 years this concept has been adapted to animals, and since the initial formative paper in 2004 (Harding *et al.*), applied ethologists have been actively investigating these judgement biases in animals as well. Mike Mendl has been a major thought leader in the field of animal cognition and cognitive bias. In Box 5.2, he describes some of the key methodological achievements in applied ethology, and directions for future research.

By investigating judgement biases in animals, we now know that things like the predictability of the environment, the complexity of the environment, social interactions, short term management, and painful husbandry practices all influence how they judge and respond to information (for reviews see Baciadonna and McElligott, 2015; Bethell, 2015; Mendl *et al.*, 2009). To date, the results of some judgement bias studies have been difficult to interpret, partly due to variation in methodology and statistical analyses, and

Box 5.2. Mike Mendl.

Some important methodological and technological achievements in applied ethology during the last 50 years:

- Development of preference and consumer demand methods to assess what animals want and don't want.
- Implementation of non-invasive methods for assessing physiological welfare indicators (e.g. salivary, faecal and urinary hormone sampling, HR and HRV recording, thermal imaging).
- Development of new approaches and methods to study and measure animal affect and welfare (e.g. anticipation behaviour, cognitive bias, qualitative behavioural assessment, vocalisation analysis).

Some promising areas for current and future development:

- Implementation of computer-vision methods for assessing behaviour as an indicator of health and welfare *in situ*
- Implementation of brain-scanning technologies (e.g. near infrared spectroscopy, functional magnetic resonance imaging) to assess brain changes in relation to affect and welfare.
- Development of metabolomic and proteomic indicators of animal stress, welfare, and health.
- Use of computational modelling to elucidate the cognitive and affective processes that underpin behaviour and decision-making.
- Use of robots / drones as data-collection tools in the animal environment, and even to interact with animals and hence help explore how (social) behaviour is organised.

partly to results contrary to the *a priori* hypotheses. Additionally, studying judgement bias requires elaborate training paradigms so it can currently be used only in experimental settings, rather than being used for ongoing assessment. While ethologists are working on more practical ways to administer the test, the current training processes tend to be labour intensive, and complex, and the repeatability of results is influenced by the rate of animal learning. This highlights that while we're constantly evolving the ways we assess the animals we interact with, being conclusive about whether or not we're enhancing or reducing their quality of life remains difficult.

Attention bias is another growing technique that has potential to assess behaviour and affect. Research in humans demonstrates that a person's focus towards a potentially threatening stimulus can be modulated by their affective state. Individuals experiencing a more negative affective state are more easily distracted by negative stimuli, and the distraction lasts longer than for individuals in a more positive state. Furthermore, when compared with individuals in a more positive affective state, those experiencing a negative short-term affective state (such as anxiety) show a greater bias towards negative stimuli (faster reaction times and are distracted longer).

In contrast, individuals in a longer-term depressive-like state will show a reduced bias towards negative stimuli (being slower to react/observe the stimulus; Dalgleish and Watts, 1990). These examples from human psychology are also being adapted to assess animals. Animals' responses to potential threats have now been demonstrated to be modulated by how we manage them. In cage-housed starlings, prevention of bathing resulted in a heightened response to conspecific alarm calls (an indicator of threat), suggesting that by restricting this natural behaviour the anxiety of the birds increased (Brilot and Bateson, 2012). In Rhesus macaques, a short term negative experience also affected attentional biases, and following a veterinary check the macaques detected threatening stimuli more readily (Bethell *et al.*, 2012). In both of these examples, the training, testing and response phases were relatively simple, and required minimal labour, indicating that attention bias may be a practical measure that is useful outside of controlled research trials. Current research also supports the existence of affect-modulated attentional biases in sheep (Lee *et al.*, unpublished data), and so this may be an effective way to assess our impact on farm animals as well.

Measuring other simple cognitive processes may also give us important insights into animal behaviour. The startle reflex is a ubiquitous, cross-species response to abrupt and intense stimulation. The amplitude of the response varies according to the internal state of the organism, including its level of fear or anxiety (Lang *et al.*, 2000), and it is generally potentiated in negative states. Previous research has demonstrated an increased startle response in rats experiencing heightened anxiety (Risbrough and Geyer, 2005). Work to date has also identified affect-induced differences in the startle response of pigs (Statham *et al.*, 2013) and sheep (Doyle *et al.*, 2015). The startle response provides a potential method to assess affective states while animals are in groups, which doesn't add additional stressors of isolation, and doesn't require training or significant infrastructure.

Based on the major expansion of this area in recent years, studies on these cognitive responses are likely to increase in the future. The use of well-recognised phenomena in human psychology is likely to continue. Future research may well focus on the concept of memory bias for example, where the recall of memory, and subsequent behavioural response to this, can be affected by the valence experienced at the time of learning.

5.2.4 The role of qualitative assessment

Qualitative assessment of animal behaviour provides a way to identify incidental behaviours, subtle details of movement and posture, and aspects of the context in which the behaviour occurs (Wemelsfelder, 1997). There is a large body of work that supports this method of quantifying behaviour. Qualitative assessment involves general observation of an animal and the generation of open-ended descriptive terms describing the animal being observed. These descriptions are well correlated between observers, even in observers from a broad range of backgrounds, and have been shown to relate to objective behavioural and physiological measures in a variety of environments.

Qualitative measures may be able to replace laborious behavioural observations with simpler subjective assessments that can provide similar outcomes. As such they represent a useful approach in difficult situations, such as during on-farm welfare assessment where time available for evaluation is very limited.

5.3 Novel technologies

Many technological advances are now considered part of our daily lives and work. Possibly as a result of habituation, we tend to underestimate the value or leap in progress that technologies have allowed us to achieve. Furthermore, most of us are slow to adopt and realise the potential of new technological discoveries. Nevertheless, technologies are becoming mainstream faster than ever before, from conception to market, and many are now affordable and practical. As we will cover in this section, computer processing, data storage, video camera systems, automated and remote technologies, and also statistical techniques and biomedical devices have all contributed to making our jobs easier, while also unravelling the complexity of understanding behaviour and the potential application of this knowledge. Most of these advances have made data collection, analysis and interpretation easier or more refined, creating new opportunities that we thought never possible before. All of this has happened in the past 50 years.

5.3.1 Novel methods of observation

Traditionally the science of applied ethology relied on direct, live observations of animal behaviour. Video camera systems brought about the ability to collect more data, allowing more animals to be observed at once, or over an extended period of time. These systems are constantly being improved, with lighter-weight, higher resolution, increased data storage, wireless data transfer, flexible placement, and various other features. Infrared cameras allow animals to be observed at night, a useful feature when studying nocturnal species, and a wide range of time-lapse or automatic recording cameras (e.g. that turn on following detection of animal movement or approach) are now commercially available. Nonetheless, live or video-based observations of animal behaviour remain notoriously time consuming, as any ethologist can attest. However, these lengthy manual observations of behaviour are now being replaced by a layer of automation.

Software and other data analysis packages are now available that can automatically record behaviour (e.g. Noldus observation software), or estimate behavioural change based on parameters such as pixel changes on pictures or video images. ImageJ for example is freely available from the (United States) National Institute for Health and can perform a number of visual analyses. This enables the measurement of distances, movements or other activity changes from a series of picture frames.

New technologies now allow for the automatic, and often continuous, recording of animal behaviour. Accelerometer electronic data loggers (also marketed as 'pedometer' or 'activity tracker') permit the long-term monitoring of various movements to infer the time spent lying down, the number of steps taken, or any number of other movements.

The data can be interpreted as measures of comfort (if the animals spend more time lying down in one situation rather than another, or vary their postural changes), pain (lameness, injuries), or social interactions (displacement around particular resources, preferential interactions). The most advanced devices are able to analyse data patterns and determine specific behaviours, based on the acceleration in a three dimensional representation and the length of the acceleration vector, from which a mathematical model can infer a particular movement and the frequency of movements, which differ between behaviours (Cornou and Lundbye-Christensen, 2008). In addition to measuring activity, numerous other behaviours can now be remotely measured, thanks to pressure mats (used for lying behaviour, lameness detection), rumination collars, and ingested or implanted monitors equipped with sensors to assess pH, internal temperature, heart rate, and an ever-increasing number of behavioural and physiological variables.

Another technological field focuses on location-based technologies, allowing us to track animal movements in real-time, over an extended period and from virtually anywhere in the world. Global Positioning Systems (GPS) is probably the most well-known, but a wider range of techniques are appearing on the market, with varying location accuracy and autonomy (battery life, storage capacity). Most of these systems calculate location based on a triangulation method between different points based on time and distance travelled. Other devices are constructed as a network of devices which can interact with one another. This is the case for proximity data loggers, which record when animals come within a set distance of one-another. Proximity data loggers placed on a few animals in a group can tell us about social relationships, which are notoriously difficult to study considering the complexity of social organisation in most social species, and also the large groups that many animals are now kept in. It allows us to examine the social network, the frequency of relationships, and in some cases even patterns of social avoidance and attraction between particular individuals, with minimal effort compared to manual observations.

Precision livestock farming has now developed into a field of its own, incorporating technological advances into production systems (Berckmans, 2014). For instance, a computer system combining cameras and algorithms can allow continuous monitoring and analysis of chickens' spatial distribution in a shed, signalling to farmers when particular areas have low or high densities of birds, possibly indicative of problems. Robotic milking, automatic electronic feeding systems, and the computer monitoring of environmental parameters and husbandry in intensive housing systems have now been widely adopted. These allow caretakers to monitor changes in the animal's behaviour such as food and water intake or estrus behaviour; but it also represents a rich source of (often untapped) data for researchers. The data can be used to assess whether there is competition, early detection of illness, or another problem which leads to a change in individual or group behaviour on a systemic, long-term basis. Dawkins and colleagues (2012) used optical flow patterns based on automatic image analysis to determine behavioural changes in flocks of 35,000 chickens. The responses were found to be linked with welfare outcomes (mortality, hock burn, lameness) and could potentially be used to improve management of animals in large groups.

Radio frequency identification (RFID) tags are the foundation of many precision farming systems, through applying an electronic chip to the animal (ear-tag, leg band, subcutaneous microchip implant) which activates an electronic mechanism or is recorded when it comes within range of a detection device (antenna, plate, scanner). For instance, RFID forms the basis of most automatic electronic feeding systems and has recently allowed tracking the movement of individual chickens in large poultry flocks in sheds or to an outdoor range area (Richards *et al.*, 2011). Such knowledge was nearly impossible to gather before, as these groups can number several thousand individuals. Increasing requirements for traceability of livestock have also contributed to interest in the use of RFID, however tag loss remains a concern both for use of precision livestock systems and traceability of animals.

5.3.2 The ups and downs of 'big data'

The main advantage of technologies is that it makes data collection easier, quicker, and more precise than manual methods; this represents an unprecedented revolution in human history. Whereas earlier studies were restricted to a limited number of animals due to the time and labour involved, it is now technically possible to collect data from hundreds or thousands of animals with relatively little effort. The RFID tags discussed previously are a good illustration, allowing the movements of thousands of animals to be recorded simultaneously and continuously over months, or even years. A novel challenge is in finding methods to manage and analyse these large amounts of data (discussed further below).

Data are also more readily available, as sharing becomes easier, faster and far-reaching. In this age when cats have invaded the Internet, popular video online sharing websites such as YouTube represent a rich source of largely untapped behavioural data. The fascination many people have for animals, along with the revolution in communication technologies, makes sharing of self-captured videos via online media a logical source of data. Indeed, a few behavioural studies have already looked at online social media archives as a way to collect data, and coined the term 'video mining' (Rault *et al.*, 2013). This new approach can allow us to study species or situations for which it is difficult to obtain sufficient sample sizes (e.g. companion animals, endangered species in zoos), behaviour which occurs rarely (but is salient enough for people to record), or behaviours that typically occur only in the home environment (e.g. tail chasing in dogs; Burn, 2011).

Another interesting approach used recently in the medical sciences is to engage non-experts. The idea in this case is to use a larger number of non-experts to screen the fast growing body of scientific literature. The goal is to summarise and group recent findings on specific diseases, demonstrating that a large enough number of observers can compensate for their lower expertise. This approach is also used in applied ethology courses for instance by asking students to observe a large number of animals in zoo settings, conduct comparative studies across species, or in approaches to qualitative behavioural assessment.

5.3.3 Problem-solving and refinement

The use of video cameras may be a given nowadays, but it has allowed great progress by avoiding observer interference and the numerous tricks required to circumvent it, such as position of an angled mirror, a one-way screen or hides. This is a major refinement, as any good ethologist wants to study the behaviour of an animal (seemingly) unaware of the observer's presence, or at least undisturbed.

Other technologies allow us to detect or record phenomenon that are beyond the five-senses (sight, sound, touch, smell, taste) of human ability, which previously were considered beyond our ability to study. Hence, technologies such as the recording of ultrasonic vocalisation have greatly increased our knowledge of animal communication (Burman *et al.*, 2007). Software allow us to perform highly complex analyses, such as sound spectrography, analyses of electrical frequencies such as power spectra for brain EEG, or social network analysis (Makagon *et al.*, 2012). These methods have provided new insights into the mechanistic explanation of behaviour and development of new assessment methods. For example, Manteuffel and colleagues (2004) have developed a system of automatic recording and analysis of vocalisations, used on pig farms to detect welfare problems.

Other technological advances allow us to test animal responses in a non-invasive, remote manner. Wireless technologies have been a major step: allowing data recording to continue as animals freely move around. Although animals may need to remain within a range of the receiver device, this can vary from a few centimetres (in the case of RFID tags), or several meters with amplifiers, to large distances with satellite-reliant technologies such as GPS. For example, researchers are now able to count whales from space using satellite images (Fretwell *et al.*, 2014)!

Infrared lasers can be used to monitor movement and quantify activity, with the device recording each time an animal cuts the laser beam. This offers a simple, low labour and cost-effective alternative to manually having to record animal movement based on video watching. A laser can also be used to measure nociception in response to heat, using a computer-controlled carbon dioxide-laser beam applied to the surface of the skin. The application of this technique is wide ranging, including investigating pain sensitivity threshold following various husbandry procedures, inflammation or other thermal changes. Thermographic cameras (also known as infrared or thermal imaging cameras) have been widely used to assess changes in surface (skin) temperature reflective of arousal or injuries, by forming an image using infrared radiation.

5.3.4 Testing devices: control, repeatability and the removal of human influence

Various behavioural testing paradigms are commonly employed by applied ethologists, as a way to 'ask the animals'. Yet, most of these methods remain time-consuming or labour intensive, due to the testing procedure itself or because of the amount of training required for the animals to properly learn the tasks. Technologies have made behavioural testing easier and more accessible, particularly for assessing the motivation to work for a reward (e.g. feed, social motivation). The resistance of a weighted-door, speed of a tread mill for an operant runway test, pecking or touching a key and other fixed-ratio tasks can all be controlled, recorded or actioned through computer and electronic systems. More elaborate automatic rodent or fish mazes have also been developed to allow repeated testing, at appropriate times of the day and night in accordance with the animal's preference and on a voluntary basis at its own rhythm. These systems can now perform hundreds of tests on animals without any human involvement or interference. Handling, a procedure which is recognised to induce significant (and often overlooked) distress and concomitant behaviour changes, can be virtually eliminated by technological systems such as automated handling gates, weighing and drafting handlers for livestock or laboratory animals alike. Data can also be transmitted at a distance through the internet or mobile network, so that a computer recording data on-farm can send it directly to the farmer's (or researcher's) office for monitoring or analysis.

Everyone who has worked on social behaviour knows very well how difficult it can be to standardise conditions when two or more individuals simultaneously change behaviour in response to each other; this is a challenge when studying the nature of social interaction. Video or audio playbacks have long been used by pioneers in ethology in an attempt to standardise stimulus presentation, such as Niko Tinbergen (1951) in his landmark study on sign stimuli using three-spined stickleback models. Nevertheless, this approach is becoming more customisable with the help of technology. Through software such as anyFish (Veen *et al.*, 2013), researchers can create computer-animated stimuli of small fishes behaviour as interactive stimulus to test live animals. The field of Virtual Reality is becoming increasingly popular in animal behaviour research (Stowers *et al.*, 2014). In a recent study conducted in Spain, researchers showed that a rat and a human can interact through humanoid avatars, using virtual reality systems (Normand *et al.*, 2012). These technologies allow us to study social interactions by simulating a conspecific, and controlling that social stimulus.

Currently, animal-computer interaction is a booming field (Rault *et al.*, 2015). It focuses on the way animals interact with digital technologies, their perception of it and potential uses to accommodate behavioural needs. The time when each dog left at home during working hours will be provided with an iPad-like device to kill time or connect to their owner is perhaps not far away (Pongracz *et al.*, 2003)! These new technologies may allow us to provide cognitive enrichment, or conversely help us elucidate the animals' cognitive and mental needs through behaviourally-relevant computer interactions. The greatest potential of animal-computer interaction, or any other technology accessible to animals, is that it can provide them with greater control over their environment, a component that is critically lacking in many captive environments (Rault *et al.*, 2015).

5.3.5 Data handling and statistical approaches

Collecting data is one thing, but knowing how to best extract and analyse the information is equally important, and our technological options for this are expanding and becoming increasingly complex and powerful. Selecting the proper tools becomes crucial.

There are now countless statistical and analytic packages that researchers use depending on the type of data or research question. Readers are referred to other sources for a more extensive overview of the statistical analyses available for the analysis of animal behaviour (e.g. Dawkins, 2007; Jensen *et al.*, 1986). Modern statistical analyses have become amazingly powerful for making sense of data, provided that you ask the right questions! These models can account for the myriad of factors which are known to affect an animal's behaviour, and at speeds that were inconceivable even 20 years ago.

New software and computers are able to make sense of an immense amount of data, far beyond our metal abilities. Indeed, in this new digital age, the sacred words are 'database' or 'metadata'. We now have the capability to analyse very large datasets collected through computer systems, or centralised from numerous smaller entities. While it is wonderful to collect all these data, it can also be daunting, and in many instances these data are still under-exploited because the information contained has not been recognised or utilised to its full potential. Many researchers, having used automatic recording of behaviour have found themselves swamped with millions or more data points to organise and analyse. Fortunately, various methods exist to filter erroneous recording from datasets, sort and calculate various outputs that can be used by applied ethologists. With the help of mathematicians and computer programmers, algorithms have been developed, which are merely a procedure or formula for solving a problem. These algorithms can recognise patterns in the data, or combine data to derive new information and answer behavioural research questions. This approach is used for instance in meta-analysis, which allows combining data from multiple independent studies on a common topic to increase power and the ability to detect global effects across studies. Epidemiological studies have also probably been underused in our field compared to other fields such as medicine or psychology.

Input from biostatisticians is increasingly solicited to analyse large or complex datasets. Biostatisticians consider the nature of information available and the way it was collected, and are often particularly knowledgeable about analytic methods, for example to account for missing data or complex interactions among factors.

Modelling has been used, with more or less enthusiasm by applied ethologists who have a strong tradition to rely on empirical observations. Modelling allows the prediction or simulation of what could happen according to a set of estimated parameters (Stillman *et al.*, 2015). Modelling approaches can be particularly useful, especially as the need for animal use is becoming more stringent and less socially acceptable. It could for instance be used to test theories prior to confirming the biological validity through animal trials, increasing efficiency and reducing animal use.

5.3.6 The interpretation of behaviour

Two of Tinbergen's Four questions pertain to proximate explanations: what causes a behaviour to occur (mechanism); and how does the behaviour develop (ontogeny)? An ethologist at heart, Niko Tinbergen recognised that the study of behaviour is a holistic science, and that ethologists need to cooperate with other biological disciplines such as physiology or neurobiology.

Much has been learned about the hormonal control of behaviour, through internal and external factors, causing behaviours to be released, inhibited or primed (Bolhuis and Giraldeau, 2008). Through advances in technology, physiological and neurobiological measurements have evolved to be less invasive (e.g. wireless electroencephalogram and heart rate measurements). Hormones or their metabolites can now be measured in a wide variety of samples such as saliva, faeces, urine, and even hair samples.

The use of physiological measures has allowed researchers to validate the importance of behavioural measures and behavioural scales (e.g. is lameness in chickens painful, and if so, at which score/severity does it start to be painful?). Measures of heart rate and heart rate variability are very useful for interpreting freezing behaviour in open-field tests, or other tests of exploration or behavioural reactivity: is an animal that is not moving not stressed about the situation, or is it so stressed that it cannot move? Changes in heart rate can help us to discriminate these two phenomena, which result in a similar behaviours. Advances in biochemistry have allowed us to analyse the chemical composition of pheromones (chemical signals which induce changes in behaviour), and help us understand their significance related to mechanisms of reproduction, and sexual or maternal behaviours (Schulte *et al.*, 2007). For instance, Boissy *et al.* (1998) sampled urine from stressed cattle and found an increased latency to feed when cattle were exposed to urine of a stressed conspecific, supporting the idea that fear in cattle can be socially transmitted through urine pheromones.

Neurobiology has also progressed a long way towards elucidating functional molecules, often referred to as neurotransmitters or neurohormones, which act as messengers to organise behaviour. Affective neuroscience has determined that dopamine underlies the 'wanting', and associated seeking behaviours, whereas opioids underlie the 'liking', or reward systems along with serotonin. For instance, we have learned that maternal behaviour and drug addiction can both activate the brain 'reward' system, with pup stimulation indeed being more reinforcing than cocaine in lactating rat dams (Ferris *et al.*, 2005).

Newer techniques such as electroencephalography (EEG), functional magnetic resonance imaging (fMRI), or positron emission tomography (PET) scans now allow us to gain insight into neural pathways and how the brain works. This knowledge, along with particular behavioural patterns, can help us understand the origin or development of particular behavioural problems (see for example: Ferris *et al.*, 2005). We are now able to simultaneously measure changes in behaviour and observe activity within the brain regions involved. The work of Jaak Panksepp on the neurobiology of emotions and behaviour (Panksepp, 1998) illustrates nicely the knowledge that can be gained from a cross-disciplinary approach that includes applied ethology along with other disciplines. Panksepp has endeavoured to synthesise our knowledge of behaviour with its associated neurobiological origins and control systems (Panksepp, 1998), shedding new light on the origin and function of emotions in animals.

Genetics and genomics can also help us understand variation in fearfulness or the causes of abnormal behaviours, as well as providing insights into the consequences of selection on behaviour variables. This work is possible thanks to new laboratory techniques such as quantitative trait locus (QTL) analysis or cDNA microarray technology (Jensen *et al.*, 2008). Knock-out animal models, principally in mice, have determined the genetic control of specific behaviours, for example showing that oxytocin is required for social recognition (Ferguson *et al.*, 2000). Epigenetics has also shed some light on behaviour, particularly behavioural development such as the work from Meaney and Champagne (Weaver *et al.*, 2004), who demonstrated transgenerational effects of maternal care in rats as being determined initially by genetic factors, and subsequently shaped by their dams in early life through epigenetic changes.

5.4 Conclusions

Applied ethologists have been very ingenious, to say the least, in not shying away from learning from other disciplines or borrowing tools which may be relevant and help to advance our discipline further. We are now able to assess animal behaviour more comprehensively, and measure our impact on the animals we engage with, more than ever before. While our knowledge is expanding, we must still be very cautious in how we assess behaviours and the conclusions we draw as we try to understand animals better. Having comprehensive, objective data about how animals behave and process information can allow us to make conclusions on their welfare, and the possible experiences they may have, without having to project our emotions onto them. In Box 5.3, the noted ethologist, Marion Dawkins, summarises the advances in applied ethology, from a fringe activity to an important field with a significant role to play in sustainability and animal welfare.

Box 5.3. Marian Stamp Dawkins.

In my working life-time, I have seen animal welfare go from being seen as a 'fringe' subject, not really seen as a scientific subject at all to becoming the mainstream science it is today. We owe a lot to pioneers such as W.H. Thorpe and David Wood-Gush who had the vision to recognise the important role that ethology could play in the scientific evaluation and improvement of animal welfare and to the development of rigorous standards of data collection and hypotheses testing. We have also benefitted from technological advances in tracking and monitoring of behaviour that have given us insights into physiological mechanisms that could once only have been dreamt of. With increasing pressure on agriculture to become more 'sustainably intensive' and to feed yet more people, it is my hope that applied ethologists will continue to be in the forefront of combining scientific rigour with a determination to improve animal welfare.

Some of technological advances discussed here (although now considered basic), have greatly facilitated the study of applied behaviour. Nevertheless, most of these have not really changed the basic principles of ethological study and observation; instead they have enabled us to perform more detailed observations, to gather more data and to perform more thorough and complex analyses. In parallel, we have seen the development of less invasive physiological and neurobiological measures that provide insight into the mental states of animals. In time, with the expansion of behavioural observation techniques and the development of novel technologies, we will move closer to understanding how the animals we engage with feel, and how humans can positively influence these states.

References

Anderson, C., Yngvesson, J., Boissy, A., Uvnäs-Moberg, K. and Lidfors, L., 2015. Behavioural expression of positive anticipation for food or opportunity to play in lambs. Behavioural Processes 113: 152-158.

Baciadonna, L. and McElligott, A., 2015. The use of judgement bias to assess welfare in farm livestock. Animal Welfare 24: 81-91.

Berckmans, D., 2014. Precision livestock farming technologies for welfare management in intensive livestock systems. In: Mellor, D.J. and Bayvel, A.C.D. (eds.) Animal welfare: focusing on the future. Revue Scientifique et Technique Office Internationale des Epizooties 33: 189-196.

Bethell, E.J., 2015. A 'How-To' guide for designing judgment bias studies to assess captive animal welfare. Journal of Applied Animal Welfare Science 18: S18-S42.

Bethell, E.J., Holmes, A., MacLarnon, A. and Semple, S., 2012. Evidence that emotion mediates social attention in rhesus macaques. PloS ONE 7: e44387.

Boissy, A., Aubert, A., Desiré, L., Greiveldinger, L., Delval, E. and Veissier, I., 2011. Cognitive sciences to relate ear postures to emotions in sheep. Animal Welfare 20: 47-56.

Boissy, A., Manteuffel, G., Jensen, M.B., Moe, R.O., Sprujit, B., Keeling, L.J., Winckler, C., Forkman, B., Dimitrov, I., Langbein, J., Bakken, M., Veissier, I. and Aubert, A., 2007. Assessment of positive emotions in animals to improve their welfare. Physiology and Behavior 92: 375-397.

Boissy, A., Terlouw, C. and Le Neindre, P., 1998. Presence of cues from stressed conspecifics increase reactivity to aversive events in cattle: evidence for the existence of alarm substances in urine. Physiology and Behavior 63: 489-495.

Bolhuis, J.J. and Giraldeau, L.-C., 2008. The behaviour of animals: mechanisms, function and evolution. Blackwell Publishing, Malden, MA, USA.

Brilot, B.O. and Bateson, M., 2012. Water bathing alters threat perception in starlings. Biology Letters 8: 379-381.

Burghardt, G.M., 2005. The genesis of animal play: testing the limits. MIT Press, Cambridge, MA, USA.

Burman, O., Ilyat, A., Jones, G. and Mendl, M., 2007 Ultrasonic vocalizations as indicators of welfare for laboratory rats (*Rattus norvegicus*). Applied Animal Behaviour Science 104: 116-129.

Burn, C., 2011. A vicious cycle: a cross-sectional study of canine tail-chasing and human responses to it, using a free video-sharing website. PloS ONE 6: e26553.

Chambers, C.T. and Mogil, J.S., 2015. Ontogeny and phylogeny of facial expression of pain. Pain 156: 798.

Chapagain, D., Uvnäs-Moberg, K. and Lidfors, L.M., 2014. Investigating the motivation to play in lambs. Applied Animal Behaviour Science 160: 64-74.
Cooper, J.J. and Appleby, M.C., 2003. The value of environmental resources to domestic hens: a comparison of the work-rate for food and for nests as a function of time. Animal Welfare 12: 39-52.
Cornou, C. and Lundbye-Christensen, S., 2008. Classifying sows' activity types from acceleration patterns: an application of the Multi-Process Kalman Filter. Applied Animal Behaviour Science 111: 262-273.
Dalgleish, T. and Watts, F.N., 1990. Biases of attention and memory in disorders of anxiety and depression. Clinical Psychology Review 10: 589-604.
Dalla Costa, E., Minero, M., Lebelt, D., Stucke, D., Canali, E. and Leach, M.C., 2014. Development of the Horse Grimace Scale (HGS) as a pain assessment tool in horses undergoing routine castration. PloS ONE 9: e92281.
Dawkins, M.S., 1983. Battery hens name their price: consumer demand theory and the measurement of ethological 'needs'. Animal Behaviour 31: 1195-1205.
Dawkins, M., 2007. Observing animal behaviour: design and analysis of quantitative data. Oxford University Press, Oxford, UK, 176 pp.
Dawkins, M., Cain, R. and Roberts, S., 2012. Optical flow, flock behaviour and chicken welfare. Animal Behaviour 84: 219-223.
Doyle, R.E., Freire, R., Cowling, A., Knott, S. and Lee, C., 2014. Performance of sheep in a spatial maze is impeded by negative stimuli. Applied Animal Behaviour Science 151: 36-42.
Doyle, R.E., Lee, C., McGill, D.M. and Mendl, M., 2015. Evaluating pharmacological models of high and low anxiety in sheep. PeerJ e1510.
Duncan, I.J.H. and Petherick, J.C., 1991. The implications of cognitive processes for animal welfare. Journal of Animal Science 69: 5017-5022.
Duve, L.R., Weary, D.M., Halekoh, U. and Jensen, M.B., 2012. The effects of social contact and milk allowance on responses to handling, play, and social behavior in young dairy calves. Journal of Dairy Science 95: 6571-6581.
Ferguson, J.N., Young, L.J., Hearn, E.F., Matzuk, M.M., Insel, T.R. and Winslow, J.T., 2000. Social amnesia in mice lacking the oxytocin gene. Nature Genetics 25: 284-288.
Ferris, C., Kulkarni, P., Sullivan Jr., J.M., Harder, J.A., Messenger, T.L. and Febo, M., 2005. Pup suckling is more rewarding than cocaine: evidence from functional magnetic resonance imaging and three-dimensional computational analysis. Journal of Neuroscience 25: 149-156.
Fretwell, P., Staniland, I. and Forcada, J., 2014. Whales from space: counting southern right whales by satellite. PLoS ONE 9: e88655.
Gaillard, C., Meagher, R.K., Von Keyserlingk, M.A., Weary, D.M. and McElligott, A., 2014. Social housing improves dairy calves' performance in two cognitive tests. PloS ONE 9: e90205.
Greiveldinger, L., Veissier, I. and Boissy, A., 2009. Behavioural and physiological responses of lambs to controllable vs uncontrollable aversive events. Psychoneuroendocrinology 34: 805-814.
Harding, E.J., Paul, E.S. and Mendl, M., 2004. Cognitive bias and affective state. Nature 427: 312.
Held, S.D. and Špinka, M., 2011. Animal play and animal welfare. Animal Behaviour 81: 891-899.
Jensen, P., Algers, B. and Ekesbo, I., 1986. Methods of sampling and analysis of data in farm animal ethology. Tierhaltung/Animal Management. Springer Basel AG, Basel, Switzerland.
Jensen, P., Buitenhuis, B., Kjaer, J., Zanella, A., Mormede, P. and Pizzari, T., 2008. Genetics and genomics of animal behaviour and welfare – challenges and possibilities. Applied Animal Behaviour Science 113: 383-403.

Kirkden, R.D. and Pajor, E.A., 2006. Using preference, motivation and aversion tests to ask scientific questions about animals' feelings. Applied Animal Behaviour Science 100: 29-47.
Lang, P.J., Davis, M. and Ohman, A., 2000. Fear and anxiety: animal models and human cognitive psychophysiology. Journal of Affective Disorders 61: 137-159.
Lonardi, C., Scollo, A., Normando, S., Brscic, M. and Gottardo, F., 2015. Can novel methods be useful for pain assessment of castrated piglets? Animal 9: 871-877.
Makagon, M., McCowan, B. and Mench, J., 2012. How can social network analysis contribute to social behavior research in applied ethology? Applied Animal Behaviour Science 138: 152-161.
Manteuffel, G., Puppe, B. and Schon, P., 2004. Vocalization of farm animals as a measure of welfare. Applied Animal Behaviour Science 88: 163-182.
McLennan, K., Rebelo, C., Holmes, M., Corke, M. and Constantino-Casas, F., 2014. The facial expression of sheep with footrot and its relationship wtih lameness, total lesion scores and faecal cortisol. In: Estevez, I., Manteca, X., Marin, R.H. and Averós, X. (eds.) Proceedings of the 48th Congres of the International Society for Applied Ethology. Wageningen Academic Publishers, Wageningen, the Netherlands, p. 111.
Meagher, R.K., Daros, R.R., Costa, J.H., Von Keyserlingk, M.A., Hötzel, M.J. and Weary, D.M., 2015. Effects of degree and timing of social housing on reversal learning and response to novel objects in dairy calves. PloS ONE 10: e0132828.
Mendl, M., 1999. Performing under pressure: stress and cognitive function. Applied Animal Behaviour Science 65: 221-244.
Mendl, M., Burman, O.H.P., Richard, M.A. and Paul, E.S., 2009. Cognitive bias as an indicator of animal emotion and welfare: emerging evidence and underlying mechanisms. Applied Animal Behaviour Science 118: 161-181.
Mintline, E.M., Stewart, M., Rogers, A.R., Cox, N.R., Verkerk, G.A., Stookey, J.M., Webster, J.R. and Tucker, C.B., 2013. Play behavior as an indicator of animal welfare: disbudding in dairy calves. Applied Animal Behaviour Science 144: 22-30.
Mintline, E.M., Wood, S.L., De Passillé, A.M., Rushen, J. and Tucker, C.B., 2012. Assessing calf play behavior in an arena test. Applied Animal Behaviour Science 141: 101-107.
Normand, J.-M., Sanchez-Vives, M., Waechter, C., Giannopoulos, E., Grosswindhager, B., Spanlang, B., Giger, C., Klinker, G., Srinivasan, M. and Slater, M., 2012. Becoming into the rat world: enabling real-time interaction between rat and human each at their own scale. PLoS ONE 7: e48331.
Panksepp, J., 1998. Affective neuroscience: the foundations of human and animal emotions. Oxford University Press, Oxford, UK.
Panksepp, J. and Burgdorf, J., 2003. 'Laughing' rats and the evolutionary antecedents of human joy? Physiology and Behavior 79: 533-547.
Paul, E.S., Harding, E.J. and Mendl, M., 2005. Measuring emotional processes in animals: the utility of a cognitive approach. Neuroscience and Biobehavioral Reviews 29: 469-491.
Pongracz, P., Miklosi, A., Doka, A. and Csanyi, V., 2003. Successful application of video-projected human images for signalling to dogs. Ethology 109: 809-821.
Rault, J.-L., Elmore, M., Biehl, D., Russell, M. and Garner, J., 2013. The world is a natural laboratory, and social media is the new petri dish. Ethology 119: 1-4.
Rault, J.-L., Webber, S. and Carter, M., 2015 Cross-disciplinary perspectives on animal welfare science and animal-computer interaction. The Second International Congress on Animal Human Computer Interaction, Malaysia. Available at: http://tinyurl.com/j9sfqft.

Reefmann, N., Wechsler, B. and Gygax, L., 2009. Behavioural and physiological assessment of positive and negative emotion in sheep. Animal Behaviour 78: 651-659.
Richards, G.J., Wilkins, L.J., Knowles, T.G., Booth, F., Toscano, M.J., Nicol, C.J. and Brown, S.N., 2011. Continuous monitoring of pop hole usage by commercially housed free-range hens throughout the production cycle. Veterinary Record 169: 338.
Risbrough, V.B. and Geyer, M.A., 2005. Anxiogenic treatments do not increase fear-potentiated startle in mice. Biological Psychiatry 57: 33-43.
Rushen, J. and De Passillé, A.M., 2012. Automated measurement of acceleration can detect effects of age, dehorning and weaning on locomotor play of calves. Applied Animal Behaviour Science 139: 169-174.
Schulte, B., Freeman, E., Goodwin, T., Hollister-Smith, J. and Rasmussen, L., 2007. Honest signalling through chemicals by elephants with applications for care and conservation. Applied Animal Behaviour Science 102: 344-363.
Shettleworth, S.J., 2010. Cognition, evolution and behavior. Oxford University Press, New York, NY, USA.
Spruijt, B.M., Bos, R.V.D. and Pijlman, F.T.A., 2001. A concept of welfare based on reward evaluating mechanisms in the brain: anticipatory behaviour as an indicator for the state of reward systems. Applied Animal Behaviour Science 72: 145-171.
Statham, P.T.E., Campbell, N.W., Hannuna, S.L., Jones, S.M., Paul, E.S., Colborne, G.R., Browne, W.J. and Mendl, M.T., 2013. Development of an automated measure of 'defence cascade' in pigs. In: Behaviour 2013: joint meeting of the 33rd International Ethological Conference (IEC) and the Association for the Study of Animal Behaviour (ASAB). Newcastle-Gateshead, UK.
Stillman, R., Railsback, S., Giske, J., Berger, U. and Grimm, V., 2015. Making predictions in a changing world: the benefits of individual-based ecology. BioScience 65: 140-150.
Stowers, J., Fuhrmann, A., Hofbauer, M, Streinzer, M, Schmid, A., Dickinson, M. and Straw, A., 2014. Reverse engineering animal vision with virtual reality and genetics. IEEE Computer Society 0018-9162, 14: 38-45.
Sutherland, M.A., Worth, G.M., Schütz, K.E. and Stewart, M., 2014. Rearing substrate and space allowance influences locomotor play behaviour of dairy calves in an arena test. Applied Animal Behaviour Science 154: 8-14.
Tinbergen, N., 1951. The study of instinct. Oxford University Press, New York, NY, USA.
Veen, T., Ingley, S., Cui, R., Simpson, J., Asl, M., Zhang, J., Butkowski, T., Li, W., Hash, C., Johnson, J., Yan, W. and Rosenthal, G., 2013. anyFish: an open-source software to generate animated fish models for behavioural studies. Evolutionary Ecology Research 15: 361-375.
Vick, S., Descovich, K., Farningham, D., Leach, M., Flecknell, P. and Buchanan-Smith, H., 2015. Facing up to pain: can facial expression indicate pain states in rhesus macaques? In: 34th International Ethological Conference, Cairns, Australia. Available at: http://tinyurl.com/hvs6oxj.
Weaver, I., Cervoni, N., Champagne, F., D'Alessio, A., Sharma, S., Seckl, J., Dymov, S., Szyf, M. and Meaney, M., 2004. Epigenetic programming by maternal behaviour. Nature Neuroscience 7: 847-854.
Wemelsfelder, F., 1997. The scientific validity of subjective concepts in models of animal welfare. Applied Animal Behaviour Science 53: 75-88.

6. Applied animal behaviour science and animal welfare: seeking the best balance between our science and its application

A.B. Lawrence
Animal and Veterinary Sciences, SRUC, West Mains Road, Edinburgh EH9 3JG, United Kingdom; The Roslin Institute, University of Edinburgh, Edinburgh, EH25 9RG, United Kingdom; alistair.lawrence@sruc.ac.uk

Abstract

Applied animal behaviour science has been closely connected with animal welfare since the formation of the Society for Veterinary Ethology (SVE) in the mid-1960s. It is important to understand the implications of this 'special relationship' both in terms of the positives and opportunities it brings, but also its possible risks and adverse effects on our science. I have used two 'frameworks' in this paper to help understand the relationship. The first is 'Pasteur's Quadrant' proposed by Donald Stokes as an alternative view on science and its application (Stokes, 1997). Stokes rejects the idea that science can only be divided into 'basic' and 'applied' categories and also that the relationship between these is a simple linear progression. He introduced the idea of 'use-inspired basic research' which he exemplified through the work of Louis Pasteur the 'father of microbiology'. The proposition that we can seek fundamental understanding within the context of applying that knowledge, seems especially relevant to our field and interestingly seems to be the direction that science funders are progressing towards. I use three examples of strategic research in applied animal behaviour science, each of which presents a mix of basic science and its application to understanding the animals' 'perspective'. My second 'framework' is provided by the UK research excellence exercise (REF 2014), which I use to revisit the question of the impact of applied animal behaviour science in improving animal welfare. As a part of REF 2014 academic institutions were required to submit Impact Studies, with the aim of demonstrating how a peer reviewed paper progressed towards a measurable impact outside of the science community. I found 27 Impact Studies in the REF database that involve both behavioural science and animal welfare. The results are heartening in terms of the range and extent of the impacts claimed by the studies. I suggest that 'REF-like' exercises conducted in other countries with a tradition of applied animal behaviour science would yield similar results. These two approaches give grounds for optimism for the future based on evidence of strategic science developments within the field, and also that our science is having impact in improving animal welfare. I conclude by emphasising the importance of maintaining the best balance possible between development of our science and its application, something that as a science community we should work on at both a local and global level.

Keywords: applied animal behaviour science, animal welfare, 'Pasteur's quadrant', strategic research, research impact

DOI 10.3920/978-90-8686-828-5_6, © Wageningen Academic Publishers 2016

6.1 Introduction

Applied animal behaviour science is a set of disciplines (e.g. ethology, psychology and neuroscience) applied to the study of animal behaviour in practical contexts (Lawrence, 2008). There are a number of potential applications of animal behaviour science (e.g. the management of animals, pest control), yet the consideration of animal welfare predominates and overshadows all other potential applications. As other chapters in this book will confirm, the origins of what we now regard as applied animal behaviour science came largely as a response to the publishing of '*Animal Machines*' (Harrison, 1964) and the resulting Brambell Report (Command paper 2836, 1965, referred to as the Brambell report because F.W.R Brambell chaired the committee), on the welfare of intensively farmed livestock. As described in Chapter 1 (Newberry and Sandilands, 2016), a small group of veterinarians felt the need to form a society (the Society for Veterinary Ethology (SVE)) in response to the growing interest in the welfare of farmed animals, no doubt influenced by the emphasis Brambell placed on the importance of behaviour as a means of assessing animal welfare (Command paper 2836, 1965). Early meetings of the SVE (e.g. British Veterinary Journal, 1975) tended to have an even balance between welfare and non-welfare related papers, and the early work of pioneers in the field such as David Wood-Gush was often directed at understanding the behaviour of domesticated animals in its own right, probably reflecting the interest then current in behavioural science and the influence of key scientists such as Tinbergen and Lorenz (Lawrence, 2008).

However by the 1980s the pattern had changed and from then to the present day the vast majority of papers and posters given at SVE (and the International Society for Applied Ethology (ISAE) as SVE became) have had a clear link to animal welfare. The sustained application of applied animal behaviour science to central scientific issues raised in the animal welfare debate has required substantial investment and reflects the level of funding (mainly from public sources) that has been directed towards addressing animal welfare issues in a number of countries since the 1980s. Lawrence (2008) illustrated this trend by describing the increases in output of refereed papers in the field and especially in papers that refer to the application of behavioural science to animal welfare.

It is therefore the case that much of the current work in applied animal behaviour science directed towards animal welfare issues is dependent on the extent of public opinion and hence political interest or concern being directed at animal welfare. This amongst other considerations underlines the importance of understanding the implications of this close relationship in terms of both the positives and opportunities it brings, and its possible risks and adverse effects on our science. Hence it is the implications of this 'special relationship' between applied animal behaviour science and animal welfare which is the subject of this chapter.

6.2 The balance between science and its application

Behind the question of how we should understand and deal with the consequences for applied animal behaviour science, of the strong focus on animal welfare issues, lie complex questions about scientific discovery and the balance between what can be referred to as 'basic' and 'applied' research. In '*Pasteur's Quadrant*', (see Stokes, 1997 and Figure 6.1) it is proposed that the division between basic and applied research originates from an influential report on science discovery and application (Bush, 1945, cited from Stokes, 1997). Stokes (1997) suggests that this dichotomy between basic and applied research prevailed during the rapid expansion of scientific research that occurred in the aftermath of World War II, and had a substantial impact on public policy underlying decisions on research funding, for example shaping US government policy on research funding in the post-World War II period (M.H. Armacost in the Preface to Stokes, 1997). In the UK, the 'Barnes Report' (commissioned by the UK government and published in 1988), proposed that a substantial amount of agricultural research previously funded from public funds should be classified as 'near market' and either funded by industry or terminated (Read, 1989). This led to a contraction of the UK agriculture research base, the after effects of which continue to the present time, and also to publicly funded research being more tightly tied to specific (short-term) policy interest, as opposed to longer-term strategic issues (Murphy-Bokern, 2008).

Underlying the decision over who should pay for applied research is an economic debate about the role of the State in the application of science and what is now referred to as innovation. The position taken by the Barnes report was that as the private sector benefited from near-market research then it should pay. However this view has been challenged partly on the basis of evidence that the private sector often only invests once the State has made high-risk research investments (Mazzucato, 2013) and also because the inter-relationships between the private sector and the State in research is complex

Figure 6.1. Pasteur's Quadrant redrawn from Stokes (1997, p. 73). Stokes used the work of Niels Bohr (the early 20th century atomic physicist) to exemplify what he referred to as Pure basic research (referred to in the text as basic research); the work of Thomas Edison (inventor) to exemplify Pure applied research (referred to as applied research) and the work of Louis Pasteur (the father of microbiology) to exemplify Use-inspired basic research (referred to as strategic research).

and easily perturbed (Read, 1989). More recently in a number of countries including the UK there has been a growth in what are referred to as 'public-private partnerships' where there is a shared funding of research and where industry is permitted to 'pay' part or all of their share through 'in-kind' contributions (e.g. access to animals or data; see Innovate UK as an example of this approach).

As an alternative view, Stokes (1997) proposed that we can recognise three categories of research (Figure 6.1):

1. Pure basic research (Stokes exemplified this category of research through the work of Niels Bohr, the early 20th century atomic physicist);
2. Pure applied research (exemplified by the work of Thomas Edison, inventor);
3. Use-inspired basic research (exemplified by the work of Louis Pasteur the 'father of microbiology').

What is different about Stokes' (1997) argument is that he sees a resolution to the supposed tension between pursuit of science and application through what he calls 'use-inspired basic science'. Another term used in this context is 'strategic research' that also implies basic research conducted in the context of discovery of uses and applications, and I will use this term mainly for convenience. Relevant to the future development of applied animal behaviour science, Stokes suggests that, as in the case of Pasteur, it is possible to carry out what might otherwise be regarded as basic science but with a focus on an applied outcome (the 'best of both worlds'). Furthermore, Stokes proposes that the relationship between basic research and application is rarely linear and we need to see science and its application in a more interactive and dynamic way (Figure 6.2 for Stokes's interpretation of a 'dynamic' interaction between science and application). It could be

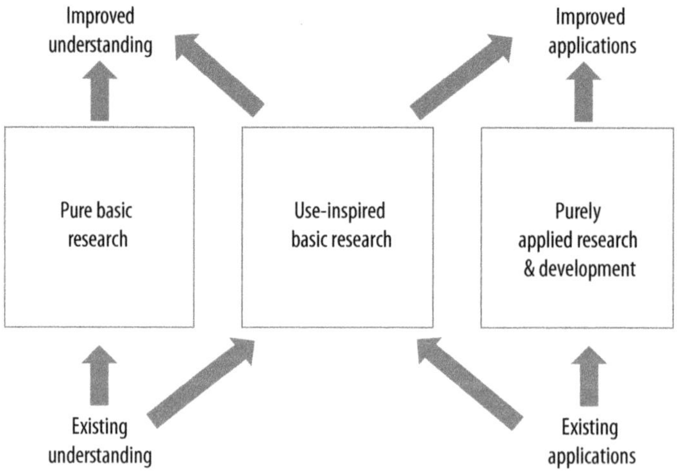

Figure 6.2. A 'dynamic model' of interactions between different categories of science redrawn from Stokes (1997, p. 88) as an alternative to the 'traditional' linear model of progression from basic science to its application.

argued that Stokes's proposition is in the process of being re-invented, as government funding of research (including basic research) is to an extent required to have 'research impact' (more on this later).

Hence, consideration of the influence of animal welfare on the development of applied animal behaviour science can be seen in terms of the best balance between scientific enquiry in its own right versus the application of our science, but also about how we promote research that sits in the 'sweet spot' occupied by the likes of Pasteur. In the remainder of this chapter I will explore these considerations from the perspective both of developing our science and of improving its application, and the interactions between these two objectives.

6.3 The impact of animal welfare on the development of applied animal behaviour science

6.3.1 The 'positive side'

Arguably the most important positive resulting from the focus on animal welfare has been the science and innovation it has led to with respect to assessing the 'animals' perspective'. The application of behavioural science to understanding of animals' inner psychological or mental state has had something of a chequered history in science (see Lawrence, 2008, pp. 3-4 for a brief resumé). It follows from the Brambell Report and the Five Freedoms (formulated by the UK Farm Animal Welfare Council; FAWC, 1979) that animal welfare is about both physical and mental well-being and hence any science directed at animal welfare needs to address both of these elements. The position of many applied animal behaviour scientists was most clearly laid down by Duncan (1993) in a paper entitled 'Animal welfare is to do with what animals feel', and despite misgivings within the wider field of animal behaviour (e.g. Kennedy, 1992; MacPhail, 1998) applied animal behaviour scientists have set about establishing scientific approaches to understanding animals' mental states. I want to highlight three examples of approaches which appear to display the characteristics that Stokes had in mind when he coined the category of use-inspired basic research; these are intended as examples and not an exhaustive list and others I am sure could produce their own examples to fit the strategic research category.

1. *Preference testing and derivatives.* The first of these approaches to emerge was championed by Marian Dawkins (1980) in her book '*Animal Suffering: The Science of Animal Welfare*', which promoted the use of animal 'preference testing' and study of animals' 'choices' as a method of objectively assessing animals' motivational priorities and by inference, their experiences. In further refinements of preference testing, methods were developed for estimating the 'cost' animals were 'willing to pay' to access different rewards once preferences had been established (e.g. estimating cost based on the willingness of the animal to press a device such as a key or panel (Dawkins, 1990)). This approach can be used to derive what are referred to as 'demand functions' for behaviours that display the characteristics of being 'elastic' (where the animal 'demands' less of a resource as the 'price' increases) or 'inelastic' (where the animal's demand for a resource is relatively unaffected by the price (Matthews and Ladewig,

1994)). The link to the economic analysis of supply and demand (which analyses human demand in relation to price) resulted in the approach often being referred to as the 'consumer demand' approach to assessment of behavioural needs (Dawkins, 1990). A study using the consumer demand approach to examine the motivation of farmed mink for access to water was published in Nature (Mason *et al.*, 2001), and was the first applied animal behaviour science paper to achieve this distinction.

Both preference testing and consumer demand have also been used in the wider behavioural sciences; preference testing has been used to assess habitat preferences (e.g. Morris *et al.*, 2001) and consumer demand approaches have been used to study the characteristics of motivational states believed to underlie behavioural expressions (e.g. Hogan and Roper, 1978). However in its application to animal welfare, Dawkins and others have assumed that preference and the 'cost paid' equate to animals' emotional states (or feelings) and to what animals 'want' or 'don't want' (Dawkins, 2008).

More recent development of preference testing involves the study of the associations between animals preferences and so-called 'welfare indicators'. Research on welfare indicators has arisen as a result of the growing interest in assessing welfare under practical conditions (e.g. on farms; Botreau *et al.*, 2007), where welfare indicators ideally should be reliable (e.g. show small variance between observers or assessors), valid (be a validated measure of the animals' experience) and also practical (taking account of constraints such as time availability). Of these three requirements the validation of welfare indicators is the most challenging, and it can be argued that current welfare indicators are selected more on the basis of practical considerations rather than for validity (see FAWC, 2005, Appendix A). Christine Nicol has argued that one approach to validation of welfare indicators is to study their association with animals' environmental preferences. Nicol *et al.* (2009) found that whilst some putative welfare indicators were associated with birds' positive choices (whatever those choices were), other indicators were not associated with environmental preferences, which suggests that this approach may have the capability of discriminating between indicators in terms of their relevance to the animals' experience. Current work (Davies *et al.*, 2015) is exploring the associations between animal decisions (which lead to expressed preferences) and so-called 'somatic markers' (e.g. heart rate), adding to our knowledge of the biology that links animal preferences with potential welfare indicators.

2. *Cognitive bias and other approaches developed from human psychology.* My second example also illustrates how approaches developed in one field (here studies of human cognition and emotion) were innovatively 're-configured' for use in applied animal behaviour science for assessing animals' 'mental state'. Alain Boissy, Isabelle Veissier and co-workers (e.g. Désiré *et al.*, 2002), proposed that animal emotions can be studied through presenting to animals stimuli with characteristics known to induce specific emotional states in humans (i.e. the study of cognitive processing on emotional expression). For example Boissy *et al.* (2011) showed that ear posture in sheep could be consistently related to different emotional contexts where the emotional context was defined by its 'cognitive characteristics' (e.g. the suddenness or unfamiliarity of the triggering stimulus).

An alternative approach proposed by Mike Mendl, Liz Paul and co-workers was based on the reverse process, where an underlying affective (emotional) state can influence how we process information about our environment (e.g. depressed or anxious people tend to judge ambiguous stimuli more negatively (Mathews and Mackintosh, 1998)). In the first study of what is now referred to as the 'cognitive bias' approach, rats were tested for their response to ambiguous stimuli having experienced different housing prior to the testing; the results (again published in Nature) indicated that rats housed in 'unpredictable housing' (predicted from previous work to induce mild depressive symptoms) showed a negative 'pessimistic' bias when responding to ambiguous stimuli (Harding *et al.*, 2004). Since its first use, cognitive bias has been widely taken up in applied animal behaviour science. In a Web of Science® search I was able to find over 20 experimental studies with more than five citations in studies ranging from primates to honey bees. The study on bees (Bateson *et al.*, 2011) which demonstrated that a negative event (being shaken) resulted in more 'pessimistic choices' in response to ambiguity, raised the interesting and somewhat perplexing question of how to interpret 'pessimistic' or 'optimistic' biases in 'lower animals' (Mendl *et al.*, 2011).
It seems likely that human cognitive studies will foster both further innovation in applied animal behaviour science and also theoretical developments. For example, along similar lines to cognitive bias, the 'response to reward loss' approach, also based on human cognitive studies, argues that animals in a negative emotional state will show more enhanced responses when experiencing reward loss (Burman *et al.*, 2008). In addition Mendl and Paul have developed conceptual frameworks to understand and interpret results from work on animal emotions (Mendl *et al.*, 2010; Paul *et al.*, 2005) providing an example of strategic research 'feeding back' to basic research.

3. *Qualitative behavioural assessment.* My last example of science and innovation in applied animal behaviour science in the context of assessing animals' mental state is qualitative behavioural assessment (or QBA), developed by Françoise Wemelsfelder and colleagues. I had personal involvement in the development of QBA, working with Françoise during the early 1990s at what was then the Scottish Agricultural College (SAC). QBA arose from Wemelsfelder's position (part philosophical and part biological) that it can be legitimate to study animal behaviour from a qualitative perspective, and indeed that it may be essential to do so, in order to capture the subjective aspects relating to mental state that are of concern in animal welfare (Wemelsfelder, 2012). The result of this thinking led Wemelsfelder to develop an approach to the recording of animal behaviour that focuses on expressive quality of the behaviour as opposed to the physical characteristics that are normally used in behavioural data collection. An innovative step in the approach was provided by the collaboration between Wemelsfelder and Tony Hunter. Hunter was a bio-statistician who specialised in sensory food testing and suggested an approach to qualitative assessment of animal behaviour through free-choice profiling (FCP). FCP allows observers to develop and use their own qualitative terminology which, combined with a statistical analysis (Generalised Procrustes Analysis), provides an approach to collect and analyse qualitative descriptions of animal behaviour (Wemelsfelder *et al.*, 2001). Early tests of the approach, conducted using videos of individual pigs interacting with a person in their pen, with observations being made by observers using their own

> qualitative terms, demonstrated significant agreement between observers, suggesting that observers' assessments of the pig were reliable (Wemelsfelder *et al.*, 2001). Since then there has been research looking further at issues of observer reliability (Wemelsfelder *et al.*, 2012) and validity (Rutherford *et al.*, 2012; Stockman *et al.*, 2011), and considerable interest in the application of QBA to on-farm assessment of welfare (Wemelsfelder and Mullan, 2014). QBA was also incorporated into the WelfareQuality® assessment protocols for the assessment of positive welfare and has been used in a number of studies testing the efficacy of the WelfareQuality® protocol (Vries *et al.*, 2013).

However, QBA has raised tensions in applied animal behaviour science (acknowledged by Wemelsfelder, 2012). By openly acknowledging the qualitative nature of the approach, QBA can be seen as being 'more about the observer than the animal' and at risk of being anthropomorphic and hence unreliable. Despite these concerns my view is that there is a growing acceptance of the reliability, validity and utility of QBA within our field. This is founded upon results from peer reviewed papers, the experience of using the approach and also perhaps by the realisation that, as with the distinction between basic and applied science discussed earlier, the distinction between objective and subjective science is more blurred than we may realise or acknowledge.

6.3.2 The 'negative side'

I have just argued that welfare provides plenty of scope for strategic research and I have given examples of where, in my view, applied animal behaviour science has risen to this challenge. On the opposite (negative) side it can be argued that the focus on welfare issues has curtailed development of applied animal behaviour science, as we focus too much on the application and not sufficiently on the development of our science. Certainly one criticism that the applied animal behaviour and welfare 'community' has faced in the UK, is that our science is too often derivative and lacking international excellence, albeit perhaps unfairly given my previous section!

One obvious problem is the availability of funding to support basic/ strategic science in our field (which links back to science funding policy in relation to basic and applied research). Examples from my own research experience where funding has influenced the balance between basic and applied research includes work we carried out on stereotypic behaviour in closely confined sows. The Brambell Committee had voiced their concern over such confinement (Command paper 2836, 1965) and it became something of a hot topic in research in the 1980s and 1990s (Rushen *et al.*, 1993). A series of research papers from different groups pointed to the combination of food restriction (and elevated hunger and foraging motivation) and the thwarting effects of close confinement on that foraging motivation as strong contributing factors in the development of stereotypies in sows (Lawrence and Terlouw, 1993). However, the research did not provide much illumination on more fundamental questions about the relationship between stereotypic behaviour and welfare. Nevertheless, when the UK phased out the use of close confinement systems for pregnant sows in 1999, UK research on sow stereotypies ceased, despite these fundamental questions remaining. Another example comes from our work aimed at developing indicators of stress in farrowing sows. We had proposed that opioid-inhibition of oxytocin could be used as a sensitive indicator of stress in the farrowing sow (based on Lawrence *et al.*, 1992). However when we analysed blood samples from sows farrowing in different environments, we found no effect of environment on circulating oxytocin and vasopressin (Lawrence *et al.*, 1995). We did however find that the opioid antagonist naloxone increased circulating levels of both oxytocin and vasopressin, suggesting that a degree of opioid inhibition of these reproductive hormones is part of a normal parturition (Jarvis *et al.*, 2000) (Figure 6.3). However our funding at the time, which was more directed at applied questions, did not allow us to investigate the more basic question of the function of this opioid-inhibition (e.g. whether it played a role in preventing depletion of the finite stores of pituitary oxytocin (Russell *et al.*, 2003)). Of course work ceasing in one research group or even in one country does not preclude others working in the area; regarding the links between stereotypies and welfare, other scientists found new animal models to address these questions (e.g. the work of Joe Garner and Georgia Mason (2002)). Furthermore, the finding that stereotypies were related to food restriction in some species, helped to trigger a considerable body of research on the welfare impacts of food restriction which still continues to this day (D'Eath *et al.*, 2009; Dunn *et al.*, 2013). However, funding can clearly constrain individuals or groups from pursuing continuity of research on specific issues, due to changing funding policies and strategies.

From a UK perspective, over the last 10 years there have been rather dramatic changes in the funding of applied animal behaviour and welfare research which are undoubtedly causing adjustments in the type of work being funded. Previously government departments responsible for welfare policy funded a considerable amount of welfare-related research (much of it involving elements of animal behaviour); the research ranged across the spectrum from more basic to more applied and policy related. This source of funding has

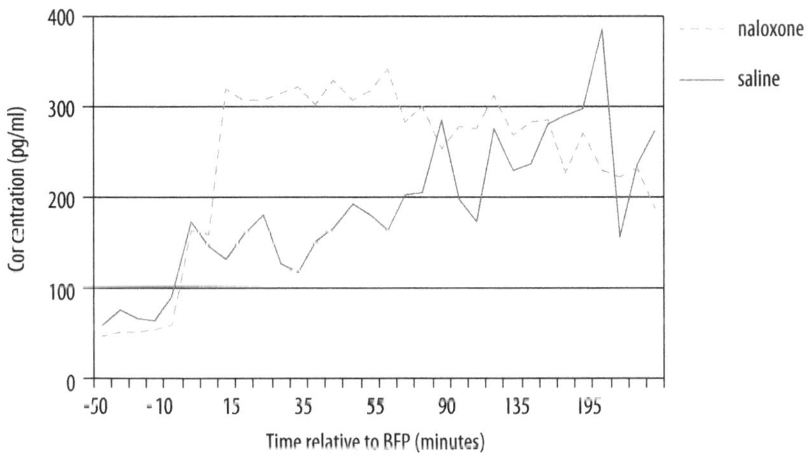

Figure 6.3. The pattern of plasma oxytocin (ng/ml) in sows as a response to administration of naloxone or saline. BFP = birth of the first piglet. There was a significant increase in oxytocin in the early stages of parturition in sows receiving naloxone (redrawn from Jarvis et al., *2000).*

now almost completely ceased. Today in the UK applied animal behaviour and welfare research has two main types of funding available: (a) 'Research council' funding: this is highly competitive and for research to be funded requires that it is deemed to be of high international excellence; this therefore tends towards the funding of more basic research although there is now also a requirement to illustrate how the research will have impact (e.g. see BBSRC, 2012); (b) Innovation funding: this is at the applied end of the spectrum and in the UK is now intended to be led by commercial interests, and likely success in implementation (e.g. product development) is a key element for gaining funding (e.g. Innovate, UK). In the UK, therefore, the major difference to the past is the lack of funding for strategic research to bridge between basic and applied research, similar to the work I described above as examples of strategic research. From what I know of other countries it seems that they vary in the extent to which they have funding to cover strategic research. In my view there would be benefit in a formal analysis of why these important changes have and are taking place as they are likely to have an impact both on the amount and type of welfare related research activity and also on impact in relation to improving animal welfare.

Funding considerations aside, we can also question whether the focus on welfare issues reduces the extent of scientific enquiry. For example, are researchers more focused on addressing practical issues rather than trying to understand underlying principles or to 'break through' with new scientific discoveries? This concern was raised by Dawkins (1997) and Rushen (2003), both pointing to the risks of succumbing to political pressure to provide quick answers, and hence avoiding addressing important and complex issues such as how we validate welfare indicators (see discussion above). In the examples given above this was not the case and all three areas described have generated results and theoretical writings which I think, by any measure, make significant and deep contributions to animal biology and the understanding of how to assess animals' mental states (Dawkins, 1990; Mendl *et al.*, 2010; Wemelsfelder, 2012). But it is a question for the field in general of whether there should be more encouragement of theoretical developments.

We can also question whether a focus on welfare leads researchers to intrinsic biases in the conduct and reporting of our science. For example, changes to housing are described as enrichments prior to there being any evidence that they have a positive relevance for the animal. In the other direction, treatments may only be imposed that are likely to occur in practice, thus not presenting the animal with true positive controls and giving little indication of the full spectrum of the animals' responses.

In summary, with respect to the influence of animal welfare on the development of applied animal behaviour science, I suggest that: (a) animal welfare provides many opportunities for basic science and its innovation; (b) it is possible to find examples from our field that fit into 'Pasteur's cell' of being excellent strategic research and I have given three examples (others I am sure could produce their own); (c) my examples illustrate that more than one scientist and often more than one group is responsible for a breakthrough or step-change, and that the process through which the science is developed and then used is dynamic and rarely linear (as suggested by Stokes (1997)); (d) it is also possible that the focus on welfare has negative impacts on the development of applied animal behaviour science because: (1) there are funding constraints (e.g. lack of funding for strategic science); (2) there is insufficient attention given to theoretical developments in the field; and (3) there are inherent biases in the way we approach our science that stem directly or indirectly from our focus on animal welfare.

6.4 The impact of applied animal behaviour science on animal welfare

Given the close relationship between applied animal behaviour science and animal welfare it is perhaps not surprising that the field has had an interest and even a concern about the impact of our science in terms of improving animal welfare with a number of authors debating the extent to which applied animal behaviour science has made a difference to animal welfare (Dawkins, 1997; Millman *et al.*, 2004; Rushen, 2003). As these authors pointed out, the question of assessing the impact of research is complex.

The topic of research impact is currently of considerable policy interest. As pointed out above, research councils in the UK have introduced the requirement that research be accompanied by statements of 'Pathways to Impact', and this emphasis on research impact is even more keenly emphasised in innovation funding platforms (e.g. Innovate UK). In Scotland, Scottish Government funding of agricultural, food and rural research is now strongly influenced by the concept of research impact, where the funded research is aimed at building 'a platform of knowledge that strengthens policy and contributes to the delivery of national outcomes' (RESAS, 2015). In reporting research activities to the Scottish Government's Rural and Environment Science and Analytical Services Division (RESAS), we not only cover traditional metrics (e.g. numbers of papers, amounts of external funding) but also our interactions with policy makers, with industry and with other stakeholders, including the public. Value is placed on these 'knowledge, transfer, exchange' activities where there is evidence that they have had impact (e.g. where scientific knowledge has been incorporated into policy or used by industry).

Furthermore, funding to universities in the UK is partly distributed on the basis of an assessment of research quality. Previously this was achieved by assessing the quality of individual outputs (usually peer reviewed papers). However in the current round an additional requirement was to provide impact case studies that demonstrate how a piece of science has achieved impact (see Research Excellence Framework (REF), 2014). The format was for impact studies to start with a peer reviewed piece of work (the initial paper which for REF 2014 had to have been published no earlier than 1993 and which had to be of a stated quality (2* or above on the REF Assessment criteria)); other papers could be quoted in support of the study. The impact case study had to demonstrate how the science led to a specific and measurable impact. However, note that this presumes the application of science as a linear progression as opposed to the more dynamic process suggested by Stokes (1997). The case study also had to be backed-up with corroborating sources, such as letters from policy contacts or commercial partners.

These impact case studies are now publicly available on the REF web-site (search REF impact case studies) and they provide an opportunity to explore the impact of our science in improving animal welfare. It is necessary to point out that the specific scores given to impact statements are not available. However, each submitting institute will have gone through an extensive, competitive screening process and for an impact statement to be found on the REF database suggests that at least the institute concerned (often assisted by external advisors) believed it to have had measurable impact. What can also be found are the references underpinning the study and where available an Altmetric score, which provides a record of all online discussions and shares of the research (http://www.altmetric.com/audience/researchers/).

I have searched the REF database using various appropriate terms and the results are shown in Table 6.1. I found 27 impact case studies that appear to fall within the scope of applied animal behaviour science and animal welfare. We could ask whether this number is more or less than we would expect. I found this hard to judge. It is a very small proportion of the over 6,000 submitted impact case studies but the UK applied animal behaviour and welfare community is also very small (I suggest less than 100 'full time equivalent' staff). The distribution is perhaps not surprising, with over 50% of the studies coming from farm animals (where historically funding has been allocated); nonetheless there is also work on laboratory, companion and captive animals. Looking at the areas covered, I classified eight (>30%) as being on pain related issues, five (~20%) on aspects of housing, and smaller numbers on handling (3), transport and slaughter (3), human-animal interactions (2), well-being (1) and applications of behaviour recording (including automation of behaviour recording (4). Some Impact Studies overlapped between areas but I have not tried to capture this.

I have also provided an indication of the impact being claimed, with brief quotes from the REF (2014) web-site (Table 6.1). From this collation I make the following observations.

- This seems a useful resource with excellent search functions and could be used within our field to provide examples and perhaps spread best practice on making our science more impactful. For example, to assess the impact of research more fully and efficiently, data on impact (e.g. numbers of animals affected) needs to be collected during the process rather than retrospectively.
- There is a very broad and impressive range of impacts claimed here: influencing UK, European Union (EU), and other (e.g. commercial, charity) policy making, including forming legislation; changing husbandry practice (e.g. handling of rodents, pain relief in various species, management of lameness in cows); changes to housing (e.g. aerial perches for hens, water provision for ducks, loose farrowing for sows); policy and practice relating to transport and slaughter; use of behavioural recording (e.g. cognitive testing in rodents, oestrous detection in cows) and public engagement (e.g. the Living Links exhibit).
- This range of impacts illustrates the many and interacting routes that science can follow on its path to impact. For example the study 'Minimising heat and other stresses during animal transportation' (Edinburgh) has informed scientific opinion as expressed by the European Food Safety Authority, EU and UK legislation, and codes of practice followed by hauliers and vehicle design.

Table 6.1. Relevant impact studies from the UK Research Assessment Exercise (REF 2014) downloaded from http://www.ref.ac.uk/; key words used were 'animal welfare', 'animal behaviour' and 'animal cognition' (searched together and separately). Shown here is a sample of 13 studies selected from among 27 Impact Studies that lie within the scope of applied animal behaviour science, the submitting institute and if available the named scientist(s) and a quote taken from the study to indicate the impact being claimed (phrases in italics added for clarity).

Impact study, organization (scientist)	Quotes on proposed impact
Transforming the welfare of commercially-reared poultry – Oxford (Dawkins)	'findings in relation to stocking densities for broiler chickens influenced the 2007 EU Broiler Directive ... has had a major impact on the industry ... For ducks, research examined the provision of water ... and identified solutions'.
Improving animal welfare through effective pain assessment and alleviation in laboratory rodents and rabbits – Newcastle	'research changed policy and practice relating to the provision of pain relief to rodents and rabbits. This has impacted on up to 35 million animals worldwide during the REF period'.
Pain research improves welfare of fish – Edinburgh (Braithwaite and Gentle)	'UK, EU and RSPCA animal welfare policies and guidelines have been informed by the research'; (numbers affected) ... millions of farmed fish in the EU'.
Whole-house gassing improves the welfare of birds requiring culling during a major disease outbreak and is now adopted by Defra – Edinburgh (Sparks and Sandilands)	'Policy Impact: Developed a humane culling method for large numbers of poultry in preparation for an outbreak of a notifiable disease such as avian flu ... adopted by Defra'.
Aerial perches improve the welfare of laying hens and are now recommended by the European Union – Edinburgh (Sparks, Sandilands and Green)	'research informed welfare guidelines impacting upon housing of around 200 million laying birds in the EU ... work has been adopted in EC regulations, and they are pushing all EU member states to ensure all their producers install aerial perches over slatted surfaces.'
Minimising heat and other stresses during animal transportation improves animal welfare and has driven EU legislation – Edinburgh (Mitchell)	'~60 billion animals are transported world-wide each year...definition of optimum transport environments has underpinned improved transport vehicle design and ... development of regulations'.
Loose-farrowing systems Improve the welfare of the sow whilst protecting the welfare of the piglet and have superseded the farrowing crate, now banned in three countries – Edinburgh & Newcastle (Baxter, Jarvis. Lawrence, Roehe, Edwards)	'This body of research has provided a fundamental platform of robust scientific evidence about the welfare detriments of crated sows whilst also working on solutions to improve piglet survival that are translated directly to the industry, thus advancing the welfare of both the sow and piglets.'
Effect of electronic training aids (E-collars) on pet welfare – Lincoln	'has stimulated debate and action by both anti-collar campaigners and the manufacturing industry; e.g. industry bodies are now working with Government to produce guidelines to reduce risks identified'.
Improvements in egg quality and hen welfare have enhanced productivity in the egg industry – Glasgow	'research on the long-term health and welfare implications of infrared beak trimming influenced UK policy debate, preventing a ban on beak trimming (due to be enacted in 2011) that would have exposed 35 million laying hens to potential pecking injury or death, as well as costing the industry an estimated £4.82-£12.3 million annually'.
Advancing clinical assessment of acute pain in companion animals – Glasgow	'researchers have been instrumental in developing international pain guidelines with the World Small Animal Veterinary Association, which represent more than 180,000 veterinarians worldwide...has promoted awareness of pain management in companion animals'.

Table 6.1. Continued.

Improved dairy-cattle welfare and reduced financial losses result from Bristol research into lameness – Bristol	'Dairy-cow herds in the UK and overseas, together with the dairy farming industry, are benefiting from strategic animal-husbandry changes and lameness-control programmes underpinned by research undertaken at the University of Bristol since 1997'.
Welfare of millions of laying hens across Europe transformed by the introduction of enriched cages – Bristol	'Hundreds of millions of laying hens in the European Community are now kept in enriched cages ... Research undertaken at Bristol University provided much of the evidence base for the full implementation of the relevant European directive'.
Electronic monitoring of dairy herds increases efficiency and reduces costs for UK and EU farmers - Strathclyde	'decision support platform combining artificial intelligence with low power wireless sensor technology, which was capable of alerting farm staff to animal conditions (changes in behaviour) ... more than 250 farms in the UK and Europe have adopted the technology'.

- A number of the studies illustrate the length of time it takes for science to evolve and have a measurable impact (Dawkins, 1997; Lawrence, 2008). For example the initial paper for the study 'Welfare of millions of laying hens across Europe transformed' (Bristol) was published at the cut-off date of 1993 (Sherwin and Nicol, 1993; Table 6.1 and REF, 2015 for similar examples). Interestingly there are also examples that show a much faster route to application; for example the technology based study 'Electronic monitoring of dairy herds' (Strathclyde) has an initial paper published in 2009 (Sasloglou *et al.*, 2009).
- There is some continuity here between the examples of applied animal behaviour science-led approaches to welfare assessment (see above) and these impact studies. For example, the studies 'Transforming the welfare of commercially-reared poultry' (Oxford) and 'Welfare of millions of laying hens across Europe transformed' (Bristol) both refer to the use of preference and choice testing (Jones *et al.*, 2009; Sherwin and Nicol, 1993).
- There are also examples where the science has come from outside the normal range of applied animal behaviour science. For example, the technology driven applications on automating behavioural recording or cognitive testing ('Actual Analytics Ltd: automated processing of video data' and 'Touch screen based cognitive testing for rats and mice'); also the 'Living Links to Human Evolution' Research Centre', which stems primarily from research into primate behaviour and ecology.
- Lastly, on a cautionary note, this exercise does not seem to fully address the difficulty of assessing science impact. For example, the contribution of individual papers (within the context of the wider literature and other information) to a specific impact remains a complex issue and (in my view) difficult to define. Also given that this is a competitive process the submitters of the Impact Statements are unlikely to be unbiased when assessing the actual impact of their work.

In summary, the question of how impactful applied animal behaviour science has been to improving animal welfare now has wider relevance given the increasing interest and attention being given to science impact. The UK research assessment exercise (REF) has provided us with a process and data for reviewing some of our science impacts. The results of my survey of the impact case studies submitted to the REF are, I suggest, heartening in terms of their range (across species and issues) and their extent (e.g. as assessed through influence on policy, forming of legislation, economic return on investment). Although this is UK based, I would suggest that similar exercises conducted in other countries with a tradition of applied animal behaviour science would yield similar results. The submission of impact case studies from other disciplines points to the need for applied animal behaviour science to continue to seek cross-discipline collaborations, including with commercial science and technology groups. It will be interesting to see whether in future more of the strategic research in applied animal behaviour science (such as my earlier examples) will find its way to having meaningful impact, thus justifying the label of 'use-inspired basic research' (Stokes, 1997).

6.5 Conclusions

In this paper I have applied two 'frameworks' to explore the relationship between applied animal behaviour science and animal welfare. I used 'Pasteur's Quadrant' provided by Stokes (1997) to discuss the relationships between science and its application, and the UK REF exercise to assess the impact of our science on improving animal welfare. 'Pasteur's Quadrant' seems a particularly relevant approach for a field such as applied animal behaviour science, which has such a close alignment to an applied issue (animal welfare) and where there is a need to consider the balance between developing the science and its application. Previous concerns about how impactful our research is (e.g. Dawkins, 1997) have lacked easily accessible evidence with which to assess impact. The REF database (with some obvious limitations) helps to fill that gap.

Using these frameworks I have emerged reasonably optimistic for the future. My optimism is based on two foundations: (1) the examples of 'use-inspired basic' or 'strategic' research (of which I am sure there are others than the three I chose to highlight), that indicate the scientific strengths within the field; (2) the examples of science impact that I collated from the UK's REF database. Again we can assume that this is just a sample of such impacts that would be substantially increased if this type of exercise was extended to other countries.

I hope that this chapter has helped to illustrate the importance of finding the best possible balance between development of the science and its application to animal welfare. With too little science development, we risk running out of novel developments to apply, and with the likely increasing need to apply to basic science funders, there is a potential drying up of research funding. With too little application we similarly run the risk of losing research funding (with the increased emphasis on impact), and more importantly not addressing the many pressing animal welfare issues in the field. Of these two risks my personal inclination is to be concerned more with there being sufficient science development in our field, not least because winning funds to pursue basic or strategic science is challenging.

I would suggest that as a science community we should consider more what can be done to maintain the best balance between developing our science and its application to animal welfare issues, and to encouraging more 'use-inspired basic' or 'strategic' research. This could happen at a global level (e.g. with ISAE taking the lead) or with more regional attempts to boost discussion on the balance between developing our science and its application. One starting point could be to collate further examples of strategic research and impacts to use as examples for ourselves, and in wider representation of the contribution of our field to science and animal welfare.

Acknowledgements

I would like to thank Professor Ian Crute for introducing me to 'Pasteur's Quadrant'. I also acknowledge the influence of Professor John Oldham who has long promoted and supported the aspiration to conduct 'excellent science within an applied context'. I would like to acknowledge funding support for this chapter from the Scottish Government's Rural and Environment Science and Analytical Services Division (RESAS), and the BBSRC Strategic funding to The Roslin Institute.

References

Bateson, M., Desire, S., Gartside, S. and Wright, G., 2011. Agitated honeybees exhibit pessimistic cognitive biases. Current Biology 21: 1070-1073.

Biotechnology and Biological Sciences Research Council (BBSRC), 2012. BBSRC policy on maximising the impact of research. BBSRC policy on impact. Available at: http://www.bbsrc.ac.uk/documents/bbsrc-impact-policy-pdf.

Boissy, A., Aubert, A., Desire, L., Greiveldinger, L., Delval, E. and Veissier, I., 2011. Cognitive sciences to relate ear postures to emotions in sheep. Animal Welfare 20: 47-56.

Botreau, R., Bracke, M., Perny, P., Butterworth, A., Capdeville, J., Reenen, C. and Veissier, I., 2007. Aggregation of measures to produce an overall assessment of animal welfare. Part 2: analysis of constraints. Animal 1: 1188-1197.

British Veterinary Journal, 1975. Proceedings of the Society for Veterinary Ethology (SVE). British Veterinary Journal 133: 95-101.

Burman, O.H., Parker, R.M., Paul, E.S. and Mendl, M., 2008. Sensitivity to reward loss as an indicator of animal emotion and welfare. Biology Letters 4: 330-333.

Bush, V., 1945. Science – the endless frontier: a report to the president on a program for postwar scientific research. National Science Foundation, Washington, DC, USA.

Command Paper 2836., 1965. Report of the technical committee to enquire into the welfare of animals kept under intensive livestock husbandry systems. Her Majesty's Stationery Office, London, UK.

Davies, A.C., Radford, A.N., Pettersson, I.C., Yang, F.P. and Nicol, C.J., 2015. Elevated arousal at time of decision-making is not the arbiter of risk avoidance in chickens. Scientific Reports 5: 8200.

Dawkins, M.S., 1980. Animal suffering: the science of animal welfare. Oxford University Press, Oxford, UK.

Dawkins, M.S., 1990. From an animal's point of view: motivation, fitness, and animal welfare. Behavioral and Brain Sciences 13: 1-61.
Dawkins, M.S., 1997. DGM Wood-Gush Memorial lecture: why has there not been more progress in animal welfare research? Applied Animal Behaviour Science 53: 59-73.
Dawkins, M., 2008. The science of animal suffering. Ethology 114: 937-945.
D'Eath, R.B., Tolkamp, B.J., Kyriazakis, I. and Lawrence, A.B., 2009. 'Freedom from hunger'and preventing obesity: the animal welfare implications of reducing food quantity or quality. Animal Behaviour 77: 275-288.
Désiré, L., Boissy, A. and Veissier, I., 2002. Emotions in farm animals: a new approach to animal welfare in applied ethology. Behavioural Processes 60: 165-180.
Duncan, I.J.H., 1993. Welfare is to do with what animals feel. Journal of Agricultural and Environmental Ethics 6, Suppl. 2: 8-14.
Dunn, I.C., Wilson, P.W., Smulders, T.V., Sandilands, V., D'Eath, R.B. and Boswell, T., 2013. Hypothalamic agouti-related protein expression is affected by both acute and chronic experience of food restriction and re-feeding in chickens. Journal of Neuroendocrinology 25: 920-928.
Farm Animal Welfare Council, 1979. Farm Animal Welfare Council, press statement. Available at: http://tinyurl.com/gm7svar.
Farm Animal Welfare Council, 2005. Report on the welfare implications of farm assurance schemes. Farm Animal Welfare Council, London, UK.
Garner, J.P. and Mason, G.J., 2002. Evidence for a relationship between cage stereotypies and behavioural disinhibition in laboratory rodents. Behavioural Brain Research 136: 83-92.
Harding, E.J., Paul, E.S. and Mendl, M., 2004. Animal behaviour: cognitive bias and affective state. Nature 427: 312.
Harrison, R., 1964. Animal machines. Stuart, London, UK.
Hogan, J.A. and Roper, T.J., 1978. A comparison of the properties of different reinforcers. Advances for the Study of Behaviour 8: 156-255.
Jarvis, S., Lawrence, A.B., McLean, K.A., Chirnside, J., Deans, L.A., Calvert, S.K., Gilbert, C.L., Goode, J.A. and Forsling, M.L., 2000. The effect of opioid antagonism and environmental restriction on plasma oxytocin and vasopressin concentrations in parturient gilts. Journal of Endocrinology 166: 39-44.
Jones, T.A., Waitt, C.D. and Dawkins, M.S., 2009. Water off a duck's back: showers and troughs match ponds for improving duck welfare. Applied Animal Behaviour Science 116: 52-57.
Kennedy, J.S., 1992. The new anthropomorphism. Cambridge University Press, New York, NY, USA.
Lawrence, A., 2008. Applied animal behaviour science: past, present and future prospects. Applied Animal Behaviour Science 115: 1-24.
Lawrence, A.B. and Terlouw, E.M., 1993. A review of behavioral factors involved in the development and continued performance of stereotypic behaviors in pigs. Journal of Animal Science 71: 2815-2825.
Lawrence, A.B., Petherick, J.C., McLean, K., Gilbert, C.L., Chapman, C. and Russell, J.A., 1992. Naloxone prevents interruption of parturition and increases plasma oxytocin following environmental disturbance in parturient sows. Physiology and Behaviour 52: 917-923.

Lawrence, A.B., Petherick, J.C., McLean, K.A., Deans, L.A., Chirnside, J., Vaughan, A., Gilbert, C., Forsling, M.L. and Russell, J.A. 1995 The effects of chronic environmental stress on parturition and on oxytocin and vasopressin secretion in the pig. Animal Reproduction Science 38: 251-264.

MacPhail, E.M., 1998. The evolution of consciousness. Oxford University Press, Oxford, UK.

Mason, G., Cooper, J. and Clarebrough, C., 2001. Frustrations of fur-farmed mink. Nature 410: 35-36.

Mathews, A. and Mackintosh, B., 1998. A cognitive model of selective processing in anxiety. Cognitive Therapy and Research 22: 539-560.

Matthews, L.R. and Ladewig, J., 1994. Environmental requirements of pigs measured by behavioural demand functions. Animal Behaviour 47: 713-719.

Mazzucato, M., 2013. The entrepreneurial state: de-bunking public vs. private sector myths. Anthem Press, London, UK.

Mendl, M., Burman, O.H. and Paul, E.S., 2010. An integrative and functional framework for the study of animal emotion and mood. Proceedings of the Royal Society B 277: 2895-2904.

Mendl, M., Paul, E.S. and Chittka, L., 2011. Animal behaviour: emotion in invertebrates? Current Biology 21: 463-465.

Millman, S.T., Duncan, I.J.H., Stauffacher, M. and Stookey, J.M., 2004. The impact of applied ethologists and the International Society for Applied Ethology in improving animal welfare. Applied Animal Behaviour Science 86: 299-311.

Morris, A.J., Whittingham, M.J., Bradbury, R.B., Wilson, J.D., Krykos, A., Buckingham, D.L. and Evans, A.D., 2001. Foraging habitat selection by yellowhammers (Emberiza citrinella) nesting in agriculturally contrasting regions in lowland England. Biological Conservation 101: 197-210.

Murphy-Bokern, D., 2008. Securing food supplies up to 2050: the challenges faced by the UK – Environment, Food and Rural Affairs Committee Contents. Available at: http://tinyurl.com/h2uyq8p.

Newberry, R.C. and Sandilands, V., 2016. Pioneers of applied ethology. In: Brown, J., Seddon, Y. and Appleby, M. (eds.) Animals and Us - 50 years and more of applied ethology. Wageningen Academic Publishers, Wageningen, the Netherlands, pp. 51-75.

Nicol, C.J., Caplen, G., Edgar, J. and Browne, W.J., 2009. Associations between welfare indicators and environmental choice in laying hens. Animal Behaviour 78: 413-424.

Paul, E.S., Harding, E.J. and Mendl, M., 2005. Measuring emotional processes in animals: the utility of a cognitive approach. Neuroscience Biobehavioural Reviews 29: 469-491.

Read N., 1989. The 'near market' concept applied to UK agricultural research. Science and Public Policy 16: 233-238.

Scottish Government's Rural and Environmental Science and Analytical Services Division (RESAS), 2015. Strategic research 2011-2016. Available at: http://tinyurl.com/hgwyycu.

Research Excellence Framework, 2014. Available at: http://www.ref.ac.uk.

Rushen, J., 2003. Changing concepts of farm animal welfare: bridging the gap between applied and basic research. Applied Animal Behaviour Science 81: 199-214.

Rushen, J., Lawrence, A.B. and Terlouw, E.M.C., 1993. The motivational basis of stereotypies. In: Lawrence, A.B. and Rushen, J. (eds.) Stereotypic animal behaviour: fundamentals and applications to animal welfare. CAB International, Wallingford, UK, pp. 41-64.

Russell, J.A., Leng, G. and Douglas, A.J., 2003. The magnocellular oxytocin system, the fount of maternity: adaptations in pregnancy. Frontiers in Neuroendocrinology 24: 27-61.

Rutherford, K.M., Donald, R.D., Lawrence, A.B. and Wemelsfelder, F., 2012. Qualitative behavioural assessment of emotionality in pigs. Applied Animal Behaviour Science 139: 218-224.

Sasloglou, K., Glover, I.A., Goh, H.G., Kwong, K.H. Gilroy, M.P, Tachtatzis, C., Michie, C. and Andonovic, I., 2009. Antenna and base-station diversity for WSN livestock monitoring. Wireless Sensor Network 1: 383-396.

Sherwin, C.M. and Nicol, C.J., 1993. Factors influencing floor laying by hens in modified cages. Applied Animal Behaviour Science 36: 211-222.

Stokes, D.E., 1997. Pasteur's quadrant: basic science and technological innovation. Brookings Institution Press, Washington, DC, USA.

Stockman, C.A., Collins, T., Barnes, A.L., Miller, D., Wickham, S.L., Beatty, D.T., Blache, D., Wemelsfelder, F. and Fleming, P.A., 2011. Qualitative behavioural assessment and quantitative physiological measurement of cattle naïve and habituated to road transport. Animal Production Science 51: 240-249.

Vries, M., Bokkers, E., Schaik, V.G., Botreau, R., Engel, B., Dijkstra, T. and Boer, I., 2013. Evaluating results of the Welfare Quality multi-criteria evaluation model for classification of dairy cattle welfare at the herd level. Journal of Dairy Science 96: 6264-6273.

Wemeslfelder, F., 2012. A science of friendly pigs: carving out a conceptual space for addressing animals as sentient beings. In: Birke, L. and Hockenhull, J. (eds.) Crossing boundaries. Brill, Leiden, the Netherlands, pp. 223-249.

Wemelsfelder, F., Hunter, T., Mendl, M. and Lawrence, A., 2001. Assessing the 'whole animal': a free choice profiling approach. Animal Behaviour 62: 209-220.

Wemelsfelder, F., Hunter, A.E., Paul, E.S. and Lawrence, A.B., 2012. Assessing pig body language: agreement and consistency between pig farmers, veterinarians, and animal activists. Journal of Animal Science 90: 3652-3665.

Wemelsfelder, F. and Mullan, S., 2014. Applying ethological and health indicators to practical animal welfare assessment. Scientific and Technical Review of the Office International des Epizooties 33: 111-120.

Part 3. Applied ethology around the globe

7. Applied ethology in the EU: development of animal welfare standards and actions

B. Algers
Professor Emeritus in Animal Hygiene, Department of Animal Environment and Health, Swedish University of Agricultural Sciences, P.O. Box 234, 532 23 Skara, Sweden; bo.algers@slu.se

Abstract

Why does the European Union bother about animal welfare? How did applied ethology come to play such an important role when animal welfare was to be measured and assessed? And was it disputed? Is it still disputed? A breakthrough in the EU for animal welfare and also for applied ethology, was the 1997 Amsterdam Treaty, stating that the EU's member states are supposed to pay full regard to the welfare requirements of animals. However, there were differences of opinion on the priority to be given to animal welfare in the political sphere. As a result, the EU took the initiative to further develop animal welfare regulations in Europe. The development of farming practices, with increasing scientific evidence of poor welfare associated with many housing, management and genetic factors caused more and more concern among citizens but also in production chains and retailers. A new trend in animal welfare research developed, involving animal based measures of welfare, making use of positive emotions as measures of good animal welfare and employing a multidisciplinary approach to the subject of animal welfare. The WelfareQuality® project aimed to accommodate societal concerns and market demands, to develop reliable on-farm monitoring systems, product information systems, and practical species-specific strategies to improve animal welfare. Internationally, a consensus is being sought by traders as well as competent authorities on animal welfare regarding what consideration should be given to animal protection and welfare in global trade. This chapter presents the scenario for when animal welfare became an issue for politicians, how applied ethology became a recognised science in the EU and why applied ethology plays a major role in the assessment of animal welfare. It also presents EU activities in developing standards, methodologies and awareness raising activities where applied ethology plays a key role.

Keywords: animal health, regulations, strategies, EFSA, EU Commission

7.1 Introduction

Why does the European Union bother about animal welfare? How did applied ethology come to play such an important role when animal welfare was to be measured and assessed? And was it disputed? Is it still disputed? This chapter presents the scenario for when animal welfare became an issue for politicians, how applied ethology became a recognised science in the EU and why applied ethology plays a major role in the

DOI 10.3920/978-90-8686-828-5_7, © Wageningen Academic Publishers 2016

assessment of animal welfare. It also presents EU activities in developing standards, methodologies and awareness raising activities where applied ethology plays a key role.

7.2 When it all started – the scene

'Ethology – it is similar to theology!' a professor and dean in my faculty said in the mid-1970s when the issue of animal welfare and the use of ethological methods was brought up for discussion. The keeping of animals, in particular animals that we raise for food production, had undergone dramatic changes in Europe after the Second World War. As industrial practices were applied to farming, the efficiency of producing milk, meat and eggs was radically increased. Different genetic lines in cattle were selected either for milk or for meat production and, in poultry, for egg or meat production. Fast growing pigs with increasing feed efficiency were developed. Such genetic changes, however, also had some unfavourable and unwanted effects on animal welfare (Rauw *et al.*, 1998).

In parallel with these genetic changes, the housing of animals kept for food production was substantially changed. To shelter animals from extreme climatic conditions, to exclude predators and to make it easier to feed animals in a more efficient way, indoor housing was developed. Indoor production required more sophisticated ventilation and manure handling techniques, as well as automated feed distribution. This also made it possible to reduce labour costs and lowered the cost of commodities such as meat, milk and eggs. During this process, many biological aspects of the animals were overlooked, for example that most farmed animals naturally live in stable social groups, and spend much of their time in foraging and exploratory behaviour. Feed was now delivered as concentrate, which reduced eating times dramatically. Sows are motivated to perform nest building, and poultry to perform dust bathing, roosting and nesting behaviour, but no space or material for these behaviours was provided. It soon became evident that overlooking such biological aspects would result in abnormal behaviours and compromise the welfare of the animals.

After the book '*Animal Machines*' was published by Ruth Harrison in 1964, concerns about the welfare of farm animals increased in Europe and, as a result of the outcry of the British public regarding the information in the book, the British Government appointed a committee to look into the subject. In 1965, the committee, chaired by Professor Rogers Brambell, presented their 85 page 'Report of the Technical Committee to Enquire into the Welfare of Animals Kept under Intensive Livestock Husbandry Systems' also known as the Brambell Report (Command Paper 2836, 1965). At the same time, reports from district veterinarians in Sweden showed changes in disease patterns and the distribution of environmentally evoked diseases in these systems. Using epidemiological methods these changes were confirmed, and their occurrence was associated with the new housing methods for farm animals (Ekesbo, 1973).

My boss and predecessor, Professor Ingvar Ekesbo, who produced these epidemiological studies, had contacted two colleagues, Eric Fabricius, professor in ethology at Stockholm University, and Lennart Levi, professor of psychosocial medicine at the Department of

Stress Research, Karolinska Institute, to discuss the meaning of these changed disease patterns. He took the initiative to use applied ethology as a tool to research the recent development of environmentally evoked diseases. Stress research on humans had already made considerable progress, and the tools of physiological and behaviour measurements were brought together to help in understanding the causal mechanisms linking changes in farm animal housing and disease.

However, the fight for research money was tough, as always, and there is an impression, looking back, that one way for those who wanted money to research other areas was to make applied ethology seem ridiculous. Ethological methods were rather simple and did not require expensive equipment, so how could this be good science? This was perhaps why, at least in Sweden, it was argued at faculty meetings that ethology was more about beliefs than a proper part of natural sciences. Veterinarians, who were the only profession in the faculty at the time, were not taught evolutionary biology or ethology as part of their education and many did not understand the importance of motivations of animals and how and why such motivational systems developed. It is true that applied ethology was not rocket science at its beginnings; however, as more money and researchers became involved, it started to deliver answers to questions related to animal welfare concerns.

To study the movement pattern of a dairy cow getting up or lying down may sound rather simple and even idiotic unless one realises that without the ability to balance the weight of the body over the front legs at rising, the cow is prone to trample her teats. This substantially increases the risk of getting mastitis, which causes pain to the cow, less milk for the farmer and increased use of antibiotics, risking an increase of resistant bacteria. For many decades, stanchions for dairy cows did not allow proper standing up, as saving building space was a priority in farm building design. But as applied ethology developed as a tool to explain important aspects of farming such as the connection between housing design, trampled teats and mastitis, attitudes towards the discipline changed.

7.3 Why is applied ethology important in studies of animal welfare?

Interestingly, one of the most important tools for the veterinary clinician is the use of behavioural signs or symptoms when trying to make a diagnosis in a case. This is evident from textbooks in veterinary medicine. For example, the symptoms of contagious bovine pleuropneumonia (CBPP) are described as 'sudden onset of high fever, fall in milk yield, anorexia and rumination ceases. There is severe depression and animals stand apart or lag behind a travelling group. Coughing, at first only on exercise, and chest pain are evident, affected animals being disinclined to move ... Respirations are shallow, rapid and accompanied by expiratory grunting' (Radostits *et al.*, 1994, pp. 779).

From an animal welfare point of view, it is important to know what effect different injuries and diseases have on the animal's feelings as well as on its capacity for a future good life. Injury is usually associated with pain, which can be regarded as a negative feedback signal that guides the animal to keep the injured area free from pressure or manipulation, thus improving the healing process. An injured or diseased leg or foot will eventually lead to pain and thus keep the animal from using its leg normally. The observation of lameness is thus a symptom, and is indicative of welfare problems.

In broiler rearing, diseases related to leg weakness are frequent (e.g. Sanotra and Berg, 2003). Lameness was identified in 18 and 30% of broilers in Sweden and Denmark respectively. The principal causes are usually non-infectious skeletal abnormalities or infectious factors. The diseases ultimately result in lameness, affecting the standing and walking patterns of the birds (Vestergaard and Sanotra, 1999). Research has shown that the innervation and pain receptors in joints in poultry are very similar to those of mammals (Gentle, 1992), which suggests that lame birds will feel pain similar to that of cattle and pigs. Interestingly, lame broilers have been shown to consume more feed with analgesics added than broilers that are not lame (Danbury *et al.*, 2000), showing that at some level the birds recognise the benefit of analgesics and are able to self-medicate. Further, the walking pattern of broilers that eat feed with analgesics is more similar to that of birds that are not lame (McGeown *et al.*, 1999), indicating that the changes in walking pattern reflect a state of pain associated with the leg diseases.

Lameness is one of the most important animal welfare issues for dairy cattle (EFSA, 2009a). Lameness affects the animal's welfare not only because of its relation to pain but also because it reduces the animal's ability to compete for resources. For example Hassall *et al.* (1993) showed that lame cows entered the milking parlour later than non-lame cows. By studying the incidence of lameness valuable information is obtained about the animals experience of pain, which obviously is a welfare issue, but also to what extent the lameness leads to deficiencies in relation to the animal being able to access various resources such as feed, good lying space etc. Hence, studies in applied ethology looking at lameness will generate important information relating to pain, but also to the broader assessment of the animals welfare status.

Similarly, a disease in which the animal has to use substantial resources to cope, e.g. through its immune system, can reduce the animal's activity levels. An obvious example is weaning diarrhoea in pigs. Pigs with diarrhoea have poorer weight gain (Pajor *et al.*, 1991), and are therefore less able to fight for resources. Pigs with poorer weight gain after weaning also show less interactive and more submissive behaviour (Algers *et al.*, 1990). Changes in social behaviour may then be a potential indicator of other underlying welfare related causes.

7.4 When animal welfare became a significant topic in Europe

In 1949, the Council of Europe (CoE) was founded. It was set up to promote democracy and protect human rights and the rule of law in Europe, and has its office in Strasbourg, Germany. Today, it comprises 47 European states. The CoE decides on internationally binding rules or Conventions, within areas such as human rights and freedom of the press, and also animal welfare. Ruth Harrison's book (1964) and the report of the Brambell Committee (1965) together with public debate resulted in increased interest in animal welfare legislation in several CoE member states. In 1971, after a request from the General Assembly in the Council of Europe, the Committee of Ministers decided that an *ad hoc* committee should be formed by delegates and experts from the member states for the task of elaborating a farm animal welfare convention. Such a convention came into force in 1976 providing general principles for the keeping of animals in modern intensive housing systems (CoE No 87, 1976).

A Standing Committee of the CoE was established in 1979 which over the years has produced recommendations for cattle (1988), poultry (1986, 1995), pigs (1986, 2005), sheep and goats (1992), calves (1993), ratites (1997), ducks, geese and fur animals (1999), turkeys (2001) and fish (2006). A convention for slaughter was issued in 1979 and revised in 1991. Already in 1968 the Council of Europe had adopted a convention on minimum requirements for the transport of animals. That document referred mostly to the health of the animals and their handling, while later (1987-1990), more recommendations for transport of various species were made. The welfare recommendations pay regard to nutrition, health, freedom of movement, physical comfort, social contacts, normal behaviour and protection against physical and psychological stressors. All of these conventions were developed as a result of more and more studies of applied ethology being published, together with increased insights into how this science relates to animal welfare. Hence, the concern for animal behaviour was legally recognised, and regarded as relevant to the welfare of the animals.

The three conventions on housing, transport and slaughter were approved by 46 of 47 member states in the Council of Europe, as well as the EU. In doing so, these countries had to translate the recommendations into their national legislation. The conventions are very different from each other in their design. Conventions regarding transport, slaughter, pet animals and especially experimental animals contain detailed regulations, whereas the farm animal convention has the character of a framework law. The more open farm animal convention allowed the standing committee to further elaborate detailed rules with the same legal status as the convention itself. The Committee then invited multiple inter-governmental and non-governmental organisations to each appoint an expert for consultation during the sessions. This was done by the ISAE, the World Society for the Protection of Animals, together with Eurogroup for Animal Welfare, the European Confederation of Agriculture and the Federation of Veterinarians of Europe.

During the meetings, each country was usually represented by the appropriate bureaucrat from their competent authority, sometimes with the aid of an academic expert. When my predecessor, Ingvar Ekesbo, retired after a long career chairing the standing committee and playing a significant role in the development of its work, I was asked by the Swedish competent authority to join the meetings as their expert. New to such experiences I travelled to Strasbourg with curiousness and fear. What giants were we to meet? How could a small country such as Sweden get our arguments through? And how do the microphones and earphones work? On which channel is the translation to English? Amazingly, all spoke English except the French! And even more amazingly, the French spoke English in the corridors and during lunch!

I found that much of the work was done during breaks and evenings, when teaming up with representatives from other countries was an important part of the work. Often, the Nordic country representatives had some wine together in a hotel room preparing proposals. Sometimes, the push from some of us to go further in developing CoE regulations was resisted by the European Commission. Their representative, Andrea Gavinelli, had to ask for a pause in the negotiations, explaining to us as new EU member states (with just a slight indication of Italian temper), that some of our ambitions were not in line with the EU Commission's view of what should be done.

It was not until later that I reflected on the fact that in this forum, the contribution of applied ethology to the assessment of animal welfare seemed to be undisputed. This contribution was not self-evident later, in the expert panel of the European Food Safety Authority, as we shall see below.

A breakthrough in the EU for animal welfare and also for applied ethology, was the 1997 Amsterdam Treaty, stating that the EU's member states should pay full regard to the welfare requirements of animals. In various surveys, the citizens of the EU had called for better animal protection standards to improve food safety and quality. In a white paper (EC, 1999), the European Commission explained the relationships between food safety and animal protection through the entire food processing chain. However, there were differences of opinion on the priority to be given to animal welfare in the political sphere. As a result, the EU took the initiative to further develop animal welfare regulations in Europe.

Already in 1995 the European Commission's Directorate-General for Health and Community (DG SANCO) had formed the Scientific Veterinary Committee (SVC, later the Scientific Committee on Animal Health and Animal Welfare, SCAHAW) with the goal of using scientific experts to review the scientific literature on housing and management of farm animals, and to identify the main animal welfare concerns in the EU. At this stage, the initiative of developing animal welfare regulations was turning from the CoE to the now-enlarged EU.

7.5 Animal welfare assessment and management: separate responsibilities

The BSE crisis came to play an important role in the development of applied ethology as a tool for assessing animal welfare in the EU. Before the crisis, practises of intensive farming were still largely ignored by the public who rarely reflected on the link between 'the bright red clean meat which appears on white plastic trays wrapped in cling film on supermarket shelves – and the animals, stunned, killed, bled, eviscerated and sawn up on production lines in the slaughterhouse' (Chambers, 1999, pp. 97, in Vos, 2000). However, the BSE crisis resulted in changes in public opinion, awareness and regulation of food safety, and opened the door for research on animal welfare assessment.

It was in 1986 that 'mad cow disease' or Bovine Spongiform Encephalopathy (BSE) was first identified in cattle. Ten years later the British government announced that the disease had jumped the species barrier to humans; and that the strange new human brain disease called new-variant Creutzfeldt-Jakob disease (vCJD) was probably linked to BSE. It seemed possible that victims had caught the disease by eating meat from BSE-infected cattle. As the BSE-crisis exploded in the media, trust in the food industry as well as in the competent authorities responsible for food safety plunged.

Food safety is a key issue in the EU and the BSE-crisis meant that public confidence in the EU food supply chain was at stake (Neyer, 2000). To re-establish confidence, the EU created the European Food Safety Authority (EFSA), which started its activities in Brussels in 2002 and in 2004 moved to Parma, Italy. The purpose of EFSA is to serve as an independent risk assessor, to carry out risk assessments for food safety, animal health and animal welfare, mostly on request from the European Commission but also on request from member states or on its own initiative. A clear distinction is made between EFSA's scientifically based risk assessment and the Commission's risk management, which takes ethical, socio-economic, cultural and religious aspects into consideration.

With the increasingly large body of scientific evidence on how animal housing and management affects animal welfare, together with the increased awareness and concern for these issues in society, risk managers were obliged to take action to assure that food production was in accordance with societal demands, including in relation to animal welfare. Such risk managers were found at different levels, such as the European Commission, national competent authorities, farmer or producer associations, and individual farmers. For the EU, the Commission took action, producing the Community Action Plan on the Protection and Welfare of Animals 2006-2010 (EC, 2010). It established objectives, not only to promote high animal welfare standards in the EU, but also to provide greater coordination in identifying future needs, and ensuring a more consistent and coordinated approach to animal protection and welfare. The action plan also implied using resources where they were most needed and gave the most efficient output. In order to achieve this, a risk based strategy seemed much more efficient. With this approach, new opportunities emerged for applied ethology research to determine how various risk factors affect animal welfare, as well as to understand the effects caused by different levels of exposure.

7.6 Applied ethology and animal welfare risk assessment

In 2004, when EFSA set out to develop an animal welfare risk assessment methodology, scientific methods already existed for risk assessments of food safety (governed by the Codex Alimentarius, e.g. CAC, 2002) and of contagious disease spreading between countries and regions (governed by the OIE, 2004a,b).

The objective of risk assessment is to assist risk managers to prioritise resources in the handling of a defined problem. However, methods for scientifically assessing the risk of impaired animal welfare were lacking. The EFSA definition of animal welfare includes how the animal feels, but also how well the animal copes with its environment and the conditions in which it lives. Hence, the task at EFSA was to develop a method for the assessment of risks to poor welfare in farm animals.

Risk assessment includes four steps: Hazard Identification, Hazard Characterisation, Exposure Assessment and Risk Characterisation, and has been defined as the 'evaluation of the likelihood of an adverse effect' (EC, 2000). Thus, risk is a function of the severity and probability of the adverse effect occurring due to exposure to an identified hazard.

Risk assessments for animal welfare must of course start from a usable definition of animal welfare. For this purpose, it has been suggested that poor animal welfare is a violation of behavioural needs. Already in 1969 Thorpe suggested that animals have behavioural needs apart from physiological needs that behaviour serves. Thus, behavioural studies in the context of animal housing and management became crucial parts of animal welfare consideration for the EU. Both external and internal factors are of importance (Jensen and Toates, 1993). For example if an animal is subjected to strong cold winds, it will seek shelter (e.g. Houseal and Olson, 1995). Hence an external factor, weather, affects the animal so that it will carry out locomotory behaviour to seek shelter. Internal factors, such as hormonal changes in relation to farrowing in sows, will affect the sow to conduct a series of motor behaviours to seek a nest site and build a farrowing nest (Algers and Uvnäs-Moberg, 2007).

In 2005, the EFSA used its mandate from the European Commission to issue a scientific opinion on the animal health and welfare aspects of intensive calf farming systems, and began developing a method for the assessment of risk to poor welfare. I was given the task to lead the working group and initially to suggest to EFSA whom to take onboard as a risk assessor. After some consultation with colleagues, I approached professor Frans Smulders at the Department of Farm Animal and Public Health in Veterinary Medicine, University of Veterinary Medicine, Vienna. This turned out to be the beginning of a beautiful friendship. Frans was able, patiently and pedagogically, to explain the secrets of risk assessment to all of us ignorant animal health and welfare experts. Furthermore, he turned out to be a very friendly person who several times opened his beautiful flat in Vienna to our working group meetings, and for whom a good meal with some good wine should always end a good meeting.

To be used in a risk assessment, it is important to define the actual risk or problem associated with each hazard. Is it a hazard that will cause disease? Will it cause pain? Will it contribute to poor growth? Will it cause behavioural frustration? The resulting poor welfare may be extreme and prolonged. Therefore, taking a discussion of what the adverse effects of a hazard may be as a point of departure, can facilitate discussion on the factors that should be considered as hazardous to animal welfare, and which scientific studies would be needed for an assessment. Interestingly, the adverse effects described in various EFSA reports on animal welfare risk assessment, such as 'malaise', 'fear', 'anxiety', 'stress', 'pain', 'frustration' are characteristics that are often studied using applied ethology.

If pain, fear and anxiety are thought to be adverse effects on welfare, then hazards that may cause those effects should be listed, whereas if poor growth, disease, etc. are considered of importance, factors that may cause such effects should be considered as hazards. All such effects have usually been incorporated in the animal welfare risk assessments made to date. Risk assessment of animal welfare is not a tool that may be used only in relation to one specific definition of welfare. However, the interpretation of welfare should be kept transparent, as otherwise it will be left to the reader, and probably cause confusion or disagreement as different interpretations will be possible.

The calf welfare working group provided the EFSA with its report in 2006 (EFSA, 2006). This was the first ever systematic animal welfare risk assessment, which after adoption at EFSA was published and sent to the European Commission. Although very simplistic, not least to facilitate for risk managers to get an overall perspective of the various risks to poor welfare and their seriousness, the calf welfare report inspired further elaboration of the assessment methodology, engaging many of the health and welfare experts in the EFSA AHAW panel and many others who assisted in the various working groups of the panel.

7.7 EFSA and applied ethology

With the evolving challenges faced by the EU with food safety scares and emerging societal animal welfare concerns, momentum was building and animal welfare became a hot topic. Although we have seen how ethology played a strong role in helping address these challenges, I found myself still having to fight a corner, standing up for the importance of animal behaviour research!

'Frustration – I am frustrated, but what does it mean?' a member of the EFSA expert panel on Animal Health and Animal Welfare burst out, obviously in frustration. In 2007 the European Commission had given EFSA a mandate to report on the main animal health and welfare issues of sows, boars and piglets and now the panel had a meeting discussing a draft report from its working group. The comment illustrated well the conflict between experts that had their career in infectious diseases and those that had their career in animal housing and welfare with expertise in behaviour. At the time, I felt that the discussion more reflected some panel members' lack of expertise in applied ethology and welfare research and methods, rather than being a constructive discussion.

Some parts of the report from the working group were heavily disputed by some panel members, so most of the time of two panel meetings was used to try to settle differences in opinion of how the available scientific literature should be judged. Finally, the differences of opinion boiled down to the wording of the suggested conclusions and recommendations on the need for farrowing sows to have space for nest building, and on whether piglet mortality was as low in loose housing systems as in crates. Being the chairman of the working group, it was my task to explain the elaborations of the group to the panel. We were rather far apart and a difference of opinion threatened the outcome of the panel's decision on adopting the report. A final attempt to find wording that could satisfy both the working group producing the report and the opposing members of the AHAW panel was made by me and the person most opposed to the text during a break in the meeting.

It turned out that we were able to find wording that I and my strongest opponent could agree upon, and this wording was adopted by a relieved AHAW panel. Shortly afterwards, to our great surprise, the executive leader of EFSA sent a message of congratulations to the meeting and offered wine for our lunch break instead of the usual soft drinks! The disagreement on judgement of the various data on the issue was problematic for EFSA, as its aim is to provide pure science – even though science can only be judged by humans and no human is without values. The day ended with a joint dinner at one of the better Parma restaurants with speeches and fraternisation.

Only two years later, however, EFSA published an opinion with a minority report on the welfare of tied dairy cows. Here, the different opinions among the panel experts were not possible to merge into a statement that could be agreed upon. The minority were all the panel experts with a career in behavioural studies and applied ethology, and the issue was on how to judge the effect on welfare of cattle kept in tie-stalls (EFSA, 2009b).

Maybe the task for EFSA to assess animal health and animal welfare is too wide for one expert panel as mandates vary widely, from assessing the risk for the spread of Blue tongue virus in the EU to the effects of space allowance and floor types on the welfare of pigs. Clearly, EFSA and the quality of its reports could benefit from not merging scientists with too diverse expertise onto the same expert panels.

7.8 The EU forges ahead – Actions and Welfare Quality

In January 2006 the EU adopted its first action plan for animal welfare. This was in response to increasing societal demands as presented by various NGOs and in an EU barometer report as well as in the EU parliament where citizens expressed a high interest in animal welfare. The goals of the action plan included upgrading existing minimum standards for animal protection and welfare, giving a high priority to promoting policy-oriented research on animal protection and welfare, and introducing standardised animal welfare indicators for scientific assessment of welfare in production systems, livestock herds and slaughterhouses. Regulations were lacking for dairy cows and handling of animals in connection with slaughter. Already, a great deal of scientific literature present that could serve as a basis for the regulations. There were also no rules governing animal welfare for fish, such as how densely fish should be kept in the tanks and pens of fish farms, or requirements on slaughter methods. It was argued that more research was needed to provide guidance on how fish are affected by stress (see e.g. EFSA, 2009c).

The development of farming practices, with increasing scientific evidence of poor welfare associated with many housing, management and genetic factors, has caused increasing concern among citizens but also in production chains and retailers (Miele *et al.*, 2005). The EU funded the WelfareQuality® project was developed to respond to such concerns, with the goal of delivering reliable, transparent, science-based, on-farm welfare assessment systems for the main farmed species (Blokhuis *et al.*, 2010). The project aimed to accommodate societal concerns and market demands, to develop reliable on-farm monitoring systems, product information that was accessible to producers and consumers, and practical species-specific strategies to improve animal welfare. Forty-four institutes and universities, representing 20 countries participated in this integrated research project concerning cattle, pigs and chickens. The final report was delivered in 2009, including animal welfare assessment protocols for the three species.

In its declaration of principles for good animal welfare the WelfareQuality® project outlined twelve welfare criteria that should be monitored to assess the quality of life of animals concerned. They not only cover absence of negative states, such as feelings of hunger, thirst, pain or anxiety, but also positive states such as thermal comfort, resting posture and positive emotional state. As outcomes of the WelfareQuality® project, not only were animal welfare assessment protocols produced and tested, but broad ranging societal involvement in the project was achieved including the general public, social scientists, producers, animal breeders, retailers, certification bodies, NGOs and politicians, as well as (of course!) many applied ethologists (Blokhuis *et al.*, 2010).

The WelfareQuality® project is one of the most obvious examples of the new trend in animal welfare research involving animal based measures of welfare, making use of positive emotions as measures of good welfare and employing a multidisciplinary approach to the subject. It also shows how the EU was prepared to devote a great deal of attention not only to the risk of spread of contagious diseases among farm animals but also to the public concern for animal welfare and animal behaviour.

7.9 Other EU initiatives on animal welfare

With the vast and continuously growing body of applied behavioural research, which was also being put together by numerous SVC/SACHAW/EFSA reports, there was a growing need to implement some of the knowledge and practices gained. These initiatives present great opportunities for research in applied ethology. Some of these more recent EU activities and projects are described here.

A project with similar objectives to WelfareQuality® is the WELANIMAL project. It aims to evaluate whether life conditions of animals raised in central and southeast Europe meet the EU requirements for animal welfare, by sampling farm data on a large scale, evaluating the effects of different social, economic, cultural and religious approaches on welfare, and encouraging the development of 'Humane animal production systems' as part of a platform that covers common ethical values of consumers (WELANIMAL, 2010).

In 2008 a European Animal Welfare Platform (EAWP) was created bringing together representatives of major stakeholders and groups in Europe with interests in animal production, ranging from producers and retailers to scientists and NGOs. The EAWP aimed to facilitate the exchange of knowledge, experience and expertise between partners. It also disseminates information to a much broader stakeholder group (EAWP, 2010).

Internationally, a consensus is being sought by retailers as well as competent authorities on animal welfare regarding what consideration should be given to animal protection and welfare in global trade. Better Training for Safer Food (BTSF) is a Commission training initiative covering food and feed law, animal health and welfare and plant health rules. It trains government staff involved in official controls in these areas. It aims to keep participants up-to-date with EU law in these areas and should help to ensure more harmonised and efficient controls (DG SANCO, 2010). BTSF has its roots in the emergence of food safety as an important concern of EU citizens. Food emergencies, globalisation of the food trade and a lack of fully harmonised implementation of food law and official controls, together with increased consumer awareness, were factors that supported the development of the Internal Market. Many EU Member States already had a long tradition of providing training for their control staff, although such national-level training had not been uniform across the EU. With the adoption of Regulation 882/2004 in April 2004, provision of such training became compulsory for all Member States. Concrete activities within BTSF began in 2006 with the implementation of training programmes. In these, utilisation of research on applied ethology studying the effects of housing and management has been of vital importance.

At one BTSF course in Italy, we visited a facility where pig transport vehicles stopped to unload their cargo so that the pigs could rest before further transport. The unloading was done with workers in the wrong position in relation to the pigs, onto a ramp that was unstable (which made the pigs hesitant to walk), going into a 170 degrees sharp turn (which is far from optimal), and with the help of a few kicks by the workers (which is unacceptable). Filming it all gave me excellent material to show to my students. But the question remains – how far have we come with implementing the knowledge that applied animal behaviour studies have given us? Is it perhaps in increasing awareness and promoting practical implementations where researchers in applied ethology should find new tasks?

Many research projects are now taking a broadened approach to address societal concerns, consumer behaviour and industry initiatives. Large projects include the EU funded Q-PorkChains project which addresses the issue of animal welfare and provides learning resources on the topic that encompasses primary production, quality control, chain management, product development and product quality as well as consumer attitudes and behaviour (www.porktraining.org).

Although much research has been produced and multiple initiatives, as described here, have been taken to increase awareness and improve legislation, the impact of all this work in reality has been painfully slow. In 2010, the EU's action plan on Animal Welfare was evaluated by the rapporteur of the Committee on Agriculture and Rural Development, Marit Paulsen. It was acknowledged that the first action plan had been implemented satisfactorily, although it was stated that more could have been done in relation to animal transport and the issue of developing satellite systems to monitor such transport. Further, in the report's explanatory statement it was noted that 'many pig farmers in Europe are violating the provisions of Directive 2008/120/EC laying down minimal standards for the protection of pigs' (see discussion below on 'The EU and applied ethology today').

In Paulsen's report a new action plan for the forthcoming years 2011-2015 was suggested, focusing on:

- a general European animal welfare law;
- a European centre for animal welfare and animal health;
- better enforcement of existing legislation;
- the link between animal health and public health;
- indicators and new techniques.

It was argued that animal welfare standards vary a great deal among member states, which also prevents equitable competition between producers. A common law should fulfil the requirements for a good life in accordance with the nature of each animal species. It will be a great challenge to develop such a law that can meet the demands from animal welfare interests, market and producer interests, while also being usable as a proper steering document. A risk based system for monitoring will also be needed to give credibility to such a law, and certainly there will be an ongoing need for expertise in animal behaviour and welfare to achieve these objectives.

7.10 Non-regulatory approaches to animal welfare

While the EU commission does indeed engage in the question about animal welfare and how we, as humans, should treat the animals we use for food, in order to implement these regulations and disseminate knowledge, non-regulatory activities are also needed.

In recent years there has been an increase in non-regulatory approaches to increase animal welfare. Some schemes have been developed by producer organisations or independent NGOs and some by retailers (Veissier *et al.*, 2008). However, consumers may not always have the appropriate knowledge to be able to judge such schemes. Sorensen and Fraser (2010) argue that there is a need for involving panels of producers and non-producers for selection of welfare criteria that have a wide acceptance by diverse citizens and set minimum levels, while scientists should be involved to assure fairness, objectivity and efficiency. They are of the opinion that this would result in better acceptance, be more responsive to change and lead to better outcomes than regulations stipulating the features of the physical environment. However, such measures would also demand better knowledge and understanding of animal welfare among people in general. Hence, there is a need for knowledge in applied ethology not only among regulators but also among producers, NGOs, retailers and consumers.

In 2010, DG SANCO (the EU Directorate General for Health and Food Safety) launched an initiative to organise the 1st International Conference on Animal Welfare Education with the theme 'Everyone is responsible,' to raise awareness among EU citizens and to stimulate further activities to increase knowledge about animal welfare. Various materials for teaching and training in animal welfare were presented and ethical aspects and methodologies for teaching animal welfare were discussed. Clearly, new approaches have to be developed to allow for the citizen and the consumer to be able to understand messages about animal welfare presented by retailers or producer organisations, and to act as a responsible citizens and consumers. Therefore, today, there is a great opportunity for people well acquainted with applied ethology and with pedagogical skills to help transfer ethological knowledge into citizen awareness.

Citizens nowadays have much less understanding of the conditions of food producing animals; in parallel, the relation to pet animals has developed into looking at the pet as a family member. What may be called 'Disneyfication' of animals, together with over-idealistic advertisements from the food industry on the conditions of farm animals, has led consumers and people working in the food industry to have different pictures of what good animal housing and management should be. If consumer and industry expectations of requirements for good animal welfare differ, this can lead to increasing conflict, something from which society does not gain and which in the long run is not sustainable (Figure 7.1). Therefore, there is a continued role for applied behaviour research in providing objective knowledge on farm animal and their specific needs.

The welfare of humans and the welfare of animals are closely linked. A secure supply of animal food products for people depends on the health and productivity of animals, and these in turn depend on the care and nutrition that animals receive. Modern animal production has raised a wide range of ethical issues, including concern for animal welfare, which have to be considered together with environmental sustainability and secure access to food. Awareness and competence building are key factors in this context.

The current challenge is to raise awareness about animal production and welfare in all members of society, as a means to achieve a more consistent vision of what is good animal welfare, as a basis for industry to develop production schemes, and to guide consumer habits for purchasing food. As few people gain their knowledge about animals and animal

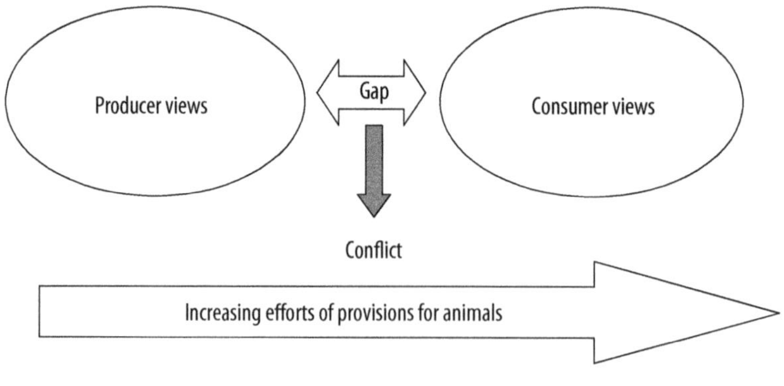

Figure 7.1. Non-sustainable situation of appreciation of good animal welfare in consumer and industry groups.

welfare through life experiences, such matters need to be included in school education and other education programmes. The goal should be to maximise the overlap between producer and consumer expectations (Figure 7.2), and to use animal welfare science together with socio-economic analyses to position the overlap on the scale of increasing efforts of provisions for animals.

The progress of these developments will depend on existing and future national, regional or EU-funded activities related to:
- developing teaching material, courses and training activities;
- making such material available, e.g. through stimulating the use of freeware;
- creating a peer review system of such material; and
- promoting the translation of such material and modifying it for regional applicability.

Here, the use of Open Educational Resources (OER), Worked Based Learning (WBL) and IT resources may be the future tools for scientists and teachers in applied ethology and animal welfare, bridging the boundaries between academia and industry, between legislators and implementers, between producers and consumers, and between politicians and citizens (Algers, 2015).

7.11 The EU and applied ethology today

'Piglets are naughty and sometimes bite each other's tails ... but a rich environment keeps the pigs happy and helps to stop bad behaviour'. This is a quote from the EU Farmland website (www.farmland-thegame.eu) intended to educate school children on the conditions of farm animals. In spite of being heavily criticised for its poor scientific quality and low pedagogical standard (Algers and Ljung, 2015), the site is still up and provides educational material which is not based on science but rather gives children prejudices about animals. When it comes to increasing knowledge among the citizens in the EU there is still a long way to go.

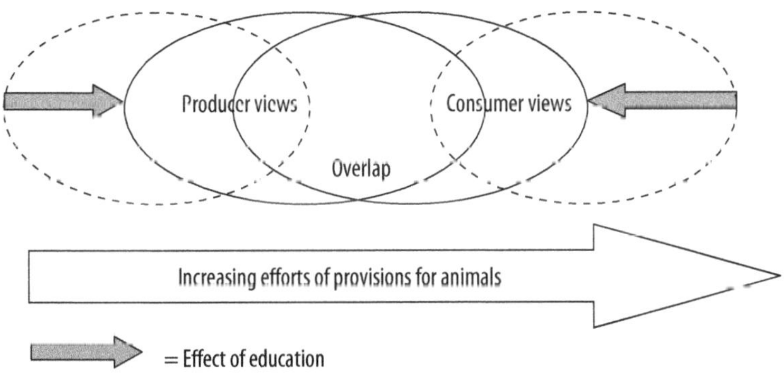

Figure 7.2. Effects of education in animal welfare and sustainable situation of appreciation of good animal welfare in consumer and industry groups.

The question of tail docking in pigs in the EU is an ongoing problem, despite the fact that it should have been solved long ago. In the Council Directive 2008/120/EC it is clearly stated that routine tail docking in pigs is prohibited and enrichment materials for the pigs must be provided, which is in line with the high animal welfare standards that the European Union aims for. This directive is in force in all member states. But the provision of enrichment material is often lacking and the habit of tail docking remains widespread as two reports by EFSA have shown (Lerner and Algers, 2013). Only Sweden, Finland and Lithuania manage their production without routine tail docking.

The EU launched the EuWelnet in 2013, evaluating the feasibility and usefulness of a network that could assist the competent authorities and stakeholders in implementing EU legislation on animal welfare. The concept is that a network of institutions, with recognised knowledge on animal welfare and independent from private interests, could supply technical assistance, promote innovation, disseminate scientific and applied research, and take a lead in education and vocational training. One of the tasks of the EUWelnet was to develop and test an e-learning tool with the objective of increasing the understanding of pigs motivations for exploratory behaviour and tail biting, raising awareness of the potential consequences of tail biting, and assessing the suitability of available enrichment material.

An online training package on enrichment and tail docking of pigs was produced in seven languages, with the aim of improving consistency of professional judgements regarding legislation compliance on farms. In a training exercise, 158 official inspectors, certification scheme assessors and advisors from 16 EU countries completed a test before, and after, viewing the training package. Training significantly increased the proportion of respondents correctly identifying that a farm with no tail lesions should stop tail docking, improved their recognition of enrichments that, by virtue of their type or use by pigs may be insufficient, as well as their identification of risk factors for tail biting and recognition of when routine tail docking was occurring (Hothershall *et al.*, in press).

I was part of the working group and had the pleasure to host the first meeting in Skara, Sweden. It was a joy to see a Swedish farmer, Peter Eriksson, showing pictures of his farm to us all and describing how, with the regular supply of straw to his pigs, he managed his farm without any tail docking. As mentioned above, the e-earning tool was shown to be successful, which gives some hope for the future. Indeed, when the EU elections approached in spring 2014, the tail docking issue was in most Swedish media and all political candidates were posing with pigs, emphasising the importance of enforcement of EU legislation.

Another EU initiative is now underway which could result in changes to tail docking practices. FareWellDock (http://farewelldock.eu) is a three-year research project which is part of the Animal Health and Welfare (ANIHWA) ERA-net initiative. This initiative aims to increase cooperation and coordination of national research programmes on the health and welfare of farm animals. The general aim of the FareWellDock project is to supply necessary information for quantitative risk assessment and stimulate the movement towards a non-docking policy in the EU. Here, applied behaviour science again plays a key role in understanding how to provide acceptable animal welfare in the pig industry.

One might argue that research and training activities on why and how to provide enrichment materials for pigs and stop routine tail docking seems like a luxury where governmental funding should not be spent. However, the issue is not only an animal welfare problem affecting most pigs in the EU but also a question of living up to an agreed minimum standard where today farmers that meet the agreed standards have to compete unfavourably with those that do not. This is a threat to one of the corner stones of the EU – free and fair trade (Edman, 2014). Hence, the knowledge, awareness and practical implementation of applied ethological science towards the elimination of tail docking is not a luxury but rather a necessity.

7.12 The way forward

Looking back, it is evident that research in applied ethology has been important for the provision of a scientific basis for development of animal welfare regulations in Europe and in the EU. However, it is also obvious that the insights that this research brings – like the stone of Sisyphus – roll back down the hill once in a while. However, it is obvious that applied ethology is a corner stone in animal welfare assessment and that the EU to a large extent relies on our field to provide insights in how the treatment that we provide our animals affects their welfare. But clearly there is still a large gap between the existing knowledge and the implementation of it. Looking forward, it seems to be high time for the ISAE to invest in a closer relationship with the EU, and consider suggestions for how this gap could be closed.

References

Algers, A., 2015. Open learning in life sciences. Studies of open educational resources in animal welfare and work-based learning in food science. Thesis, Chalmers University of Technology and University of Gothenburg. Available at: http://hdl.handle.net/2077/40580.

Algers, A. and Ljung, M., 2015. Peer reviewing of OER in a contested domain – an activity theoretical analysis. Journal of Interactive Online Learning 13(4): 21-38.

Algers, B., Jensen, P. and Steinwall, L., 1990. Behaviour and weight changes at weaning and regrouping of pigs in relation to teat quality. Applied Animal Behaviour Science Journal 26: 143-155.

Algers, B. and Uvnäs-Moberg, K., 2007. Maternal behaviour in pigs. Hormones and Behaviour Journal 52: 78-85.

Blokhuis, H.J., Veissier, I., Miele, M. and Jones, B., 2010. The Welfare Quality® project and beyond: safeguarding farm animal well-being. Acta Agriculturae Scandinavica, Section A Animal Science 60: 129-140.

Codex Alimentarius Commission (CAC), 2002. Principles and guidelines for the conduct of microbiological risk assessment. Document CAC/GL 30.

Command Paper 2836, 1965. Report of the technical committee to enquire into the welfare of animals kept under intensive livestock husbandry systems. HMSO, London, UK.

Council of Europe (CoE), 1976. Council of Europe No 87, 1976. European Convention for the protection of animals kept for farming purposes. CoE, Strasbourg, Germany.
Danbury, T.C., Weeks, C.A., Chambers, J.P., Waterman-Pearson, A.E. and Kestin, S.C., 2000. Self-selection of the analgesic drug carprofen by lame broiler chickens. Veterinary Record 146: 307-311.
DG SANCO, 2010. Available at: http://tinyurl.com/jfjeje9.
Edman, F., 2014. Do the member states of the European Union comply with the legal requirements for pigs regarding manipulable material and tail docking? Institutionen för husdjurens miljö och hälsa, BSc. thesis, SLU, Skara, Sweden. Available at: http://stud.epsilon.slu.se/7178.
European Animal Welfare Platform (EAWP), 2010. Available at: http://tinyurl.com/zambrps.
European Commission (EC), 1999. White paper on food safety. COM (1999) 719 final, January 12, 2000. Brussels, Belgium.
European Commission (EC), 2000. First report on the harmonisation of risk assessment procedures. Part 1: the report of the Scientific Steering Committee's Working Group on Harmonisation of Risk Assessment Procedures in the Scientific Committees advising the European Commission in the area of human and environmental health. European Commission, Brussels, Belgium.
European Commission (EC), 2010. Communication from the Commission to the European parliament and the Council on a Community Action Plan on the protection and welfare of animals 2006-2010. Available at: http://tinyurl.com/z9hb5vz.
European Food Safety Authority (EFSA), 2006. Opinion of the Scientific Panel on Animal Health and Welfare (AHAW) on a request from the Commission related with the risks of poor welfare in intensive calf farming systems (Question No. EFSA-Q-2005-024). EFSA Journal 366.
European Food Safety Authority (EFSA), 2009a. Scientific opinion on welfare of dairy cows in relation to behaviour, fear and pain based on a risk assessment with special reference to the impact of housing, feeding, management and genetic selection. EFSA Journal 1139: 1-66.
European Food Safety Authority (EFSA), 2009b. Scientific opinion on the overall effects of farming systems on dairy cow welfare and disease. Scientific Opinion of the Panel on Animal Health and Animal Welfare (Question No EFSA-Q-2006-113). EFSA Journal 1143: 1-38.
European Food Safety Authority (EFSA), 2009c. Scientific opinion general approach to fish welfare and to the concept of sentience in fish. Scientific Opinion of the Panel on Animal Health and Welfare (Question No EFSA-Q-2008-708). EFSA Journal 954: 1-27.
Ekesbo, I., 1973. Animal health, behaviour and disease prevention in different environments in modern Swedish animal husbandry. Veterinary Record 93(2): 36-39.
Gentle, M., 1992. Ankle joint receptors of the domestic fowl. Neuroscience 49: 991-1000.
Hassall, S.A., Ward, W.R. and Murray, R.D., 1993. Effects of lameness on the behaviour of cows during the summer. Veterinary Record 132: 578-580.
Harrison, R., 1964. Animal machines: the new factory farming industry. Stuart, London, UK.
Houseal, G.A. and Olson, B.E., 1995. Cattle use of microclimates on a northern latitude winter range. Canadian Journal of Animal Science 75: 501-507.
Jensen, P. and Toates, F.M., 1993. Who needs 'behavioural needs?' Motivational aspects of the needs of animals. Applied Animal Behaviour Science Journal 37: 161-181.
Lerner, H. and Algers, B., 2013. Tail docking in the EU: a case of routine violation of an EU directive. In: Röcklinsberg, H. and Sandin, P. (eds.) The ethics of consumption. EurSafe 2013, September 11-14, 2013. Wageningen Academic Publishers, Wageningen, the Netherlands, pp. 374-378.

McGeown, D., Danbury, T.C., Waterman-Pearson, A.E. and Kestin, S.C., 1999. Effect of carprofen on lameness in broiler chickens. Veterinary Record 144: 668-671.

Miele, M., Murdoch, J. and Roe, E., 2005. Animals and ambivalence: governing farm animal welfare in the European food sector. In: Higgins, V. and Lawrence, G. (eds.) Agricultural governance: globalization and the new politics of regulation. Routledge, London, UK, pp. 169-185.

Neyer, J., 2000. The regulation of risks and the power of the people: lessons from the BSE crisis. European Integration online Papers: 4(6). Available at: http://tinyurl.com/hytojpq.

Office International des Epizooties (OIE), 2004a. Handbook on import risk analysis for animals and animal products. Vol. 1. Introduction and qualitative risk analysis. OIE, Paris, France.

Office International des Epizooties (OIE), 2004b. Handbook on import risk analysis for animals and animal products. Vol. 2. Quantitative Risk Assessment. OIE, Paris, France.

Pajor, E.A., Fraser, D. and Kramer, D.L., 1991. Consumption of solid food by sucking pigs – individual variation and relation to weight gain. Applied Animal Behaviour Science Journal 32: 139-155.

Radostits, O.M., Blood, D.C. and Gay, C.C., 1994. Veterinary medicine. Ballière Tindall, London, UK.

Rauw, W.M., Kanis, E., Noordhuizen-Stassen, E.N. and Grommers, F.J., 1998. Undesirable side effects of selection for high production efficiency in farm animals: a review. Livestock Production Science 56: 15-33.

Sanotra, G.S. and Berg, C., 2003. Investigation of lameness in the commercial production of broiler chickens in Sweden. Swedish University of Agricultural Sciences, Sweden.

Sorensen, J.T. and Fraser, D., 2010. On-farm welfare assessment for regulatory purposes: issues and possible solutions. Livestock Science 131: 1-7.

Thorpe, W.H., 1969. Welfare of domestic animals. Nature 244: 18-20.

Veissier, I., Butterworth, A., Bock, B. and Roe, E., 2008. European approaches to ensure good animal welfare. Applied Animal Behaviour Science Journal 113: 279-297.

Vestergaard, K. and Sanotra, G.S., 1999. Relationship between leg disorders and changes in the behaviour of broiler chickens. Veterinary Record 144: 205-209.

WELANIMAL, 2010. Available at: http://tinyurl.com/zm7cqow.

Vos, E., 2000. EU food safety regulation in the aftermath of the BSE crisis. Journal of Consumer Policy 23: 227-255.

8. North American perspectives

T.M. Widowski[1], J.A. Mench[2], J.C. Swanson[3] and W.R. Stricklin[4]*
[1]Department of Animal Biosciences and Campbell Centre for the Study of Animal Welfare, University of Guelph, 50 Stone Road East, Guelph, ON, N1G 2W1, Canada; twidowsk@uoguelph.ca
[2]Department of Animal Science, University of California-Davis, One Shields Avenue, Davis, CA 95616, USA
[3]Department of Animal Science, Michigan State University, 474 S. Shaw Lane, East Lansing, MI 48824, USA
[4]Department of Animal and Avian Science, University of Maryland, College Park, Building 142, College Park, MD 20742, USA

Abstract

The discipline of applied ethology has a rich history in North America. Canada and the United States have each made special contributions to the growth of the discipline and to the advancement of animal welfare science. Early pioneers included faculty from departments of biology and zoology who engaged in work using domestic species to study behaviour, along with other scientists who later became engaged in studying the behaviour of livestock and poultry to extend the scope of behavioural research into production and other managed environments. The focus of this chapter is to recount the early development of the field in North America, and its evolution up until the present day. Today's applied ethologists in North America have expanded the boundaries of the discipline, providing strong engagement and leadership in the continuous improvement of the welfare of animals and the development of sustainable agriculture systems.

Keywords: applied ethology, Canada, USA, history

8.1 Introduction

Applied ethology in North America (NA) has a long history that is often overlooked. While the development of applied ethology in Europe had its moorings in the veterinary sciences, much of applied ethology in North America was influenced by the zoological and biological sciences. Here, scientists often tackled questions about animal behaviour through scientifically related disciplines such as genetics, with studies carried out by scientists housed within animal, dairy and poultry husbandry departments. Early scientists such as Martin (M.W.) Schein, E.B. Hale and H.B. Graves at the Pennsylvania State University Department of Poultry Husbandry, for example, studied the behaviour of poultry (and other species) for the sake of identifying, cataloguing and understanding behaviour (e.g. Graves, 1973). Ethological principles became a central focus of their work while they also contributed important basic knowledge about domesticated species for agricultural production purposes.

DOI 10.3920/978-90-8686-828-5_8, © Wageningen Academic Publishers 2016

A 1955 paper by Schein and M.H. Fohrman in British Journal of Animal Behaviour was one of the first published with meaningful data and statistical analyses on social organisation in cattle. This research inspired later work by W. Ray Stricklin on matrilinear social dominance in beef cows (Stricklin, 1983), which demonstrated how the approach of traditional ethology could apply to domestic animals. Specifically, Stricklin used the selfish herd geometrical spacing phenomenon reported by Hamilton (1971) for wild animals as a starting point for research which focused on the importance of social dominance and kinship within groups of beef cows (Swanson, 1988), and later modelled the use of artificial animals called animats (Stricklin *et al.*, 1995).

Other scientists elevated the importance of the discipline through their contributions to textbooks, such as E.S.E Hafez's (editor) '*The Behaviour of Domestic Animals*' (1962), which provided further emphasis on the study of behaviour in domestic species. Although chapters in the original and subsequent editions of this book were mainly dedicated to addressing species-specific management issues, they also featured seminal papers such as Hale's influential 1969 chapter on traits amenable to domestication, which is still cited today.

Following these early pioneers was a dedicated legion of North American scientists who had either been trained by the foundational US scientists, trained by scientists in Europe, or had retooled their skills and research programs to include the study of animal behaviour. Their work further advanced science in areas such as genetic heritability and plasticity of behaviour (e.g. Siegel, 1979), elucidating different aspects of feeding behaviour and its management (e.g. Provenza and Balph, 1987), domestication (e.g. Price, 1984), reproductive behaviour (e.g. Ford, 1983), social and spacing behaviour (Stricklin *et al.*, 1979) and other areas important to understanding animal behaviour in its evolved and managed contexts.

Several conferences also played a role in advancing applied ethology as a discipline in North America. One of the earliest conferences held in North America to focus specifically on applied ethology was the 1982 'Maryland Conference on Animal Behavior in Agriculture' hosted by Ray Stricklin and Tom Hartsock of the University of Maryland. The meeting brought together approximately 60 people including scientists from the USA and Canada. The proceedings were published in the final edition (before the name was changed) of Applied Animal Ethology (Stricklin and Hartsock, 1984), with reviews written by prominent scientists from the USA and Canada on poultry (Paul Siegel), swine (David Fraser), sheep (Harold Gonyou) and cattle (Ray Stricklin), followed by overviews of future research needs written by individuals including Graves, Stan Curtis (see below), Frank Hurnik (see below) and Ted Friend who, as an Assistant Professor at Texas A&M University since 1977, had established a research program on farm animal behaviour and welfare, with a focus mainly on cattle but also horses and swine. It was an important publication and notification that North America was building scientific capacity and impetus in applied ethology. While the primary focus of this conference was on application of animal behaviour principles to animal agriculture, topics on animal welfare and well-being featured prominently in papers aimed at identifying future research needs.

The intent of our chapter is to highlight catalysing events in the United States and Canada that incorporated applied ethology and animal welfare science into the fabric of scientific investigation, teaching and outreach in the animal and veterinary sciences in North America. Our main focus is on farm animals. Of course it is not possible to mention all scientists who contributed to the study of applied ethology in North America, and forgiveness is asked from those whose names we have not mentioned.

8.2 The development of applied ethology in the USA

By the early 1960s US researchers were investigating genetic influences on behaviour. The work of John Paul Scott with dogs is a notable example (Scott, 1962; Scott and Fuller, 1965). The study of the behaviour of farm animals also began to receive more research attention, although some of the earliest mentions were tangential. For example, in his 1943 textbook, '*Animal Breeding Plans*', Jay L. Lush, the foundational mentor of the sub-discipline of statistical genetics included a chapter entitled, '*The Family Structure of Populations*'. Lush wrote that in a herd of cows a female may be helped or hindered depending on the social position of her mother – work which led ethology into a phase with greater ties to evolutionary biology and less description-based studies and which was later expanded upon by Hamilton (1964) as part of his research on kin selection.

Some of the earliest work on animal behaviour was not published in peer-reviewed scientific journals, in part because of the lack of journals devoted to animal behaviour. An example is the work by Wagnon (1963), which for its time was a well-presented collection of data on beef cattle. This work included data collection that started in the 1940s and that was later published in a University of California Agricultural Experiment Station Bulletin.

By the 1980s, textbooks on domestic animal behaviour authored or edited by US-based authors were beginning to be published, including those by James Craig (1981) '*The Behavior of Domestic Animals*'; Katherine Houpt (1982) '*Domestic Animal Behavior for Veterinarians and Animal Scientists*'; Michael Fox (1984) '*Farm Animals: Husbandry, Behavior and Veterinary Practice*'; Ben Hart (1985), '*The Behavior of Domestic Animals*', and Edward Price (1987), '*Farm Animal Behavior: The Veterinary Clinics of North America Food Animal Practice*'. Some of these books have been published in several editions.

Early contributors that influenced later work in applied ethology include Warner Clyde Allee, a professor of zoology at the University of Chicago. Allee's work included a focus on benefits to the individual from membership in a group or aggregation (Allee, 1931). Some of Allee's work also focused on the domestic chicken. Possibly his greatest contribution to applied animal behaviour studies came through his mentoring of graduate students. A noted student was Alphaeus M. Guhl, a professor in the Department of Zoology at Kansas State University, who increased our understanding of mating behaviour, social dynamics and individual recognition in chickens (Guhl and Ortman, 1953) and influenced the careers of both James Craig, professor in the Department of Poultry Science (later the Department of Animal Science and Industries) at Kansas State University, and Paul Siegel, professor in the Department of Poultry Science at Virginia Tech University, both of whom contributed significantly to the understanding of poultry behaviour and its relationships to genetics. As a result of Allee's excellence in mentoring graduate students an award for best graduate student presentation at the annual meeting of the Animal Behavior Society was established in his honour.

Many of the early studies were focused primarily on animal behaviour in the context of animal production with goals to resolve problems, further enhance particular behavioural attributes or to simply understand the basis for behaviour. Concurrent to the growth of work in animal behaviour was a rise in public concern about the use of animals in biomedical research and agricultural production. Questions often centred on deprivation and the ability to meet the behavioural, as well as the physiological, needs of animals maintained under these conditions. This generated waves of dynamic discussion and public pressure to answer questions concerning the welfare of animals.

8.2.1 The 'mandate' for animal welfare research in the United States

In 1982, Stan Curtis and John McGlone (who at that time had just completed his PhD research in Curtis' laboratory on swine behaviour, and who went on to establish a swine welfare research program as a faculty member at Texas Tech University), published an article in the Journal of Animal Science summarising the current status of farm animal behaviour research in North America. They obtained information via literature searches and inquiries directed towards animal behaviour researchers. They classified studies based on topic addressed, as well as whether the focus was mainly on behaviour or only secondarily on behaviour. They reported that the majority of published and ongoing work was on feeding, reproductive and social behaviour, with relatively few studies having a primary emphasis on animal welfare (only 15 of the 218 studies identified).

In the early 1980s, the focus on farm animal welfare research in the USA began to grow (Mench, 2008). In 1981, the US Department of Agriculture (USDA) provided one-time competitive grants for a number of relatively short-term projects focusing on high-profile areas like housing of laying hens, veal calves, and sows. Some of the Principal Investigators were applied ethologists but others were scientists from other disciplinary areas, mainly physiologists. This reflects the strong emphasis and capacity in the US in animal physiology at that time, and the prevailing view that welfare was best assessed by integrating behavioural measures with physiological measures. By the mid-1980s research in animal welfare was clearly gaining traction within the animal science societies. Following the 1982 'Maryland Conference' mentioned above, a symposium on 'Indices to measure animal well-being' was held at the ASAS annual meeting in 1985 (Gonyou, 1986). In 1987 the American Dairy Science Association meeting included a symposium on 'Dairy Cattle Welfare' (Albright, 1987).

Many of the people already conducting research on farm animal behaviour played key roles in developing and fostering this new emphasis. One individual who was particularly influential in setting the initial trajectory was the aforementioned Stan Curtis at the University of Illinois. Curtis had a formal background in animal science and environmental physiology. He was well-known for his intellect and strong personality, and over the course of his career his attitudes about and approaches to studying animal welfare were enigmatic. On the one hand, he was a staunch supporter of animal industries and often defended some of the more contentious housing and management practices – often publicly taking on critics of animal agriculture. At the same time he supervised PhD students studying such topics as effects of environmental enrichment on neural development in young pigs (Grandin, 1989), nest-building motivation of sows in farrowing crates (Widowski, 1988) and cognitive abilities in pigs (Croney, 1999).

By the 1990s animal welfare had clearly emerged as a research focus. An important landmark was reached in 1992 with the establishment of the Livestock Behavior Research Unit (LBRU) by the US Department of Agriculture's Agricultural Research Service (ARS) in West Lafayette, Indiana. A proposal was initiated in 1991 to establish a Center of Excellence jointly with Purdue University. The LBRU was approved through a federal appropriations bill passed by the US Congress including $450,000 to fund research. Aside from the building of the facility, funds were also appropriated for hiring scientists in four areas: applied ethology, neurophysiology, endocrinology and immunology. In 1994, ISAE member, the late Julie Morrow-Tesch became the first ARS Research Leader of the unit. In 1999, ISAE member Margaret Shea-Moore assumed responsibility as Research Leader, and in 2001, ISAE member Donald Lay assumed the position of Research Leader and remains in this position today.

It was also in 1992 that the Poultry Science Association gave its first Poultry Welfare Research Award, to James Craig. Subsequent winners of this award include geneticist Bill Muir (who pioneered group selection to reduce problems with feather pecking and cannibalism; Craig and Muir, 1993), and other ethologists, physiologists and poultry management experts based in the USA whose research programs were heavily directed towards poultry behaviour and welfare research topics, including ISAE members Inma Estevez, Patricia (Scotti) Hester, Joy Mench, and Ruth Newberry.

Several conferences helped to shape the farm animal welfare research agenda in the USA in the 1990s. One, an international meeting organised by Ray Stricklin and Joy Mench and held at the Aspen Institute in Maryland in 1991, brought together scientists from various disciplines with philosophers, social scientists and key stakeholders (e.g. animal welfare NGOs, policymakers, industry) to discuss the technological, scientific, ethical and sociopolitical dimensions of animal welfare. Ruth Harrison gave the keynote address, and the 90 participants came from the USA, Canada, France, the UK, Germany, Sweden, and The Netherlands. The proceedings were published in two special editions of the Journal of Agricultural and Environmental Ethics (Mench and Stricklin, 1993; Stricklin and Mench, 1993).

Another influential conference was held in 1998 at the University of California, Davis. It was organised by Gary Moberg and Joy Mench and focused on stress biology in relationship to animal welfare, covering topics ranging from basic stress biology to welfare considerations in aquaculture. Once again, this reflected the strong association in the USA between physiological and behavioural approaches to welfare assessment.

Gary, who was a highly regarded stress physiologist with a strong interest in integrating physiological, behavioural and cognitive aspects of stress, died unexpectedly shortly after the conference. He published his final paper in the resulting conference volume (Moberg and Mench, 2000). This paper elaborated upon his concept that animals experiencing stressors that they perceive as severe or prolonged enter a pre-pathological state before evidencing pathology. He postulated that this pre-pathological state was associated with metabolic changes that could be measured and thus act as an earlier indicator than pathology of compromised animal welfare. This idea continues to have a major influence on the conceptualisation of the relationship between stress biology and animal welfare.

By this time, due to the increasing interest among the public and the agricultural industries, it had become obvious that there was a need for a continuous funding source for animal welfare research, and it became a permanent part of USDA's competitive grants program. The USDA's animal welfare research priorities were based on the outcome of two meetings. The first, held in 1992 and attended by a diverse group of animal scientists, veterinarians, producer groups, processors, and other animal agriculture stakeholders, was the Food Animal Integrated Research Priorities (FAIR '95) meeting. This meeting, which was coordinated by the Federation of Animal Science Societies and the Forum for Animal Agriculture, had the goal of articulating unified research priorities for animal agriculture for funding by 1995, and animal welfare was featured as one of the areas of emphasis (Mench, 1993). Via a consensus process conducted after the meeting ended (FAIR'95), two overarching areas of research focus for animal welfare were identified: (1) determine scientific measures of well-being in food producing animals; (2) develop long-term management options and short-term production practices based on scientific findings about animal well-being. The USDA followed up on the FAIR '95 process by holding a conference at Purdue University focused on animal welfare (Food Animal Well-Being, 1993), with the aim of building on the FAIR '95 research priorities and also understanding factors affecting the social acceptability of various practices within animal agriculture.

Although there has been continuous USDA funding of animal welfare research for several decades, animal welfare has never been a completely independent funding category, but instead treated as a component of animal health or (more recently) agricultural engineering. The percentage of the USDA competitive grant budget devoted directly to animal welfare research is extremely small (averaging around $3.7 million per year from 2003-2012, which is only 3% of the total USDA competitive grants budget), and due to federal budgetary constraints both the allocation for animal welfare and the overall USDA competitive grants budget has remained static for many years (NRC, 2015). In all probability this has had some negative impact on faculty hiring and hence graduate student and postdoctoral training in animal welfare. Notably, there has never been a funding category specifically for animal behaviour research within USDA, unlike other disciplinary areas like physiology or genetics, which has likely constrained the growth of the discipline of applied ethology in the USA.

Regardless, there has been increasing faculty strength in animal welfare in the US throughout the last few decades. The first two faculty positions in farm animal welfare were established in the US in the mid-1980s. One was a research and teaching position in the Department of Poultry Science at the University of Maryland (held by Joy Mench, who later moved to University of California, Davis, and expanded her research program on poultry behaviour and welfare to incorporate an emphasis on abnormal behaviour and environmental enrichment for laboratory animals, companion birds, and zoo animals), and the other a Cooperative Extension (outreach) position in the School of Veterinary Medicine at the University of California, Davis (held by Carolyn Stull). This is not to say that applied ethologists and other scientists were not already making contributions to animal welfare prior to that time, just that the issue had gained enough visibility and traction by then to be featured in position titles.

Many animal science, dairy science, poultry science, or veterinary departments in the USA now have at least one faculty member with applied ethology and/or animal welfare as their primary research, teaching or outreach focus. There are also a number of established academic centers/programmes focusing on welfare and/or behaviour. In addition to the LBRU discussed above, these include Iowa State, Michigan State, Penn State, Purdue University, Washington State, University of California, Davis, Texas A&M, and University of Arkansas.

However, given the increasing public concern about the treatment of animals the National Academy of Sciences recently identified remedying limited capacity in animal welfare research as a high priority for animal agriculture in the USA in order to ensure sustainability (NRC, 2015). The Academy recommended that USDA establish a process to determine how to address this critical need, although that has not occurred as of the time this chapter was written.

8.2.2 Relationships with stakeholders

In the EU, the regulatory framework has had a major influence on the direction of farm animal welfare research. By comparison, in the USA, a lack of regulatory focus (or even a national system of Codes of Practice as in Canada), has resulted in the animal welfare research planning process being fragmented. The research priorities developed as part of the periodic FAIR process (2002, 2012) continue to be influential in shaping USDA funding, as are the research focus areas identified by USDA multi-state committees and projects (Mench, 2008). However, an important stimulus has been research funding by the animal industries, which have stepped in to fill at least part of the gap created by declining federal and state funds. Both the National Pork Producers Council and the US Poultry and Egg Association regularly provide competitive funding for animal welfare research, with emphasis areas targeted to current industry needs. Tyson Foods funded the establishment of the John W. Tyson Center of Excellence for Poultry Science at the University of Arkansas, a component of which is the Center for Animal Wellbeing. This year, Perdue Farms began providing 2-year fellowships via the Arthur Perdue Foundation for graduate students studying poultry behaviour and enrichment. Public-private partnerships are also entering the mix, with the potential for significant funding for animal welfare research. An example is the Coalition for Sustainable Egg Supply, a public-private partnership with industry and retailer participation that funded a multi-million dollar study on laying hen welfare in the context of overall housing system sustainability (Mench *et al.*, 2016). These illustrate just a few of the types and sources of industry-based funding for animal welfare research.

In the absence of regulation, market-based and voluntary animal welfare assurance programs have become the dominant model in the USA (Mench, 2008). Temple Grandin, Janice Swanson, and Joy Mench played key roles early on in the development of these kinds of programs, providing input and helping to write animal welfare standards and develop auditing programs for many retailers, food distributors, producer groups, trade associations, and welfare labelling programs. Many other applied ethologists, veterinarians, and other animal welfare researchers/extension specialists in the USA (and Canada) are now heavily involved in these ongoing efforts.

As a consequence, much animal behaviour and welfare research in the USA has been very practically oriented and focused on addressing concerns and needs relevant to the animal industries. As just one example, issues surrounding animal handling and problems with effective stunning procedures led to the development of guidelines for transportation, handling and the slaughter of livestock. In 1991 the American Meat Institute (now called the North American Meat Institute), developed and published guidelines, now entitled 'Animal Handling Guidelines & Audit Guide: A Systematic Approach to Animal Welfare', for best practices associated with the humane handling and slaughter of livestock for the red meat processing industry (Grandin, 2013). Over time assessment and audit criteria were added along with special conferences and training programs. Temple Grandin was instrumental in leading the research efforts that formed the foundation for these evidence-based guidelines, assessment and audit criteria. Grandin developed metrics that quantified both quantity and quality of animal handing, stunning efficacy and slaughter outcomes relative to achieving loss of consciousness and a humane death in beef and swine processing plants across the US. This data served as the basis for rapid improvement and setting in place audit practices and training programs for employees working with livestock, and gauging the attention given by management to animal welfare. Grandin's focus on understanding and using animal behaviour as an important tool for improving the care, handling and management of livestock marked an important advancement in improving the welfare of farm animals. Her early impacts were on ranches, farms and processing facilities in the USA and rapidly reached around the globe. She is likely the most recognised applied ethologist and the most celebrated high functioning autistic person in the world.

Although our focus here has mainly been on farm animals, applied animal behaviour has also played a key role in other areas. One of these is in laboratory animal welfare, and particularly the welfare of primates used in research. Pioneering work by ethologists and comparative psychologists like Kathryn Bayne, DVM PhD (formerly of the National Institutes of Health and currently Global Director of Association for the Assessment and Accreditation of Laboratory Animal Care International); Molly Bloomsmith, professor, Emory University; Melinda Novak, professor, University of Massachusetts; Steve Shapiro, professor, University of Texas; and German-trained veterinarian Viktor Reinhart (Attending Veterinarian at the Primate Research Center at the University of Wisconsin), was critical in establishing practices for primate housing, social management, and environmental enrichment in the USA (NRC, 1998). This research was spurred in part by a 1985 regulatory mandate (Animal Welfare Act, 1985) requiring those who keep nonhuman primates for research or exhibit to develop and follow appropriate plans for 'promoting the animals' psychological well-being'.

US ethologists have also contributed to advances in the study of zoo animal welfare, Hal Markowitz being perhaps the best known because he is widely considered to be the 'father of zoo environmental enrichment'. Markowitz, who was a professor at San Francisco State University when he published his highly influential book, '*Behavioral Enrichment in the Zoo*' (1982), inspired global efforts to study zoo animal behaviour and design creative enrichment strategies. His research still sets the standard for those working to develop cognitive enrichment methods for animals.

8.3 The development of applied ethology in Canada

The growth of applied ethology as a discipline and the roles that the scientists have played in animal welfare policy followed a bit of a different path in Canada. A search of the Canadian Journal of Animal Science prior to the 1980s produces only a few examples of papers that included behaviour in the methods. Even fewer of these focused on animal behaviour *per se* but rather used behaviour as one of a number of tools to study nutrition, reproduction and energetics. Looking back at the earliest issues of Applied Ethology/Applied Animal Behaviour Science, one also finds only a handful of authors with Canadian affiliations, but by the mid-1990s, that situation had clearly changed.

8.3.1 The 'mandate' for animal behaviour and welfare research in Canada

In 1980, the Canadian Federation of Humane Societies (CFHS) began coordinating the process of drafting voluntary codes of practice for farm animals in Canada. According to their website, CFHS had pushed the federal government for regulations for the treatment of farm animals and after 'considerable consultation' the federal government proposed (and the Minister of Agriculture agreed to pay for) the development of voluntary national codes as a compromise.

The first Codes of Practice for the Care and Handling of Chickens (meat birds and layers), Pigs and Special Fed Veal Calves were published in 1983, 1984 and 1988, respectively. Special fed veal calves were fed a milk based diet that was low in iron to produce the white muscle veal delicacy favoured by Italian chefs. Codes for mink and ranched fox soon followed and by 1998 codes had been developed for beef, dairy, sheep, deer and horses. The process for code development in Canada is unique in that it is a multi-stakeholder process involving representatives from the 'agricultural industries and their organisations, federal and provincial government, associations of animal science, representatives of the animal welfare movement, and interested individuals' (Agriculture Canada, 1989). Frank Hurnik, a professor of Animal & Poultry Science at the University of Guelph, was instrumental in the development of these early Codes. He is considered to be one of the first applied ethologists in Canada, and has been recognised as an honorary fellow of the ISAE. Hurnik (1988) firmly believed that the primary responsibility for setting and maintaining standards for the well-being of farm animals lay in the hands of the animal producer groups. Hurnik chaired several of the first code development committees, working with producers to draft several of the early codes.

In 1987, an Expert Committee on Farm Animal Welfare and Behaviour was formed to complement the code process; again this was a multi-stakeholder group comprised of farm animal commodity groups, government research and regulatory agencies and animal welfare organisations. One of the first jobs of the committee was to identify welfare concerns for each species of farm animal (Tennessen, 1988). According to Tennessen, an applied ethologist based at the Nova Scotia Agricultural College, the committee also made two general recommendations. Regarding research, it was the committee's opinion that 'government agencies and universities should re-order their priorities so that research on farm animal welfare and behaviour will have the staffing and support typical of other disciplines' (e.g. nutrition, genetics, reproductive physiology). Furthermore, in recognising the vital importance of education the committee recommended that 'all students of animal production, at the college or university level, and of veterinary medicine should receive instruction in animal welfare and ethics of livestock production'.

Two years later, a Workshop on Farm Animal Behaviour and Welfare Research was held to establish a framework for the Research Branch of Agriculture Canada (today Agriculture and Agri-food Canada, AAFC) to establish direction, scope and location for animal welfare research for each of the major animal commodities (Canada Department of Agriculture, 1989). Outcomes of the workshop included mandates for animal welfare research at four of the Agriculture Research Stations located across the country, recommendations for developing partnerships between the government researchers and universities for research, teaching and technology transfer as well as specific research priorities for the housing, transportation and management practices concerning each species of farm animals.

Whether by cause or coincidence, the fields of applied ethology and animal welfare science grew considerably across Canada from the time of the first codes through the early 1990s. Several applied ethologists, stress physiologists and housing engineers had already been conducting research in these fields at Agriculture Canada and universities including David Fraser led an active program in pig ethology at the AAFC Central Animal Experiment Station in Ottawa at that time. New positions for applied ethologists emerged across the country including positions for AAFC research scientists Ruth Newberry (recipient 2014 ISAE Creativity Award), Anne-Marie de Passillé (ISAE Fellow), Jeff Rushen (ISAE Fellow) and Dan Weary. Tina Widowski moved up from the United States to join the University of Guelph as a post-doc working with Ian Duncan. Tina later took up a faculty position at Guelph, becoming the director of the Campbell Centre for the Study of Animal Welfare in 2007 and the University Chair in Animal Welfare in 2009. Two positions emerged in Saskatchewan: Joseph Stookey took up what has become his >20 year position as Professor of animal behaviour at the Western College of Veterinary Medicine. Harold Gonyou (ISAE Fellow) returned to his Canadian homeland to take up a position at the Prairie Swine Centre in Saskatoon. Harold would soon go on to become the ISAE's first North American president and editor-in-chief of Applied Animal Behaviour Science. These Canadian applied ethologists all became influential on the international stage and made significant contributions to the field of applied ethology.

8.3.2 The role of Canadian veterinary colleges

A number of veterinary faculty with strong interests in applied ethology and animal welfare were also foundational in the establishment of the culture for animal welfare research and policy in Canada. Andrew Fraser, a founding member of the ISAE, and now an honorary fellow, had done some of his early academic training at Ontario Veterinary College (OVC, now part of the University of Guelph). After working in the UK, Andrew returned to Canada permanently in 1973, spending several years at the Western College of Veterinary Medicine (WCVM) at the University of Saskatchewan. Andrew Fraser was founder of the journal Applied Ethology and served as Editor-in-chief for many years. Harry Rowsell, a faculty member at OVC from 1953-1965, created the first animal care committee for use of animals in research and teaching in Canada, at Guelph in 1964. Rowsell also served as a department chair at WCVM for several years before moving on as professor of pathology on faculty of medicine at the University of Ottawa. Rowsell notably established the Canadian Council on Animal Care for animals in research and teaching in 1968, and in 1988 he was appointed officer of the Order of Canada for his contribution to the humane treatment of animals used in biomedical research.

David Porter, a respected veterinarian and biomedical scientist originally from Bristol, was an outspoken advocate for the humane treatment of research animals. Porter accepted a position at the Ontario Veterinary College with the understanding that a centre for animal welfare be established. Porter worked with Frank Hurnik, philosophy professor Hugh Lehman and other faculty in the veterinary and agriculture colleges, and in departments of psychology and zoology, to launch a multi-disciplinary Centre for the Study of Animal Welfare at the University of Guelph in 1989. Together they also planned first steps for the centre. One of those steps was recruiting Ian JH Duncan (ISAE Fellow) to emigrate from Scotland and become the Centre's first director. In 1995 Duncan was appointed the Colonel K.L. Campbell Chair in Animal Welfare, which was the second chair in animal welfare to be established in the world. Today, the Campbell Centre for the Study of Animal Welfare (CCSAW), renamed after major private benefactor, Mona Campbell, is a thriving hub of research and graduate training in applied ethology and animal welfare science. The centre has attracted many world-class scientists, including Georgia Mason who was appointed a Canada Research Chair in 2004. An endowed chair in Companion Animal Behaviour and Welfare was established in OVC in 2010. With a core group of 10 faculty members focusing on applied ethology and animal welfare science and dozens of associated faculty across the colleges of agriculture, veterinary science, business and social sciences, today CCSAW offers a broad array of research and graduate training opportunities in behaviour, welfare and human-animal relationships of farm, laboratory, companion and zoo animals.

The late Nora Lewis was a Canadian ethologist highly respected for both her research and university teaching program in animal behaviour. Nora trained in applied ethology, receiving an MSc (1978) and PhD (1982, training with Frank Hurnik), and later a DVM (1994) all from the University of Guelph. She became Associate Professor of Applied Ethology and Acting University Veterinarian at the University of Manitoba, where she played a significant role to establish the Animal Care and Use Program.

Andrew Luescher, a DVM with interest in behaviour, studied for a PhD in applied ethology with Frank Hurnik. After some years at Guelph working in veterinary behaviour, Andrew took up a position at Purdue University, later becoming director of Purdue's animal behaviour clinic. At the time, he was one of a handful of board certified veterinary animal behaviourists in the United States. He had a strong focus on dog behaviour, studying behavioural development in dogs, compulsive disorders and aggression.

In 1994, the Sir James Dunn Animal Welfare Centre was established at the Atlantic Veterinary College in Prince Edward Island. Caroline Hewson, a UK trained DVM, with a PhD from the University of Guelph working with Andrew Luescher, became the inaugural Research Chair in Animal Welfare (2000-2006), positioned at the Sir James Dunn Animal Welfare Centre. Canada's newest veterinary college at the University of Calgary opened in 2008 with dedicated faculty to research pain and animal welfare. Today, all four of the English speaking veterinary schools in Canada have at least one faculty position dedicated to an applied ethologist focusing on farm animals, and St Hyacinthe at Université of Montréal has ties with the AAFC Research Branch Scientists in Quebec. Thus, Canadian veterinary schools have historically been and continue to be committed to research and veterinary training in animal behaviour and animal welfare. Interestingly, most of these positions in the veterinary schools are held by faculty trained in applied ethology rather than veterinary medicine, and this trans-disciplinary culture has been important for the growth of applied ethology in Canada.

8.3.3 Relationships with stakeholders

Despite the hiring of several new applied ethologists within the AAFC Research Branch in the early 90s, there was downward trajectory of funding support by government and an increase in the expectation that the animal industries themselves would increasingly fund more animal welfare research. AAFC began requiring cost-sharing by the agri-food industry for research carried out by government scientists in all areas related to animal production including animal welfare. In 1997 the AAFC Animal Research Centre in Ottawa closed, and applied ethologists, David Fraser and Dan Weary moved to the University of British Columbia (UBC) to launch the Animal Welfare Program there. Both Fraser and Weary were awarded Natural Science and Engineering Council (NSERC) Industrial Research Chairs, which require substantial industry funding which is then matched by the federal government. A variety of farm animal commodities, allied industries (e.g. nutrition and genetics companies), veterinary and humane organisations contributed to establishment of the program at UBC. Over the next decades, Canadian government research positions in applied ethology and animal welfare science continued to wax and wane. Today, there are only a few animal welfare researchers left in AAFC with research programs focusing on pig and beef cattle welfare. As concerns about the welfare of farm animals from the broader community and the need for animal welfare research continued to grow, funds from the private sector and individual beneficiaries filled the gap, resulting in an increase in positions for applied ethologists working with farm animals at Canadian universities.

Similar to the situation of the US, animal welfare research funding in Canada today relies very heavily on engagement and investment from the animal agricultural industries rather than on government sources. Much of the provincial funding and federal funding for animal welfare research requires industry-matching funds. For example, the Dairy Farmers of Canada (DFC), Beef Cattle Research Council, Canadian Poultry Research Council (CPRC) and Swine Innovation Porc provide substantial competitive funding for animal welfare research, and these funds are currently matched by the federal government. Perhaps due to the long history and culture of having animal producers collaboratively engaged in the dialogue with scientists about animal welfare, industry support of animal welfare research in Canada is significant, including a number of industry supported research chairs at universities.

Core funding for positions, research and technology transfer for applied ethology at the Prairie Swine Centre (associated with University of Saskatchewan) has been supported by pork producer agencies in the western provinces of Canada for decades. Together with provincial dairy boards and associated dairy industries, the DFC has been funding several federally-matched research chairs for applied ethologists (e.g. at UBC and more recently at McGill University). In order to fill a gap in poultry welfare research capacity, the Egg Farmers of Canada established a research chair in poultry welfare at the University of Guelph in 2011 and in 2014, the CPRC together with Burnbrae Farms (a family-run egg producer and processor) funded an additional professorship in poultry welfare there. Also in 2014, the Anderson-Chisolm Chair in Animal Care and Welfare was established at University of Calgary veterinary college with a gift from an Alberta cattle ranching family.

While this private sector funding has resulted in considerable opportunities for applied ethologists to work directly with farmers, serve on industry advisory committees and conduct large-scale research on farms, it has also resulted in a shift from more fundamental research questions about animal welfare to solving more short-term and pressing needs of the industries. On the positive side, research results are being translated directly into policy and practice. However, it does limit the ability of applied ethologists and other researchers to ask more basic research questions about animal emotions and cognition as well as more general questions related to public good.

8.3.4 The increased role of applied ethologists in the codes of practice in Canada

Although the codes of practice for farm animals served as the backbone for Canada's farm animal care infrastructure, they had often been viewed with some scepticism as they were primarily animal industry-driven, voluntary and the standards mainly reflected the industry status quo for space allowances, housing and management practices (Bradley and MacRae, 2011). In 2005, the National Farm Animal Care Council (NFACC) was established to coordinate a new national approach to animal care in Canada (Serecon Management Consulting, 2004). While the animal commodity organisations still comprise the majority of members, several new stakeholders, including from the retail sector, have joined in the process.

Applied ethologist David Fraser has been integral in this national consultation and his influence, along with that of other applied ethologists in Canada is apparent. The code development process now includes a scientists' committee comprising at least one applied ethologist (nominated by ISAE and usually the chair), one veterinarian and one animal production scientist to reflect the different values that scientists from different fields place on various measures of animal welfare (Fraser, 2003). The scientists are tasked with writing an internationally peer-reviewed scientific report on priority animal welfare issues for the species and required to include evidence from biological function, affective states and natural living perspectives (Fraser, 2008; NFACC, 2014). The report is used to inform the drafting of the code by the multi-stakeholder Code Development Committee, which typically consists of farmers, processors, government agencies, chair of the scientist committee (usually an applied ethologist), humane society and retailers. The initial draft of the Code is then released for public comment via the NFACC web-site and these comments are then considered in the final version. This new, more inclusive, approach is resulting in some substantial changes to industry practices including pain relief during routine surgical procedures (e.g. dehorning and castration) as well as elimination of continuous confinement of sows in gestation stalls. Although not specifically legislated, the codes are increasingly used in provincial animal welfare regulations and a growing number of Canadian processors, food retailers and food service industries are requiring compliance with the codes.

8.4 The North American region

The USA and Canada have a lot in common. The two countries share the longest international border in the world and have each other as their largest mutual trading partners. This trade arrangement has also been true for applied ethologists and animal welfare scientists. Throughout the growth period of the 1980s and 1990s there was a nearly continuous flux of scientists and students between the USA and Canada. Today a number of Americans hold faculty and research scientist positions in Canada, a number of Canadians hold faculty positions in the USA and quite a few Canadians and Americans have held faculty positions in both places over the years.

In addition to research, teaching and training graduate students together, American and Canadian applied ethologists have also engaged in a number of important collaborations. One was the Bio-Ethics'87 Symposium jointly organised and hosted by Tom Hughes from the Animal Welfare Foundation of Canada and Frank Loew from Tufts University (Hughes, 1988). The conference was held in conjunction with the World Veterinary Congress in Montreal (Bryant, 1988) with the objective of encouraging active development of veterinary and applied ethology in North America. The conference brought together scientists from Canada, the USA and Europe with proceedings published in Applied Animal Behaviour Science. Another was the North American Symposium on Poultry Welfare, which was organised by Ian Duncan (by that time the first director of the Center for the Study of Animal Welfare at the University of Guelph) and Joy Mench. The symposium was modelled on the European Poultry Welfare Symposia and held in conjunction with the 1995 meeting of the Poultry Science Association in Canada, with the papers published in Poultry Science (Mench and Duncan, 1998). It attracted a very diverse North American audience, including key poultry industry representatives, helping to increase industry awareness about recent scientific advances. The first North American ISAE meeting was held in Guelph in 1994 and since then the Canada and USA regions of the ISAE have alternated its location between the countries, with organisers from both countries working together.

One final North American activity that deserves mention is the Intercollegiate Animal Welfare Judging and Assessment Contest. This extracurricular activity has become a highly popular and important experiential learning exercise in animal welfare science for undergraduate, graduate and veterinary students across North America (and recently some non-NA teams). The contest which is modelled after livestock judging teams that are highly popular in agricultural science departments in the US, was developed and launched by ISAE members Camie Heleski and Aldroaldo Zanella at Michigan State University (MSU) (Heleski *et al.*, 2003). Students analyse the welfare of various types of animals as presented in comparative scenarios and determine whether the welfare of animals in one scenario is better or worse than in the other. The contest teaches ethical reasoning, encourages objective assessment of animal welfare on the basis of scientific theory and data, promotes critical thinking, and improves communication skills. Although the contest has been mainly hosted and organised by faculty at MSU, Canadian teams have participated since its inception, and Canadian faculty have helped with developing materials and have co-hosted on a couple of occasions. In the last few years, the contest has grown to well over 100 student contestants and the American Veterinary Medical Association has assumed much of the funding and organisation of the contest as a way for veterinary students in US colleges that do not have animal welfare faculty to gain experience in animal welfare science.

8.5 The way forward

The disciplines of applied ethology and of animal welfare science in North America are currently robust but are still growing. The trans-national relationships among our American and Canadian scientists and their links to others worldwide have catalysed the growth of both academic and industry positions for scientists in this field. Scientists are engaged in issue specific and trans-disciplinary research, and increasingly serve in capacities important to setting standards of animal care and advising public policy. We expect continued growth as both public and corporate concern for animal welfare has passed the tipping point in North America, and the need for research and scientific expertise has outpaced our current capacity. We look forward to the new generation of scientists who will enhance our understanding of animal behaviour and work for the continuous improvement of the welfare of animals.

References

Agriculture Canada, 1989. Recommended code of practice for the care and handling of poultry from hatchery to slaughter plant. Agriculture Canada Publication 1757/E, Communications Branch, Ottawa, Canada, 41 pp. Available at: http://tinyurl.com/hxlyhz3.

Albright, J.L., 1987. Dairy animal welfare: current and needed research. Journal of Dairy Science 70: 2711-2731.

Allee, W.C., 1931. Co-operation among animals. American Journal of Sociology 37: 386-398.

Animal Welfare Act, 1985. Improved standards for laboratory animals act. Public Law 99-198, Food Security Act of 1985, Subtitle F- Animal Welfare. Bradley, A. and MacRae, R., 2011. Legitimacy and Canadian farm animal welfare standards development: the case of the National Farm Animal Care Council. Journal of Agricultural and Environmental Ethics 24: 19-47.

Bryant, D.G., 1988. Preface to the special issue Bio-Ethics'87. Applied Animal Behaviour Science 20: 7-8.

Canada Department of Agriculture Research Branch, 1989. Report: workshop on farm animal behaviour and welfare research. Lennoxville, Quebec, Canada.

Craig, J.V., 1981. Domestic animal behavior. Prentice Hall Inc., Englewood Cliffs, New York, NY, USA.

Craig, J.V. and Muir W.M., 1993. Selection for reduction of beak-inflicted injuries among caged hens. Poultry Science 72: 411-420.

Croney, C., 1999. Cognitive abilities of domestic pigs (*Sus scrofa*). Doctoral dissertation, Pennsylvania State University, State College, PA, USA.

Curtis, S.E. and McGlone, J.J., 1982. Status of farm animal behavior research in North America. Journal of Animal Science 54: 450-455.

FAIR '95, 1993. Symposium on food animal research priorities. Federation of American Societies of Food Animal Sciences, Champaign, IL, USA.

FAIR 2002, 1999. Symposium on food animal research priorities, Federation of American Societies of Food Animal Sciences, Baltimore, MD, USA.

FAIR 2012, 2012. Farm animal integrated research priorities. Federation of Animal Science Societies. Champaign, IL, USA. Available at: http://tinyurl.com/j6p9hxy.

Food Animal Well-Being, 1993. Conference sponsored by USDA and Purdue University Office of Agricultural Research Programs. Purdue Research Foundation, West Lafayette, IN, USA.

Ford, J.J., 1983. Postnatal differentiation of sexual preference in male pigs. Hormones and Behavior 17: 152.

Fox, M.W., 1984. Farm animals: husbandry, behavior and veterinary practice. University Park Press, Baltimore, MD, USA.

Fraser, D., 2003. Assessing animal welfare at the farm and group level: the interplay of science and values. Animal Welfare 12: 433-443.

Fraser, D., 2008. Understanding animal welfare, the science in its cultural context. UFAW Animal Welfare Series, Wiley-Blackwell, New York, NY, USA.

Gonyou, H.W., 1986. Assessment of comfort and well-being in farm animals. Journal of Animal Science 62: 1769-1775.

Grandin, T., 1989. Effect of rearing environment and environmental enrichment on behavior and neural development of young pigs. PhD thesis, University of Illinois at Urbana-Champaign, Champaign, IL, USA.
Grandin, T., 2013. Recommended animal handling guidelines and audit guide: a systematic approach to animal welfare. American Meat Institute Animal Welfare Committee. American Meat Institute Foundation, Washington, DC, USA.
Graves H.B., 1973. Early social responses in *Gallus*: a functional analysis. Science 182: 937.
Guhl, A.M. and Ortman, L.L., 1953. Visual patterns in the recognitions of individuals among chickens. The Condor 55: 287-298.
Hafez, E.S.E., 1962. The behaviour of domestic animals. Williams and Wilkins, Baltimore, MD, USA.
Hale, E.B., 1969. Domestication and the evolution of behavior. In: Hafez, E.S.E. (ed.) The behavior of domestic animals, 2nd edition. Williams and Wilkens, Baltimore, MD, USA.
Hamilton, W.D., 1964. The genetical theory of social behavior. Journal of Theoretical Biology 7: 1-52.
Hamilton, W.D., 1971. Geometry of the selfish herd. Journal of Theoretical Biology 31: 295-311.
Hart, B., 1985. The behavior of domestic animals. W.H. Freeman and Co., New York, NY, USA.
Heleski, C.R., Zanella, A.J. and Pajor, E.A., 2003. Animal welfare judging teams – a way to interface welfare science with traditional animal science curricula? Applied Animal Behaviour Science 81: 279-289.
Houpt, K., 1982. Domestic animal behavior for veterinarians and animal scientists. Iowa State University Press, Ames, IA, USA.
Hughes, T., 1988. Opening address and welcome. Applied Animal Behaviour Science 20: 1-3.
Hurnik, J.F., 1988. Welfare of farm animals. Applied Animal Behaviour Science 20: 105-117.
Lush, J., 1943. Animal breeding plans. Iowa State University Press, Ames, IA, USA.
Markowitz, H., 1982. Behavioral enrichment in the zoo. Van Nostrand Reinhold, New York, NY, USA.
Mench, J.A., 1993. Animal well-being. In: FAIR '95, symposium on food animal research priorities. Federation of American Societies of Food Animal Sciences, Champaign, IL, USA, pp. 99-100.
Mench, J.A., 2008. Farm animal welfare in the USA: farming practices, research, education, regulation, and assurance programs. Applied Animal Behaviour Science 113: 298-312.
Mench, J.A. and Duncan, I.J.H., 1998. Poultry welfare in North America: opportunities and challenges. Poultry Science 77: 1763-1765.
Mench, J.A. and Stricklin, W.R., 1993. An international conference on farm animal welfare: scientific perspectives. Journal of Agricultural and Environmental Ethics 6, Suppl. 2.
Mench, J.A., Swanson, J.C. and Arnott, C., 2016. The coalition for sustainable egg supply: a unique public-private partnership for conducting research on the sustainability of animal housing systems using a multistakeholder approach. Journal of Animal Science 94: 1296-1308.
Moberg, G.P. and Mench, J.A., 2000. The biology of animal stress. Basic principles and implications for animal welfare. CABI Publishing, Wallingford, UK.
National Farm Animal Care Council, 2014. NFACC code of practice development process. Available at: http://tinyurl.com/jkaogaa.
National Research Council (NRC), 1998. The psychological well-being of nonhuman primates. National Academies Press, Washington, DC, USA.
National Research Council (NRC), 2015. Critical role of animal science research in food security and sustainability. National Academies Press, Washington, DC, USA.

Price, E.O., 1984. Behavioral aspects of animal domestication. Quarterly Review of Biology 59: 1-32.
Price, E.O., 1987. Farm animal behavior: the veterinary clinics of North America food animal practice. W.B. Saunders, Philadelphia, PA, USA.
Provenza, F.D. and Balph, D.F., 1987. Diet learning by domestic ruminants: theory, evidence and practical implications. Applied Animal Behaviour Science 18: 211-232.
Serecon Management Consulting Inc., 2004. A discussion paper setting out a national approach to animal care. A report to agriculture and agri-food Canada. Serecon Management Consulting Inc., Edmonton, Canada.
Schein, M.W. and Forhman, M.H., 1955. Social dominance relationships in a herd of dairy cattle. British Journal of Animal Behaviour 3: 45-55.
Scott, J.P., 1962. Critical periods in behavioral development. Science 138: 949-958.
Scott, J.P. and Fuller, J.L., 1965. Genetics and the social behaviour of the dog. The University of Chicago Press, Chicago, IL, USA.
Siegel, P.B., 1979. Behaviour genetics in chickens. World's Poultry Science Journal 35: 9-19.
Stricklin, W.R., 1983. Matrilineal social dominance and spatial relationships among Angus and Hereford and cows. Journal of Animal Science 57: 1397-1405.
Stricklin, W.R., Graves, H.B. and Wison, L.L., 1979. Some theoretical and observed relationships of fixed and portable spacing in animals. Applied Animal Ethology 5: 201-214.
Stricklin, W.R. and Hartsock, T.G., 1984. Maryland conference on animal behavior in agriculture. Applied Animal Ethology 11: 295-298.
Stricklin, W.R. and Mench, J.A., 1993. An international conference on farm animal welfare: ethical, technological and sociopolitical perspectives. Journal of Agricultural and Environmental Ethics, Suppl. 1.
Stricklin, W.R., Zhou, J.Z. and Gonyou, H.W., 1995. Selfish animats and robot ethology: using artificial animals to investigate social and spatial behavior. Applied Animal Behaviour Science 44: 187-203.
Swanson, J.C., 1988. The effects of maternal relationship on social and spatial behavior of beef cows. PhD dissertation, University of Maryland, College Park, USA. Available at: http://tinyurl.com/z7xn4l4.
Tennessen, T., 1988. Improving the life of farm livestock: a Canadian point of view. Applied Animal Behaviour Science 20: 159-163.
Wagnon, K.A., 1963. Behavior of beef cows on a California range. California Agricultural Experiment Station, Bulletin 799. Available at: http://tinyurl.com/jrazr35.
Widowski, T.M., 1988. Environmental and physiological aspects of prepartum behavior in the sow. PhD thesis, University of Illinois at Urbana-Champaign, Champaign, IL, USA.

9. Australian and New Zealand perspectives

P.H. Hemsworth[1], D.J. Mellor[2], G.J. Coleman[1], N.J. Beausoleil[2], A.D. Fisher[1] and K.J. Stafford[2]*
[1]Animal Welfare Science Centre, University of Melbourne, Parkville, Vic 3010, Australia; phh@unimelb.edu.au
[2]Animal Welfare Science and Bioethics Centre, Massey University, Palmerston North 4442, New Zealand

Abstract

A feature of Australian and New Zealand applied animal behaviour science for many decades has been its multidisciplinarity. The disciplines involved in addition to applied animal behaviour include anatomy, biochemistry, nutrition, environmental and stress physiology, pathophysiology, affective neuroscience, and psychology, among others. The primary focus of this chapter is the enhanced understanding of animal behaviour that has been achieved by measuring behavioural responses along with other functional indices, and the wider insights that this understanding has provided in exploring challenges faced by animals used for human purposes. The examples provided illustrate the areas of investigation undertaken in Australia and New Zealand, often with the collaboration of researchers from other countries. First among these examples is the utilisation of behavioural and hormonal stress measures to assess animals' responses to intensive housing and painful husbandry procedures. Next, we discuss the importance of following a multidisciplinary approach in developing animal welfare assessments, using the Five Domains assessment model as a specific example. We then describe the behavioural contributions to understanding foetal and neonatal awareness, welfare and survival. The chapter concludes with a discussion of how understanding animal and human behaviour can improve human-animal relationships, particularly through the impact of human attitudes and behaviour on animal welfare.

Keywords: Australia, New Zealand, stress response, welfare assessment, human-animal relationship

9.1 Introduction

Scientists such as George Alexander, Rolf Beilharz, Ron Kilgour, David Lindsay, Justin Lynch, Glenn McBride, Victor Squires and others, were prominent in the mid-20th century in Australia and New Zealand studying neonatal, grazing, social and reproductive behaviour, and learning in animals. This critical mass of researchers, attentive to the practical use of scientific principles to solve productivity, health and welfare problems, established the study of farm animal behaviour as a legitimate scholarly endeavour, providing a secure foundation for the Australian and New Zealand scientists who emerged subsequently. Many of these pioneers were trained in disciplines other than

DOI 10.3920/978-90-8686-828-5_9,

ethology, for example psychology and physiology, and thereby demonstrated the value of multidisciplinary insights applied to the study of farm animal behaviour and its relevance to animal welfare.

More recently, Australian and New Zealand scientists have contributed to the discussion and development of conceptual frameworks used to assess animal welfare, namely, biological functioning, affective state and natural living (Fraser, 2003). As a subjective state, the welfare of an animal cannot be measured directly, but there is now an integrated and widely accepted appreciation in the published literature of what animal welfare represents, and how it can be assessed. Changes in an animal's behaviour, such as appearance, demeanour, activity/inactivity and vocalisation/silence are usually the first signs used in the field to identify a potential welfare problem. In fact, researchers focusing on animal welfare have utilised specific behaviours as well as several other indices including anatomical, biochemical, metabolic, physiological, pathophysiological, clinical and neurological parameters in their evaluations. Examples of behaviour and welfare-related topics investigated in Australia and New Zealand include the following: the causes and prevention of neonatal hypothermia, morbidity and mortality, especially in lambs; tolerance of transport by male dairy calves (called bobby calves in Australia and New Zealand); assessment of resting surfaces preferred by dairy calves; impacts of selective breeding on temperament in lambs; welfare focused nutrition for dairy cows; heat stress in dairy cows and in transported sheep, cattle and poultry; the welfare of farmed deer; humane slaughter of ungulates and poultry; the relative impacts of different traps, poisons, exclusion devices and other ways of controlling vertebrate pests; wildlife responses to different forms of identification marking; and zoo animal welfare. The outcomes of these and numerous other studies have provided evidence-based support for the development of regulations and codes of welfare in Australia, New Zealand and elsewhere.

However, the primary focus of this chapter is the enhanced understanding of animal behaviour that has been achieved by measuring behavioural responses along with other functional indices, and the wider insights that this behaviour-based understanding has provided in exploring specific challenges faced by animals used for human purposes. The examples provided are not comprehensive, rather, they are intended to illustrate the areas of investigation undertaken in Australia and New Zealand, often with the collaboration of researchers from other countries. First among these examples is the utilisation of behavioural and hormonal stress measures to assess animals' responses to intensive housing and painful husbandry procedures. Next, we discuss the importance of following a multidisciplinary approach in developing animal welfare assessments, using the Five Domains assessment model as a specific example. We then describe the behavioural contributions to understanding fetal and neonatal awareness, welfare and survival. The concluding section consists of a discussion of how understanding animal and human behaviour can improve human-animal relationships, particularly through the impact of human attitudes and behaviour on animal welfare.

9.2 Using behavioural and physiological responses and biological costs to assess welfare

In response to the debate in many industrialised countries about the confinement of farm animals that followed the 1964 publication of Ruth Harrison's book '*Animal Machines*', the British Government established the Brambell Committee (Command Paper 2836, 1965). The Brambell Committee report indicated that animal welfare should be evaluated using scientific evidence about the feelings of animals as shown by their structure, function and behaviour, and that this should occur via research into veterinary medicine, stress physiology, animal science and animal behaviour. While these recommendations have been generally well-accepted by researchers, there was considerable interest and concern in the 1980s about the relative merits of some behavioural and physiological parameters, which partly reflected the researchers area of expertise (Barnett, 1987; Rushen, 1986).

The Australian stress physiologist, the late John Barnett, was instrumental, together with scientists from other countries such as Don Broom and the late Gary Moberg, in proposing the notion that inadequate adaptation will generate welfare problems for animals and that such maladaptation offers the opportunity to assess animal welfare risks. Broom (1986, 2000) defined the welfare of an animal as, 'its state as regards its attempts to cope with its environment'. In this definition, state refers to the extent of biological activity underlying attempts to cope, including the involvement of body repair systems, immunological defences and physiological stress responses, as well as a variety of behavioural responses. As with Broom (2000) and Moberg (2000), Barnett argued that how well an animal is coping with the challenges it faces will be reflected in its biological functioning and fitness, with severe risks to welfare associated with the most extreme coping attempts, often reflected in hormonal stress responses (Barnett, 2003). To demonstrate this, Barnett used the example of stress in the Australian brown marsupial mouse (*Antechinus agilis* previously *A. stuartii*), and its consequences (Lee *et al.*, 1977). This small carnivorous marsupial has a highly synchronised and dramatic life history in which there is a total annual mortality of males prior to the birth of the next generation. This mortality is stress related and the main stressors are hormonally induced behavioural patterns: as the breeding season approaches, the males become very aggressive towards each other and spend long periods of time searching for females. Males consequently show a whole array of physiological changes, including a sustained increase in free (biologically active) glucocorticoids, indicative of very high mortality-inducing stress levels.

In parallel with this approach to welfare assessment, which was focused on biological functioning, there was growing recognition by the scientific community of the relevance of an animal's affective experiences in assessing animal welfare. Australian and New Zealand scientists contributed considerably to progressing rigorous preference testing methodologies. In these tests the animal's preference for a specific environmental option, or the motivation to perform a specific behaviour are measured, and the information is used to make inferences about animal welfare (Fraser and Matthews, 1997). Since animals do not always choose options that are in their long-term best interests, the combination of both biological functioning and affective experiences provide the most valuable basis for understanding an animal's welfare status (Matthews, 2008).

9.2.1 Assessing intensive animal housing

The Senate Select Committee on Animal Welfare in 1983 was decisive for animal welfare in Australia by including reference to animal welfare research. The Committee's recommendation that 'future trends in housing the dry sow should be away from individually-confined stall systems and that this be reflected in the Codes of Practice for the welfare of the pig' (Australian Government, 1990) stimulated funding and research on group housing of gestating sows. Consequently, Australian scientists have contributed significantly to international research aimed at understanding the effects of specific design features of intensive housing systems on farm animal welfare. In particular, a large amount of work has been conducted on sow welfare. Research commencing in the 1990s by John Barnett and colleagues examined the effects of the design features of group housing on gestating sow welfare (see reviews by Barnett *et al.*, 2001; Verdon *et al.*, 2015). The knowledge gained remains important as the international pork industry moves away from stall housing and towards the use of group housing during gestation.

This research was characterised by assessments of the extent of biological activity of the animals' underlying attempts to cope, including the involvement of body repair systems, immunological defences and physiological stress responses, as well as a variety of behavioural responses. In a series of studies, Barnett and colleagues investigated methods to reduce aggression, injuries and stress in gilts at mixing. Strategies tested included modifying pen size and shape, masking odours, sedation using pharmacological agents, grouping after dark and *ad libitum* feeding and concluded that all or some of these methods may only be effective in postponing aggression rather than reducing it (see review by Barnett *et al.*, 2001). Furthermore, the authors considered that aggression, injuries and stress arising from mixing were impediments to wider adoption of group housing, and recommended further research on mixing strategies to provide producers with the information needed to reduce its risks to animal welfare. Indeed the need to minimise welfare risks arising from mixing animals is still relevant today (Verdon *et al.*, 2015).

More recent research indicates that reducing floor space from 3.0 to 1.4 m^2/sow increases aggression and plasma cortisol concentrations, and reduces farrowing rate in sows (Hemsworth *et al.*, 2013). This research also determined that the effects of space allowance on aggression and stress were most pronounced early after grouping, highlighting the importance of floor space as the dominance hierarchy is established. In addition to floor space, early research also highlighted the importance of protecting gilts when feeding in a competitive system by the use of full body length feeding stalls in group pens to reduce aggression and stress and improve immunological responsiveness (see review by Barnett *et al.*, 2001).

This body of Australian research, which was characterised by measurement of the extent of biological activity underlying attempts to cope to determine welfare risks, contributed to the recognition that both the protection of sows whilst feeding and available floor space are important determinants of the welfare of sows in groups. Furthermore, this research highlighted the importance of the design of the housing system rather than just the housing system *per se*. These are key pieces of information to consider when housing sows in groups and this research has been applied all over the world.

9.2.2 Assessing painful husbandry procedures

The sources of expertise that have contributed to animal welfare studies are sometimes quite surprising. For example, during the late 1960s and early 1970s, methods were developed that allowed the physiological states of sheep foetuses to be studied continuously in conscious ewes for weeks or months during mid-to-late pregnancy (e.g. Mellor, 1980, 1984; Silver, 1980). The experience of using physiological and behavioural indices, such as absence of elevated cortisol levels, heart rate and escape responses, to assess adaptation to close confinement and regular handling before the ewes were presented for uterine/foetal surgery (Pearson and Mellor, 1976), provided foundational knowledge for later detailed evaluations of painful husbandry procedures led by David Mellor and conducted in Scotland and New Zealand in the 1980s and 1990s, respectively.

The Scottish studies, led by Vince Molony and Joyce Kent, focused on describing and seeking ways to minimise behavioural and cortisol stress responses of young lambs and calves to different painful husbandry procedures (Mellor and Murray, 1989; Mellor *et al.*, 2000; Molony and Kent, 1997; Molony *et al.*, 1995). Throughout the 1990s in New Zealand, complementary studies led by David Mellor and Kevin Stafford, when combined with the continuing Scottish work, enabled comparison of the relative negative impacts of 59 different methods of castration and/or tail docking of lambs, with or without the use of different forms of pain relief (Mellor and Stafford, 2000). For example, the relative cortisol responses, interpreted in terms of 'pain-induced distress', were lower with tail docking or castration using tight rubber rings than with the equivalent surgical procedure, suggesting that, in the absence of pain relief, ring use was preferable to surgery. This was contrary to the existing view that rings caused more pain than the knife, a view based on the striking contrast between ring-induced writhing activity and the relative physical immobility induced by surgery. It became apparent, however, that this behavioural comparison was flawed as a means of assessing the relative intensities of pain caused by the two methods. The character of ring-induced ischaemic pain would be quite different from the pain caused by surgical cutting, so that each method elicited different behaviours, without a common basis for comparison (Mellor *et al.*, 2000).

Also in New Zealand, studies of calf castration, tail docking, dehorning and disbudding underpinned analyses of the relative noxiousness of a range of different methods used for each procedure, again undertaken with or without pain relief (Stafford and Mellor, 2005a,b, 2006, 2011). For example, it was shown that the pain-induced cortisol response to amputation dehorning had two phases, an amputation phase and a subsequent inflammatory phase, and that local anaesthetic could prevent pain during the first phase and non-steroidal anti-inflammatory drugs could prevent it during the second phase. Pain-related behaviours were present in animals dehorned without pain relief and were absent in control animals (not dehorned) and in those where both types of pain relief were delivered before dehorning.

The practice of mulesing was developed in Australia in the 1930s to reduce the problem of blowfly strike to the hindquarters of wool-bearing sheep. The combination of Australian Merino sheep, ideal environmental conditions for fly populations and the arrival of the sheep blowfly *Lucilia cuprina* resulted in significant sheep deaths and morbidity from blowfly strike. This problem developed in an era when preventative chemical treatments such as insect growth regulators active against blowfly larvae were not available. The mulesing operation involved the cutting of strips of skin from the hindquarters of lambs using mulesing shears, with healing resulting in reduced wool growth and stretching of the skin through scar tissue. This reduced wool cover and skin wrinkle was found to provide around a 90% reduction in fly strike to the hindquarters over the life of the animal (Lee and Fisher, 2007). Mulesing was normally performed in combination with the more common practice of tail docking of lambs.

Although mulesing was originally viewed as beneficial to animal welfare (Beveridge, 1984), the obvious immediate welfare costs and perceived pain from the procedure, together with pressure from animal advocacy groups, prompted research from the late 1980s onwards involving applied ethologists to evaluate, refine, and if possible replace the procedure (e.g. Fell and Shutt, 1989). A step forward came with the development of a topical anaesthetic treatment (Lomax *et al.*, 2008). This approach reduced the behavioural responses of lambs to the procedure, with further improvements shown to be possible when combined with non-steroidal anti-inflammatory drugs (Paull *et al.*, 2007). Because of the medium term goal to replace the need for mulesing by anti-flystrike breeding solutions or non-surgical alternatives, subsequent ethological and physiological research has evaluated sheep responses to alternatives such as plastic occlusive clips and intradermal delivery of sodium lauryl sulphate to the breech (Colditz *et al.*, 2010; Hemsworth *et al.*, 2009).

Similar approaches have also been extended to the practices of tail-docking and castration of Australian sheep (e.g. Lomax *et al.*, 2010; Paull *et al.*, 2012). Particular strengths of this body of work include the extensive use of treatment-specific sham control groups and parallel studies of behavioural and cortisol responses (Fraser, 2008). Thus, evidence-based advice was available regarding the choice of method and appropriate use of pain relief when conducting these painful husbandry procedures. This evidence was utilised during the development of the related New Zealand code of welfare (Anonymous, 2005) This Code included a national commitment to progressively increasing the use of pain relieving agents for invasive husbandry procedures conducted on farms. It also provided information on the least painful methods for achieving the same objective (e.g. castration) and recommended that, if pain-relieving agents were not readily accessible (e.g. local anaesthetic), the method chosen for on-farm use should be the least painful of those conducted without pain relief (e.g. rings instead of surgery).

9.2.3 The Five Domains Model

A good example of the use of behaviour to support animal welfare assessments is the Five Domains Model. The primary purpose of the Model, formulated in New Zealand by David Mellor and Cam Reid in 1994, was to provide a coherent way to comprehensively assess and rank potential negative impacts of research, teaching and testing manipulations on sentient animals (Mellor and Reid, 1994). The Model differed from most contemporary assessment approaches. Most assessments focused mainly on the precise target of the manipulation, and neglected consideration of additional negative impacts of the wider circumstances of the animal. In 1997, the Model was made a mandatory part of the New Zealand institutional animal ethics committee approval system for animal-based science manipulations (Williams *et al.*, 2006), and it is still used today by all investigators wishing to study sentient vertebrates in New Zealand.

The Model was multidisciplinary in origin and focus. The Five Domains are 'nutrition', 'environment', 'health', 'behaviour' and 'mental state'. The first three are largely physical/functional, where the focus is on internal physiological or pathophysiological disturbances. The fourth domain, 'behaviour', focuses on responses to close confinement, restraint or otherwise unusual space availability, the presence or absence of other animals, and potentially threatening circumstances. Once such factors are assessed, the anticipated subjective, emotional or affective consequences of those disturbances, restrictions or threats are assigned to the fifth 'mental' domain, and it is these effects that are the focus of the impact assessment. Thus, from its inception the model incorporated knowledge of biological function and affective state, both being regarded as dynamically interacting elements in animals living as integrated whole entities. At the time this was uncommon, because then the biological function and affective state orientations were often regarded as competing notions for understanding the nature of animal welfare.

The Model has been updated progressively to facilitate its application beyond laboratory-based studies to include assessments in farm, companion, service, sport, wild and other animal sectors, and to also take account of the latest developments in animal welfare science (e.g. Mellor and Beausoleil, 2015; Mellor *et al.*, 2009, 2015). The six negative affects originally listed in the Model (anxiety, fear, thirst, hunger, pain and distress) were subsequently expanded and differentiated into two categories reflecting the major sources of sensory input.

The current list (Mellor and Beausoleil, 2015) is supported by both behavioural science and neuroscience evidence (e.g. Boissy *et al.*, 2007; Fraser, 2008; Gregory, 2004; Panksepp, 2005). Category 1 affects, which are mainly associated with sensory inputs generated internally, include breathlessness, thirst, hunger, nausea, pain, dizziness, weakness, malaise and sickness, whereas category 2 affects, which are associated mainly with the animal's cognitive assessment of its external circumstances, include anxiety, fear, panic, frustration, anger, helplessness, loneliness, boredom and depression. Recently, assessment of enhanced welfare has been incorporated into the Model by including consideration of the rewarding experiences of comfort, pleasure, interest and confidence. These additional category 2 affects may be generated when animals engage in, among other activities, animal-environment and animal-to-animal interactive behaviours (Mellor and Beausoleil, 2015).

Category 1 identifies *survival-critical* motivational affects, all of which are negative and drive animals to undertake specific behaviours aimed at acquiring essential life-sustaining resources, avoiding or minimising physical harm, or promoting convalescence from injury or disease. Included in category 2 are negative and positive *situation-related* affects, which are also identifiable using particular behaviours. Behavioural science, conducted over at least three decades (Fraser, 2008), and reinforced by affective neuroscience evidence (e.g. Boissy *et al.*, 2007; Gregory, 2004; Panksepp, 2005), has therefore made major contributions to the development of the Five Domains Model (Figure 9.1) and to its effective and increasing use in a wide range of contexts, for example, by the World Association of Zoos and Aquariums (WAZA) in its animal welfare strategy (Mellor *et al.*, 2015).

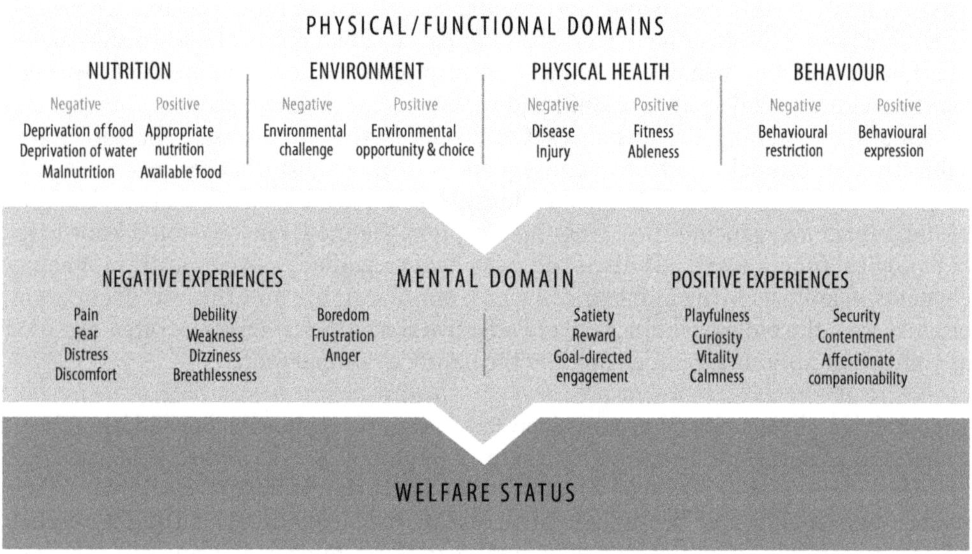

Figure 9.1. The Five Domains Model, divided into physical/functional and mental components, provides examples of how internal and external conditions give rise to negative (aversive) and positive (pleasant) subjective experiences or affects, the integrated effects of which give rise to an animal's welfare status (revised from Mellor and Beausoleil, 2015). The Model is not intended to be an accurate physical and functional representation of the body; rather it is a device to facilitate animal welfare understanding and assessment. Reproduced from Mellor et al. *(2015) with permission from the WAZA, Gland, Switzerland.*

9.3 Foetal and neonatal awareness, welfare, behaviour and survival

Mortality and morbidity rates in new-born farm animals were investigated intensively beginning in the early 1950s in Australia, New Zealand and the United Kingdom (Alexander, 1984; Hinch and Brien, 2014; Kenyon *et al.*, 2007; Mellor and Stafford, 2004). At that time and for roughly 40 years thereafter, the major driver for this research was production losses. Initially, most information was derived from clinical-pathological examination of young animals that died, but increasingly the physiological, metabolic and behavioural responses of living new-borns and their impact on subsequent survival were explored (e.g. Alexander, 1984; Fraser, 1985; Mellor, 1988; Mellor and Cockburn, 1986). The application of this knowledge contributed to a general improvement in the survival and health of new-born farm animals by the end of the 1980s (Alexander, 1984; Mellor and Stafford, 2004). However, it took at least another decade for the focus to shift towards behaviour-welfare interactions in foetal, new-born and young animals (e.g. Dwyer, 2008; Dwyer and Lawrence, 2005; Fraser and Broom, 1998; Mellor and Stafford, 2004).

This is an area where animal welfare advances resulted from knowledge acquired from animal models focused on human clinical applications. In numerous laboratories internationally, information obtained through studies of brain functionality and behaviour in chronically instrumented foetal sheep (e.g. Dawes 1988; Mellor *et al.*, 2005) provided the knowledge required for detailed consideration of the behaviour-welfare nexus in foetal, new-born and young mammals. Two examples illustrate this point.

First, a key area of interest with regard to Australian and New Zealand livestock was the extent to which foetal behavioural and physiological responses to slaughter of the dam, followed by foetal blood or tissue collection, were a cause for welfare concern (Mellor and Gregory, 2003). Detailed exploration of the literature revealed that, notwithstanding some foetal behavioural responses suggestive of arousal, neurologically mature ungulate foetuses are maintained in unconscious states by a suite of cerebrocortical inhibitors that are unique to life *in utero*, so that such foetuses are not able to experience unpleasant affects such as breathlessness and pain (Mellor and Diesch, 2006; Mellor *et al.*, 2005). Moreover, foetal hypoxia/anoxia after slaughter of the dam leads to rapid shutdown of cerebrocortical electrical activity. These and other observations led to the 2006 adoption by the World Organisation for Animal Health (OIE) of global standards for the humane management of ungulate foetuses during slaughter of their dams (Mellor, 2010; OIE, 2015). The presence of ungulate foetuses at slaughter of their dams may be unplanned, or it may be planned specifically to enable foetal blood (serum) or other tissues to be collected. The most common product is foetal calf serum used globally in tissue cultures and other scientific procedures.

Secondly, a combination of behavioural and neuroscience observations has also clarified when three different groups of mammalian young develop the survival-enhancing capacity to engage in flexible behavioural responses to unpredictable postnatal environments (Mellor and Lentle, 2015). This behavioural capacity depends on cerebrocortically-based cognitive functions which themselves require the establishment of neural connections between the cortical and the subcortical regions of the brain. Their timely establishment ensures that the required behavioural flexibility is available at the time the young first encounter unpredictable environments and need to be able to respond autonomously. This occurs after several months, days-to-weeks or minutes-to-hours in young that at birth are, respectively, neurologically exceptionally immature (e.g. marsupial joeys), moderately immature (e.g. rat pups) and mature (e.g. lambs) (Mellor and Lentle, 2015). Thus, marsupial joeys require such behavioural flexibility when they first leave the maternal pouch, rat pups to support independent exploration outside and return to the natal nest, and lambs to facilitate bonding and remaining with their mobile mothers after birth outdoors.

9.4 Human-animal relationships

With an ever-increasing body of evidence accumulating since the 1980s of the profound effects of human interactions on farm animal fear and stress responses, there has been a widespread appreciation of the importance of human-animal relationships on animal productivity and welfare. Australian researchers developed a sequential causal model that explains the pathway between stockperson beliefs, stockperson behaviour, farm animal behaviour and stress, and farm animal productivity and welfare outcomes (see reviews by Coleman and Hemsworth, 2014; Hemsworth and Coleman, 2011). This model is especially significant because it integrates human and animal factors in the causal pathway and targets outcomes that are relevant to both animal welfare and commercial profitability. The implications of human-animal relationships in other animal uses, such as in domestic and zoo settings, is also beginning to be studied and recognised.

This research in the livestock industries was characterised by its multi-disciplinary nature and international collaborations, utilising the disciplines of applied animal ethology, psychology and physiology and conducted by ISAE scientists in countries such as Australia, Austria, Canada, France, Norway, UK and USA. Early studies by two of the authors of this chapter (Paul Hemsworth and Grahame Coleman) and the late John Barnett and studies by Martin Seabrook in the late 1970s and early 1980s revealed the potential implications of human-animal interactions on farm animal productivity and welfare.

These studies led to three decades of interest among Australian researchers and many international colleagues regarding fear responses in animals, especially in commercial and experimental settings. Their work revealed variation in fear of humans by farm animals that could not be explained by farm characteristics or animal genetics. The variation in fear responses was strongly related to increased stress responses and reduced productivity. This led to the hypothesis that this variation in fear was caused by human factors, giving rise to investigations directed towards identifying the human characteristics responsible for these effects.

However, initial progress in this area was relatively slow because it was unclear what stockperson characteristics should be studied. Moreover, psychologists at this time showed little interest in this area. The focus of early research was human impacts on farm animal productivity, as fear responses to humans were shown to reduce productivity through stress. However, as the understanding of animal welfare progressively increased during the 1990s, the focus of this work shifted from the stress-productivity nexus towards the welfare-productivity implications of stockperson-induced stress, resulting in a substantial amount of research on human-livestock relationships.

The application of two well established psychological principles facilitated the development of an integrated model relating human characteristics to welfare and productivity outcomes in livestock. The first was the application of Ajzen and Fishbein's theories of Reasoned Action and Planned Behaviour, that can be used to predict human behaviour from behaviour-specific attitudes (Ajzen and Fishbein, 1980). The second was use of the principles of Cognitive Behavioural interventions (Ajzen and Fishbein, 1980; Hemsworth and Coleman, 2011) to effect behavioural change.

Ajzen and Fishbein's theory of Planned Behaviour is reproduced in Figure 9.2. Here, the antecedents of stockperson behaviour are their beliefs about the effects that their specific behaviours have on their farm behaviour and animal fear levels and productivity, and about the effectiveness and appropriateness of these behaviours.

Evidence to support the application of the Ajzen and Fishbein model came from studies in the laboratory, and correlational and intervention studies in commercial settings, of the relationships between stockperson attitudes, stockperson behaviour, animal fear responses, animal stress physiology and animal productivity, and provided evidence of causal relationships between these variables (Hemsworth and Coleman, 2011). In addition to this early focus on animal's fear responses, growing evidence showed that animals could experience positive or pleasant emotions in the presence of humans, when linked to rewarding events and associations. This work has shown that such positive relationships with humans may ameliorate animal stress during routine husbandry practices in agriculture (Hemsworth and Boivin, 2011).

The efficacy of cognitive behavioural interventions was demonstrated by the finding among dairy and pig stockpersons that cognitive-behavioural training can improve attitudes and human behaviour towards animals, with consequent reductions in animal fear and improvements in productivity (Hemsworth and Coleman, 2011). Cognitive-behavioural techniques basically involve changing a person's behaviour by first targeting both the beliefs that underlie the behaviour (attitude) and the behaviour in question, and second, maintaining these changed beliefs and behaviours. This process of inducing

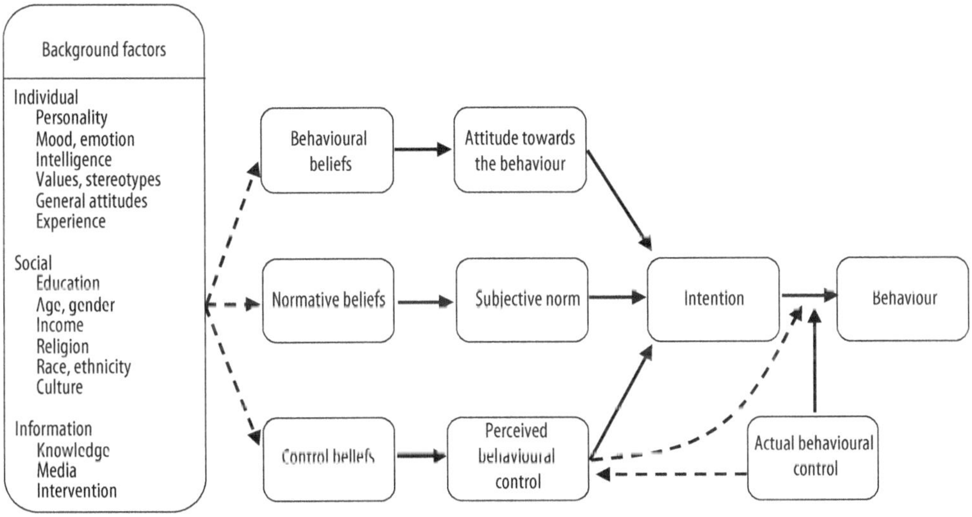

Figure 9.2. A model of the theories of reasoned action and planned behaviour (adapted from Albarracin et al. *(2005), with permission from publishers).*

behavioural change is a comprehensive procedure in which all of the personal and external factors that are relevant to the behavioural situation are explicitly targeted. This includes addressing common perceived barriers to change, addressing defensiveness about previous behaviour, changing habits, providing follow-ups to reinforce changes as well as changing the relevant attitudes and behaviour. This approach to training was also shown to be practical and effective among a wide range of stockpersons working in a variety of situations, providing strong evidence for introducing this type of training into the livestock industries.

The training programme used as an experimental tool during research in the pig industry has been commercialised for on-farm use, and is called 'ProHand' (a contraction of the 'Professional Handling of Pigs Program', Animal Welfare Science Centre, 2005). Similar research-based training programmes were also developed for on-farm application in the dairy industry and for abattoir handling of cattle, sheep and pigs in Australia. Moreover, as part of the European Union Sixth Framework programme, ProHand principles guided the development of the Welfare Quality training packages for stockpersons working in the European pig, poultry and cattle industries (Ruis *et al.*, 2010). An important characteristic of all of these programmes is that they are based on scientific research and their effectiveness in improving welfare and productivity has been demonstrated by properly designed intervention studies. There has been a progressive growth in the acceptance of these behaviour change programs as it becomes better recognised that behaviour change is different from skills training, cannot be effectively taught by conventional education approaches, and requires specific targeting of the idiosyncratic nature of an individual's beliefs that underpin his or her behaviour.

Training programmes that target the attitude and behaviour of stockpeople offer considerable opportunity for reciprocal short-term and long-term benefits to both stockpersons and animals. Improved human-animal interactions may enhance job-related characteristics, such as job satisfaction, motivation and commitment, thereby potentially improving the stockperson's job performance and career prospects (Coleman and Hemsworth, 2014). This, in turn, is also likely to reinforce a stockperson's commitment to improve animal welfare and performance. Selection of animal handlers therefore also provides an important opportunity to improve animal welfare and productivity. Research in the Australian pork industry has shown that measures of attitude, empathy, conscientiousness, work reliability and job satisfaction are useful in identifying effective animal handlers (Coleman, 2004).

This substantial body of research demonstrates that human factors contribute to farm animal welfare and productivity to an extent that has similar impact to many other inputs, such as housing. It also shows that such human factors need to be recognised as an important component of routine animal husbandry. To some this may be obvious, but to others it is less so, or even contentious. There is clearly a short-term and long-term need for livestock industries to train their personnel to care for and handle their stock effectively and humanely. Acceptance of the value of such training needs to be extended beyond the intensive livestock sectors of dairy, poultry and pigs to also include, in particular, the extensive livestock industries and post-farm gate activities in general.

Training owners or animal care staff is equally important in other animal use settings, for example, in companion animals and zoo animals where promising research-led initiatives are apparent.

9.5 Concluding remarks

A feature of Australian and New Zealand applied animal behaviour science for many decades has been its multidisciplinarity. The strengths of bringing into applied ethology the fresh and broadening perspectives of other disciplines, as well as the creative tension of operating at the margins between them, has been apparent throughout. The seminal contributions from home-based scientists having diverse disciplinary backgrounds were enhanced by a strong commitment to collaboration with many others from overseas who brought complementary perspectives and skills to their joint work. The breadth of disciplines involved is impressive. In addition to applied animal behaviour science, the others mentioned here include anatomy, metabolic and general biochemistry, nutrition, environmental and stress physiology, pathophysiology, affective neuroscience, psychology, sociology, and veterinary clinical sciences including body repair systems, immunological defences, and surgical and medical interventions. Add to this, exploration of foetal, neonatal, young and mature developmental stages and a species range that includes farm, companion, laboratory, wildlife and vertebrate pest animals, and it may be seen that, overall, Australian and New Zealand contributions have been impressively wide-ranging.

Interestingly, animal welfare science is also multidisciplinary, considers different developmental stages and provides wide species coverage, and, in a reciprocal manner, has both benefitted from and provided benefits to applied animal behaviour science. Examples of such interactions involving major inputs from applied animal behaviour science and beneficial outcomes have been noted briefly with regard to assessing intensive livestock housing, assessing and alleviating the pain caused by invasive husbandry procedures, developing a systematic and comprehensive model (the Five Domains Model) for assessment of animal welfare compromise and enhancement, and the welfare and behaviour of foetal, new-born and young mammals. Finally, discussed above is the complex and multifaceted subject of human-animal interactions. Handling studies in the laboratory and intervention studies on farms on the relationship between stockperson attitudes, stockperson behaviour, animal behaviour, stress physiology and productivity provide evidence of causal relationships between these variables. Furthermore, this research provides a strong case for introducing to livestock industries stockperson training courses that target stockperson attitudes and behaviour.

It is apparent from the examples provided here that integrating the precepts and methodologies of applied animal behaviour science and animal welfare science has brought forward refreshingly new scientific insights and the related practical satisfaction derived from their successful application to the care and management of a wide range of animals.

References

Ajzen, I. and Fishbein, M., 1980. Understanding attitudes and predicting social behaviour. Prentice-Hall Inc., Englewood Cliffs, NJ, USA.

Albarracín, D., Zanna, M.P., Johnson, B.T. and Kumkale, G.T., 2005. Attitudes: introduction and scope. In: Albarracin, D., Johnson, B.T. and Zanna, M.P. (eds.) The handbook of attitudes. Psychology Press, New York, NY, USA, pp. 3-19.

Alexander, G., 1984. Constraints to lamb survival. In: Lindsay, D.R. and Pearce, D.T. (eds.) Reproduction in sheep, Australian Academy of Science and Australian Wool Corporation. Canberra, ACT, Australia, pp. 199-209.

Anonymous, 2005. Animal welfare (painful husbandry procedures) code of welfare 2005. Ministry for Primary Industries, Wellington, New Zealand.

Australian Government, 1990. Intensive livestock production, report by the senate select committee on animal welfare. Australian Government Publishing Service, Canberra, Australia.

Barnett, J.L., 1987. Surgery may not be preferable to rubber rings for tail docking and castration of lambs. Australian Veterinary Journal 65: 131.

Barnett, J.L., 2003. Studying stress to assess animal welfare. In: Paterson, J.E. (ed.) Manipulating pig production IX, Australasian pig science association. Werribee, Australia, pp. 107-120.

Barnett, J.L., Hemsworth, P.H., Cronin, G.M., Jongman, E.C. and Hutson, G.D., 2001. A review of the welfare issues for sows and piglets in relation to housing. Australian Journal of Agricultural Research 52: 1-28.

Beveridge, W.I.B., 1984. The origin and early history of the Mules operation. Australian Veterinary Journal 61: 161-163.

Boissy, A., Manteuffel, G., Jensen, M.B., Moe, R.O., Spruijt, B., Keeling, L.J., Winckler, C., Forkman, B., Dimitrov, I., Langbein, J., Bakken, M., Veissier, I. and Arnaud, A., 2007. Assessment of positive emotions in animals to improve their welfare. Physiology and Behavior 92: 375-397.

Broom, D.M., 1986. Indicators of poor welfare. British Veterinary Journal 142: 524-526.

Broom, D.M., 2000. Coping, stress and welfare. In: Broom, D.M. (eds.) Coping with challenge: welfare in animals including humans. Daglem University Press, Berlin, Germany, pp. 1-9.

Colditz, I.G., Lee, C., Paull, D.R. and Fisher, A.D., 2010. Physiological and behavioural effects of intradermal injection of sodium lauryl sulfate as an alternative to mulesing in lambs. Australian Veterinary Journal 88: 483-489.

Coleman, G.J. and Hemsworth, P.H., 2014. Training to improve stockperson beliefs and behaviour towards livestock enhances welfare and productivity. Scientific and Technical Review of the Office International des Epizooties 33(1): 131-137.

Coleman, G.J., 2004. Personnel management in agricultural systems. In: Rollin, B.E. and Benson, J. (eds.) Maximizing well-being and minimizing suffering in farm animals. Iowa State University Press, Ames, IA, USA, pp. 167-181.

Command Paper 2836, 1965. Report of the technical committee to enquire into the welfare of animals kept under intensive livestock husbandry systems. Her Majesty's Stationery Office, London, UK.

Dawes, G.S., 1988. The 1987 James A.F. Stevenson memorial lecture. The development of fetal behavioural patterns. Canadian Journal of Physiology and Pharmacology 66: 541-548.

Dwyer, C.M. and Lawrence, A.B., 2005. A review of the behavioural and physiological adaptations of hill and lowland breeds of sheep that favour lamb survival. Applied Animal Behaviour Science Journal 92: 235-260.

Dwyer, C.M., 2008. The welfare of the neonatal lamb. Small Ruminant Research 76: 31-41.

Fell, L.R. and Shutt, D.A., 1989. Behavioural and hormonal responses to acute surgical stress in sheep. Applied Animal Behaviour Science Journal 22: 283-294.
Fraser, A.F. and Broom, D.M., 1998. Early behaviour and parental behaviour. In: Fraser, A.F. and Broom, D.M. (eds.) Farm animal behaviour and welfare. CAB International, Wallingford, UK, pp. 198-255.
Fraser, A.F., 1985. Reproductive and developmental behaviour in sheep. Elsevier Science Publishing Company, New York, NY, USA, pp. 71-430.
Fraser, D., 2003. Assessing animal welfare at the farm and group level: the interplay of science and values. Animal Welfare 12: 433-443.
Fraser, D., 2008. Understanding animal welfare: the science in its cultural context. Wiley-Blackwell, Oxford, UK.
Fraser, D., Matthews, L.R., 1997. Preference and motivation testing, in: Appleby, M.C. and Hughes, B.O. (Eds.) Animal Welfare. CAB International, Wallingford, UK, pp. 159-174.
Gregory, N.G., 2004. Physiology and behaviour of animal suffering. Blackwell Science, Oxford, UK.
Hemsworth, P.H. and Boivin, X., 2011. Human contact. In: Appleby, M.C., Mench, J.A., Olsson, I.A.S. and Hughes, B.O. (eds.) Animal welfare. CAB International, Wallingford, UK, pp. 246-262.
Hemsworth, P.H. and Coleman, G.J., 2011. 'Human-livestock interactions: the stockperson and the productivity and welfare of farmed animals'. CAB International, Wallingford, UK.
Hemsworth, P.H., Barnett, J.L., Karlen, G.M., Fisher, A.D., Butler, K.L. and Arnold, N.A., 2009. Effects of mulesing and alternative procedures to mulesing on the behaviour and physiology of lambs. Applied Animal Behaviour Science Journal 117: 20-27.
Hemsworth, P.H., Rice, M., Nash, J., Giri, K., Butler, K.L., Tilbrook, A.J. and Morrison, R.S., 2013. Effects of group size and floor space allowance on grouped sows: aggression, stress, skin injuries and reproductive performance. Journal of Animal Science 91: 4953-4964.
Hinch, G.N. and Brien, F., 2014. Lamb survival in Australian flocks: a review. Animal Production Science 54: 656-666.
Kenyon, P.R., Stafford, K.J., Jenkinson, C.M.C., Morris, S.T. and West, D.M., 2007. The body composition and metabolic status of twin- and triplet-bearing ewes and their fetuses in late pregnancy. Livestock Science 107: 103-112.
Lee, A.K., Bradley, A.J. and Braithwaite, R.W., 1977. Corticosteroid levels and male mortality in Antechinus stuartii. In: Stonehouse, B. and Gilmore, D. (eds.) 'The biology of marsupials'. MacMillan, London, UK, pp. 209-220.
Lee, C. and Fisher, A.D., 2007. A review of the welfare consequences of mulesing of sheep. Australian Veterinary Journal 85: 89-93.
Lomax, S., Dickson, H., Sheil, M. and Windsor, P.A., 2010. Topical anaesthesia alleviates short-term pain of castration and tail docking in lambs. Australian Veterinary Journal 88: 67-74.
Lomax, S., Sheil, M. and Windsor, P.A. 2008. Impact of topical anaesthesia on pain alleviation and wound healing in lambs after mulesing. Australian Veterinary Journal 86: 159-168.
Matthews, L.R., 2008. Methodologies by which to study and evaluate welfare issues facing livestock systems of production. Australian Journal of Experimental Agriculture 48: 1014-1021.
Mellor, D.J. and Beausoleil, N.J., 2015. Extending the 'Five Domains' model for animal welfare assessment to incorporate positive welfare states. Animal Welfare 24: 241-253.
Mellor, D.J. and Cockburn, F., 1986. A comparison of energy metabolism in the newborn infant, piglet and lamb. Quart. Journal of Experimental Physiology 71: 361-379.

Mellor, D.J. and Diesch, T.J., 2006. Onset of sentience: the potential for suffering in fetal and newborn farm animals. Applied Animal Behaviour Science Journal 100: 48-57.
Mellor, D.J. and Gregory, N.G., 2003. Responsiveness, behavioural arousal and awareness in fetal and newborn lambs: experimental, practical and therapeutic implications. New Zealand Veterinary Journal 51: 2-13.
Mellor, D.J. and Lentle, R.G., 2015. Survival implications of the development of behavioural responsiveness and awareness in different groups of mammalian young. New Zealand Veterinary Journal 63: 131-140.
Mellor, D.J. and Murray, L., 1989. Effects of tail docking and castration on behaviour and plasma cortisol concentrations in young lambs. Research in Veterinary Science 46: 387-391.
Mellor, D.J. and Reid, C.S.W., 1994. Concepts of animal well-being and predicting the impact of procedures on experimental animals. In: Baker, R.M., Jenkin, G. and Mellor, D.J. (eds.) Improving the well-being of animals in the research environment. Glen Osmond, Australia, pp. 3-18.
Mellor, D.J. and Stafford, K.J., 2000. Acute castration and/or tailing distress and its alleviation in lambs. New Zealand Veterinary Journal 48: 33-43.
Mellor, D.J. and Stafford, K.J., 2004. Animal welfare implications of neonatal mortality and morbidity in farm animals. Veterinary Journal 168: 118-133.
Mellor, D.J., 1980. Investigation of the fluid spaces of the sheep conceptus. In: Nathanielsz, P.W. (ed.) Animal models in fetal medicine (I), monographs in fetal physiology 2. Elsevier/North-Holland Biomedical Press, Amsterdam, the Netherlands, pp. 59-106.
Mellor, D.J., 1984. Investigation of fetal growth in sheep. In: Nathanielsz, P.W. (ed.) Animal models in fetal medicine (IV). Perinatology Press, Ithaca, NY, USA, pp. 150-173.
Mellor, D.J., 1988. Integration of perinatal events, pathophysiological changes and consequences for the newborn lamb. British Veterinary Journal 144: 552-569.
Mellor, D.J., 2010. Galloping colts, fetal feelings and reassuring regulations: putting animal welfare science into practice. Journal of Veterinary Medical Education 37: 94-102.
Mellor, D.J., Cook, C.J. and Stafford, KJ., 2000. Quantifying some responses to pain as a stressor. In: Moberg, G.P. and Mench, J.A. (eds.) The biology of animal stress: basic principles and implications for welfare. CAB International, Wallingford, Wallingford, UK, pp. 171-198.
Mellor, D.J., Diesch, T.J., Gunn, A.J. and Bennet, L., 2005. The importance of 'awareness' for understanding fetal pain. Brain Research Reviews 49: 455-471.
Mellor, D.J., Hunt, S. and Gusset, M., 2015. Caring for wildlife: the world zoo and aquarium animal welfare strategy. WAZA Executive Office, Gland, Switzerland, pp. 87.
Mellor, D.J., Patterson-Kane, E. and Stafford, K.J., 2009. Animal welfare, grading compromise and mitigating suffering. In: Mellor, D.J., Patterson-Kane, E. and Stafford (eds.) The sciences of animal welfare. Wiley-Blackwell Publishing, Oxford, UK, pp. 71-94.
Moberg, G.P., 2000. Biological response to stress: implications for animal welfare. In: Moberg, G.P. and Mench, J.A. (eds.) The biology of animal stress: basic principles and implications for welfare. CAB International, Wallingford, Oxford, UK, pp. 1-21.
Molony, V., Kent, J.E. and Robertson, I.S., 1995. Assessment of acute and chronic pain after different methods of castration in calves. Applied Animal Behaviour Science Journal 46: 33-48.
Molony, V., Kent, J.E., 1997. Assessment of acute pain in farm animals using behaviour and physiological measurements. Journal of Animal Science 75: 266-272.
Panksepp, J., 2005. Affective consciousness: core emotional feelings in animals and humans. Consciousness and Cognition 14: 30-80.

Paull, D.R., Lee, C., Colditz, I.G., Atkinson, S.J. and Fisher, A.D., 2007. The effect of a topical anaesthetic formulation, systemic flunixin and carprofen, singly or in combination, on cortisol and behavioural responses of Merino lambs to mulesing. Australian Veterinary Journal 85: 98-106.

Paull, D.R., Small, A.H., Lee, C., Palladin, P. and Colditz, I.G., 2012. Evaluating a novel analgesic strategy for ring castration of ram lambs. Veterinary Anaesthesia and Analgesia 39: 539-549.

Pearson, R.A. and Mellor, D.J., 1976. Some behavioural and physiological changes in pregnant goats and sheep during adaptation to laboratory conditions. Research in Veterinary Science 20: 215-217.

Ruis, M., Coleman, G.J., Waiblinger, S., Windschnurer, I. and Boivin, X., 2010. A multimedia-based cognitive-behavioural intervention program improves attitudes and handling behaviours of stockpeople in livestock farming. Advances in Animal Biosciences 1: 175.

Rushen, J.P., 1986. Stereotypes, aggression and the feeding schedules of tethered sows. Applied Animal Behaviour Science Journal 14: 137-147.

Silver, M., 1980. Intravascular catheterization and other chronic preparations in the mare and sow. In: Nathanielsz, P.W. (ed.) Animal models in fetal medicine (I), monographs in fetal physiology 2. Elsevier/North-Holland Biomedical Press, Amsterdam, the Netherlands, pp. 107-132.

Stafford, K.J. and Mellor, D.J. 2011. Addressing the pain associated with disbudding and dehorning in cattle. Applied Animal Behaviour Science Journal 135: 226-231.

Stafford, K.J. and Mellor, D.J., 2005a. Dehorning and disbudding distress and its alleviation in calves. Veterinary Journal 169: 337-349.

Stafford, K.J. and Mellor, D.J., 2005b. The welfare significance of the castration of cattle: a review. New Zealand Veterinary Journal 53: 271-278.

Stafford, K.J. and Mellor, D.J., 2006. The assessment of pain in cattle: a review. CAB Reviews: Perspectives in Agriculture, Veterinary Science, Nutrition and Natural Resources.

Verdon, M., Hansen, C.F., Rault, J.-L., Jongman, E., Hansen, L.U., Plush, K. and Hemsworth, P.H., 2015. Effects of group-housing on sow welfare: a review. Journal of Animal Science 93: 1999-2017.

Williams, V.M, Mellor, D.J. and Marbrook, J., 2006. Revision of a scale for assessing the severity of live animal manipulations. ALTEX 23, Special Issue: 163-169.

World Organisation for Animal Health (OIE), 2015. Management of foetuses during slaughter of their dams. In: OIE terrestrial animal health code, article 7.5.5. World Organisation for Animal Health, Paris, France.

10. The development of applied ethology in Latin America

F. Galindo[1,5], T. Tadich[2,5], R. Ungerfeld[3], M.J. Hötzel[4] and G. Miguel-Pacheco[1]*
[1]Departamento de Etología, Facultad de Medicina Veterinaria y Zootecnia, Universidad Nacional Autónoma de México, 04510 Ciudad de México, Mexico; galindof@unam.mx
[2]Departamento de Fomento de la Producción Animal, Facultad de Ciencias Veterinarias y Pecuarias, Universidad de Chile, Santa Rosa 11735, La Pintana, Santiago, 8820000, Chile
[3]Departamento de Fisiología, Facultad de Veterinaria, Lasplaces 1620, Montevideo 11600, Uruguay
[4]Laboratório de Etologia Aplicada e Bem-Estar Animal, Universidade Federal de Santa Catarina, Rod. Admar Gonzaga, 1346, Itacorubi, Florianópolis, SC 88.034-001, Brazil
[5]Centro Colaborador en Bienestar Animal y Producción Animal Sostenible para la OIE Chile-México-Uruguay

Abstract

Latin America is a diverse region of the world that is destined to play a major role in food production. This region also has a growing urban population with increasing numbers of companion animals, and major issues related to biological diversity, sustainable livestock production, and wildlife and ecosystems conservation. Considering these issues the role of applied ethology has been increasingly important in multiple areas. Although there are courses given in many countries of the region, scientific research is limited to a few groups. Still, networking and collaboration within the region has been quite strong. Some examples of this include the numerous training courses at different levels in the region and participation in international research projects. There has been an active, although fluctuating, involvement of the region in ISAE meetings, usually increasing in years when the annual ISAE conference has been hosted in Latin America. Three ISAE conferences have been held in the region, two in Brazil (2000 and 2013) and one in Mexico (2007). The World Organisation for Animal Health (OIE) agenda on animal welfare has become a good opportunity to promote research on applied ethology in the region and will be a useful tool to strengthen networking and promote collaborative research. It is important to raise awareness among decision makers and intergovernmental organisations of the work that the ISAE is doing and of the role of ethology in science-based policies. Animal behaviour studies will be a necessary tool for the development of efficient policies and legislation.

Keywords: animal behaviour, animal welfare, research, sustainability, Americas

DOI 10.3920/978-90-8686-828-5_10, © Wageningen Academic Publishers 2016

10.1 Introduction

Latin America is a large and diverse region of the world, both geographically and culturally, and is destined to play a major role as one of the main suppliers of food in the world (FAO, 2012). The livestock sector in the region has grown at an annual rate 3.7% higher than the average global growth rate 2.1%, mainly as a result of the increase in the demand for meat, with a higher demand for poultry (4.1%), followed by pork (2.67%), while demand for beef has fallen slightly (-0.2%) (FAO, 2012). In general, exports from Brazil, Argentina, Chile, Uruguay, Mexico, have grown steadily.

Due to the wide variety of food-producing animal species and husbandry systems, the constant growth of companion animal populations, the importance of working animals in rural areas, and the importance of the region's biodiversity and wildlife conservation agenda; the scope for work in animal behaviour and welfare, and its application is enormous. This chapter focuses on some general aspects of the development of applied ethology in Latin America, with emphasis on teaching and research activities in the countries where there is more information available (Mexico, Brazil, Chile and Uruguay). A general discussion on the role of the World Organization for Animal Health (OIE) is included emphasising how the OIE's regional strategy on animal welfare will help to support applied ethology work in the region.

10.2 Regional issues and the role of applied ethology

Within Latin America, livestock production has been associated with increasing greenhouse gases emissions (GHG), deforestation and loss of biodiversity (Galindo *et al.*, 2015). The global increase in the demand for animal products, combined with the environmental impact of livestock production, poses an urgent need to increase focus on sustainable livestock production systems. The intensification of efficient and sustainable livestock production systems, guaranteeing the provision of environmental services, humankind benefits from ecosystems (Macfadyen *et al.*, 2012) and addressing animal welfare issues is a major challenge for the region. In this context, animal welfare is part of a complex matrix of sustainability criteria and should be seen as another element for sustainable animal production, both from the point of view of economics and future markets, but also because of moral acceptability and social licence issues (Broom *et al.*, 2013).

Regarding the development of the science of applied ethology, it is important to emphasise that with urban populations growing rapidly, the number of companion animals has also been growing steadily. However, there is still a lack of education and legislation enforcement in terms of responsible companion animal ownership. Pet abandonment is still common in many cities in the region and overpopulation of stray dogs and cats is a common problem in disadvantaged communities, with indirect consequences on public health, animal welfare, and the environment. There is a big challenge ahead in terms of carrying out more research and outreach efforts to implement better legislation and public policy on responsible pet ownership. It is also important to address the issue of working animals, especially equids that account for several millions of animals in the region, and that are of significance for the welfare of rural populations.

Latin America includes some of the most biologically diverse countries of the world (i.e. Brazil, Mexico, Bolivia, Colombia, Ecuador, Peru, Venezuela, to name a few). In general civil society in these countries is well sensitised to topics related to biodiversity and conservation. However, there are still problems related to the illegal extraction and trade of wildlife, and inefficient management plans regarding the use of wildlife. Hence, there is a need to work on issues related to the link between animal behaviour and welfare, and conservation. These areas are often perceived as being in conflict but this is generally because animal welfare is perceived as animal protection among conservation groups. There are many opportunities to link information on animal behaviour and conservation biology that should be further investigated.

Under these scenarios, the future and potential need for science in applied animal behaviour and animal welfare in Latin America is huge. There are new mechanisms in place in the region, such as the OIE animal welfare strategy, that will allow and even demand more research and knowledge development in the context of the local needs and linking science, public policy and legislation.

10.3 Development of applied ethology in Latin America

The development of applied ethology in Latin America has varied greatly from region to region. There are groups working on the topic in most of the countries of South America and Mexico, however scientific research is limited to a few groups; the majority of groups participate mainly in teaching and outreach programmes.

Most research deals with general aspects of ethology in animal production (animal welfare assessment, animal behaviour and reproduction, animal transport and slaughter, and sustainability), and addresses research questions based on particularities of specific systems from an applied perspective, rather than attempting to generate broad views. It is important to note that, considering the many different production systems and how they differ from the prevailing systems in other countries, most research in the region has sought to respond to local, specific problems. For example, Chile has the particularity of having most livestock production concentrated in the south, and having to transport animals to the north for the last stages of fattening and slaughter. Thus, scientists have focused on studying the behavioural and physiological effects of long distance transport (by road and sea) in order to improve animal welfare during this practice (Gallo *et al.*, 2003; Tadich *et al.*, 2009). Or the positive welfare implications of introducing silvopastoral systems in tropical countries where heat stress is a problem, and were sustainable systems are imperative. Such is the case in Colombia (Navas, 2010) and Mexico (Broom *et al.*, 2013), among other countries.

While there have been some studies on aspects of agribusiness such as transport or handling of animals in slaughter plants, these are relatively rare considering the importance of these industries to the region, and have been closely related to what has been done by research groups outside of Latin America. There is a need for more studies addressing these issues from a more local perspective, for example addressing specific environmental or social features of the local systems (e.g. silvopastoral systems in the tropics).

Research concerning the ethology of native wildlife is a significant area in Latin America, and is relevant for animals in zoos, in the wild, and even in production systems. For example, in Mexico since the Department of Ethology at UNAM was opened in 1993, many zoological parks became interested in doing research on animal behaviour in order to set up programmes to improve environmental enrichment practices and linking animal behaviour studies to conservation efforts, such is the case of the Mexican wolf (*Canis lupus baiyeli*) (Pifarré *et al.*, 2012). In the case of alternative production systems with native species an example is the study of repetitive behaviours and fur chewing in chinchillas (*Chinchilla lanigera*) kept in fur-farming systems (Franchi *et al.*, 2016). Other examples include South American camelids, crocodiles, iguanas, deer, capybara, among many others. Or management programmes for the use and conservation of wildlife, focusing on research topics related to the relationships between behaviour and reproduction, stress, health, etc. One example of these are the behavioural ecology studies in the endangered Huemul deer (*Hippocamelus bisulcus*) (Corti *et al.*, 2010) or the case of the tuna-dolphin issue in the tropical pacific ocean. Studies addressing other topics such as clinical ethology (Edwards *et al.*, 2007; Heiblum *et al.*, 2007), behaviour of pets or experimental animals, or human-animal interactions in different subjects have been limited to some groups.

Students of agriculture and biological sciences have shown increasing interest in applied ethology, both at undergraduate and graduate levels. This has created an opportunity to increase the critical mass of expertise within the region, and we already see an expansion of applied ethology groups as these students start to occupy teaching and research positions in the region. International, transcontinental collaborations and more access to research funding from federal governments in recent years is also reflected in greater scientific output from the region.

Considering the geographic size of Latin America, the economic and cultural relevance of agriculture for all its countries, and the vast wild fauna and growing interest in biodiversity, there is an enormous potential for research in applied ethology. More active national and international collaborations among different groups with common or complementary interests would allow the expansion of scientific research. Within these collaborations, teams focusing on different aspects of the same production systems (e.g. environmental, social or production) and collaboration with other stakeholders (e.g. farmers, farmer associations, conservation groups) are needed to truly tackle practical issues and contribute to the economic, social and environmental sustainability of animal agriculture.

10.4 Teaching of animal behaviour and welfare

The Pan-American Council for Education in Veterinary Sciences (COPEVET) was formed in 1999 due to the need to standardise the veterinary curricula among veterinary schools present in Latin America, as well as serving other needs related to education. The COPEVET proposed a curriculum for Latin American Veterinary Schools that included Animal Welfare and Behaviour courses during the fifth semester of the career (San Martin, 2003).

In 2012, the COPEVET proposed minimum competencies for veterinarians in Latin America, in which animal welfare was included. It is proposed that veterinarians need to assure animal welfare, for which decisions need to be based on science, especially in what relates to the fundamentals of animal behaviour (Vargas-Terán, 2012).

The recommendations of COPEVET are in accordance with the Performance of Veterinary Services (PVS) scheme of the World Organization for Animal Health (OIE). The OIE published in 2012 the 'PVS recommendations on the competencies of graduating veterinarians (Day 1 graduates) to assure national veterinary services of quality' (OIE, 2012), which includes animal welfare as one of these competencies. Later on, in 2013, the OIE published the 'Veterinary education core curriculum guidelines' outlining the required competencies, in which animal welfare and ethology is included as a core course that should be taken during the first half of the career (OIE, 2013).

In Latin America there has been a rise in the number of veterinary schools in the last twenty years, with over 400 schools across the region, with approximately 25% of these in Brazil. In 2014 a survey was sent to 379 veterinary schools, and from these, 100 schools situated in 14 countries replied (Argentina, Bolivia, Brazil, Chile, Colombia, Costa Rica, Cuba, Ecuador, El Salvador, Guatemala, México, Peru, Uruguay, Venezuela). The survey included a question about the inclusion of 'Ethology and behavioural motivational systems' as part of the curricula. Seventy-two out of the 100 schools declared the subject was addressed (Taylor *et al.*, 2014). On the other hand, 98 schools declared having an animal welfare course that includes concepts of ethology, while 26 schools had courses exclusively on ethology during the undergraduate programme. Two Universities (National Autonomous University of Mexico-UNAM and Universidad de Chile) indicated that postgraduate courses in ethology were taught (Applied Ethology and Biological basis of behaviour), (Taylor *et al.*, 2014).

In Mexico, in the late 1980s, Prof. Aline S. de Aluja from the Faculty of Veterinary Medicine and Animal Science at the National Autonomous University of Mexico (FMVZ-UNAM) organised four training courses on Applied Ethology and Animal Welfare taught by Donald M. Broom (University of Cambridge, UK), Temple Grandin (University of Colorado, USA), Gerrit van Putten ('Schoonoord' Institute of Animal Production, the Netherlands), and others, directed towards academic staff that would teach the new compulsory course on Ethology and Animal Welfare (AW) in 1993 (González-Rebeles and Galindo, 2003; Hewson *et al.*, 2005). As a way to promote an academic programme in the area, that same year, the Department of Ethology and Wildlife was created at

UNAM, making it the first institution in Mexico to run a compulsory course on AW and establishing one of the first academic groups in the continent. Other veterinary schools and animal science departments (Universidad Autónoma del Estado de Morelos, Universidad Autónoma Metropolitana, Universidad de Guadalajara, Facultad de Estudios Superiores Cuautitlán-UNAM, Universidad de Guadalajara, Universidad Veracruzana, etc.) have since followed suit and included a course on the topic. Most of the vet schools in the country now teach a course on animal welfare. This has led to the formation of a Network of Universities teaching AW that has had a strong impact academically.

In the case of Chile, the first university to include ethology in the curriculum was the Universidad Mayor in 2002 (Tadich *et al.*, 2010), a private university that currently also offers a Diploma in Clinical Ethology and Animal Welfare and a Master Degree in Ethology and Animal Welfare. In 2006 the Universidad Austral de Chile developed their first course in Ethology and Animal welfare (Tadich *et al.*, 2010). In the case of Chile, currently there are 32 veterinary schools that belong to 14 Faculties/Universities. Eight of the faculties have an ethology course, usually associated with animal welfare or zoology.

Increasing concerns about animal welfare in Latin America, including Chile, combined with welfare being a multidisciplinary science, encompassing ethology among other sciences (Huertas *et al.*, 2014), could explain why ethology has been included with animal welfare in most institutions. Ethology contributes to an integral understanding of animal welfare, especially through work highlighted by the ISAE in fields such as cognition, motivation and consciousness, among others (Molento and Calderón, 2009). This is why when ethology courses do not exist, including the ethology concepts in animal welfare courses becomes essential. Probably the main concern that still exists is the shortage of trained teachers in the field of ethology, who are capable of delivering these courses at all veterinary schools.

In Brazil the number of Agricultural Sciences programmes offering animal welfare courses in their curricula has grown rapidly (Borges *et al.*, 2013; Molento, 2008; Molento and Calderon, 2009). The first undergraduate program to offer an applied animal behaviour unit was the Agronomy degree at the University of Santa Catarina, in 1981 (Machado Filho and Hötzel, 2003). A recent survey by Borges *et al.* (2013) identified animal welfare courses in 46% of the 94 surveyed veterinary medicine degrees and applied ethology in 26%. Some limitations identified were an insufficient number of undergraduate and graduate programs offering applied ethology, animal welfare courses or related contents, and insufficient educational resources available in Portuguese (Molento, 2008; Poletto and Hötzel, 2012).

In the case of Peru, the first veterinary school to consider animal behaviour and welfare in the veterinary curriculum was the Alas Peruanas University. From its beginnings in 1997, the school offered animal behaviour and animal welfare as compulsory courses. The animal welfare course followed the guidelines provided by the World Society for the Protection of Animals (now called World Animal Protection). From this point on, some universities started to introduce animal behaviour and welfare as part of other courses within their curricula. Currently, only one university still offers animal welfare and animal behaviour as a stand-alone course, but there are five schools in total that include a compulsory course in animal behaviour and welfare in their curriculum, and one school that offers this as an elective. Meanwhile 10 universities offer compulsory courses in ethics and deontology, which may include animal welfare topics.

10.5 Research in basic and applied ethology

There has been an active research group at UNAM working on Behavioural Ecology since the early 1980s led by British born, behavioural ecologist Hugh Drummond. UNAM has helped to establish a network of academic groups doing basic research on animal behaviour, with Margarita Martínez and Gabriela González-Mariscal, from the University of Tlaxcala, doing research in reproductive behaviour and physiology and Australian born Robyn Hudson at the Instituto de Biomédicas-UNAM, very active now in topics related to individual differences in development, among others. The first research groups on applied ethology started at the vet school in UNAM, thanks to the initiative of pathologist Aline S. de Aluja, who promoted the creation of a research group on Animal Behaviour and Welfare, in collaboration with researchers including: Francisco Galindo, who studied in Cambridge, UK, with Don Broom, working on social behaviour of cattle and epidemiology, sustainable livestock production, and zoo animal behaviour; Anne Sisto, trained in Texas A&M with Ted Friend, mainly working on ruminant behaviour; Marcela González, who has been collaborating with research groups in Germany on dairy cattle behaviour and stress; and, Luis F. Rodarte, trained in the Colegio de Posgraduados in Mexico working on pig behaviour and welfare and at the Faculty of Agronomy at the University of Morelos Agustín Orihuela, and who has made several research stays with Ed Price and Joy Mench in the University of California-Davis, working on social and reproductive behaviour of cattle in the tropics. In 2000, at UNAM's vet school, a working group on clinical ethology, which included veterinarians Alberto Tejeda and Moisés Heiblum, trained in Barcelona and US respectively, has been more involved in research projects.

Other veterinary schools/groups in Mexico have been active doing applied research on pig behaviour and welfare and animal transport, such as Marilú Alonso, trained in the US and Sweden, and Daniel Mota, trained in Mexico, from the Universidad Autónoma Metropolitana (UAM-Xochimilco); Genaro Miranda, trained in Spain, from the UAM-Lerma, working mainly on transport and meat quality; Angélica Terrazas and Rosalba Soto from the FES-Cuautitlán-UNAM, doing research on maternal behaviour and welfare of small ruminants; and Apolo Carrasco, Bertha Hernández, Lorena de Buen from the Universidad Veracruzana working on cattle behaviour and welfare, wildlife use and conservation. In 1996, the Mexican Society for Veterinary Ethology, now Mexican Society for Ethology and Animal Welfare (SOMEBA) was formed and has been very active in outreach activities and certification of professionals in the area of Ethology and Animal Welfare.

In Chile, at the Veterinary Faculty of the Universidad Austral de Chile most research on applied ethology has been developed by the Animal Welfare Programme led by veterinarian Carmen Gallo since the 1990s with the focus on using animal based welfare indicators in beef cattle, dairy cattle and sheep (Gallo *et al.*, 2010). On the other hand, ethology is also an important research field at the 'Applied Research in Wildlife Programme' formed in 2011 by Paulo Corti, Claudio Verdugo, Gerardo Acosta from the same Faculty with the participation of biologists working on behavioural ecology. Their research is focused in endangered native species such as the pudu (*Pudu puda*) and huemul deers (*Hippocamelus bisulcus*) More recently other universities in Chile have carried out research on applied ethology such as the Universidad de Chile where Tamara Tadich, who was trained in Chile and Scotland, runs the animal welfare programme and is working on topics related to equine behaviour, stereotypic behaviours and behavioural measures of welfare in other farmed species. Rigoberto Solis, also at the Universidad de Chile, is working on behaviour and physiology of insects, arachnids and amphibians. The Veterinary Faculty of the Universidad Mayor has formed an Ethology and Animal Welfare Unit where Beatriz Zapata and Cristian Ugaz, trained in UNAM-Mexico are currently doing research on pig behaviour and environmental enrichment in zoo animals, and the Pontificia Universidad Católica de Chile where Jessica Gimpel, trained in the UK, is working with laboratory animals. In 2011 the Chilean Association of Veterinary Clinical Ethology was formed with the objective of bringing together veterinarians working on the field.

The Faculty of Veterinary Medicine in Montevideo, Uruguay is one of the most active institutions in South America doing basic and applied research on animal behaviour. Rodolfo Ungerfeld, trained in Sweden has been doing basic and applied research on social and reproductive behaviour and the physiology of domestic and wild ruminants and participates as an editor of international journals. Within the same Faculty, Stella Huertas, is running an Animal Welfare Programme and is doing research on applied ethology in beef and dairy cattle and equines. This programme has focused on the use of ethology as a tool for improving animal welfare.

Two groups started working with applied ethology in Brazil in the 1980s led by Mateus Paranhos da Costa at Universidade Estadual Paulista (UNESP) and Luiz Carlos Pinheiro Machado Filho at Universidade Federal de Santa Catarina (Tadich *et al.*, 2010). The research performed by the animal ethology and ecology group (ETCO) at the Universidade Estadual Paulista, led by Mateus Paranhos da Costa, looks at cattle and pig behaviour, mainly associated with animal welfare and production systems (Paranhos da Costa *et al.*, 2012). The UFSC group, LETA has done work with human-animal relationships in cattle and pigs (Hötzel *et al.*, 2005), and the interactions between social hierarchy and access to resources like shade and water in cattle (Coimbra *et al.*, 2012). A third group, led by Selene Siqueira da Cunha Nogueira and Sérgio Luiz Gama Nogueira Filho is at the Universidade Estadual de Santa Cruz in Ilheus, Bahia. They have done work with behavioural studies applied to the use and conservation of wild animals, social behaviour, animal communication and control of vertebrate pests (Nogueira and Nogueira, 2011). Also active in animal welfare is the group of Carla Forte Maiolino Molento in the Universidade Federal do Paraná. Carla has done work with broiler, fish and dog welfare, is interested in the teaching of animal welfare, and has been very active organising several national meetings on animal welfare and ethics through the Federal

Council of Veterinarians, CFMV; also one important contribution of Carla Molento has been the translation into Portuguese of several texts and books on animal welfare and ethics (Broom and Fraser, 2010). More recently Adroaldo Zanella, trained in the UK, joined the Faculty of Veterinary Medicine at the Universidade de São Paulo where he has coordinated the EU-Animal Welfare Indicators project (AWIN) and has formed a group doing basic and applied research on animal welfare assessment in multiple species.

In recent years several of these institutions (i.e. UNESP-Brazil, UFSC, UNAM-México, among others) have implemented graduate programmes on Applied Ethology and Animal Welfare with a wide variety of research projects and funding sources (National Councils for Science, NGO's, European Union, among others).

Networking and collaboration within the region has been quite strong. Some examples of this include the numerous training courses at different levels in the region and participation in regional research projects. Participants in the Welfare Quality project (EU) included Mateus Paranhos da Costa (Universidade Estadual Paulista, Brazil), Stella Huertas (Universidad de la República, Uruguay), Claus Köbrich (Universidad de Chile) and Francisco Galindo (Universidad Nacional Autónoma de México). Another collaborative project is the joint World Organisation for Animal Health (OIE) Collaborating Centre on Animal Welfare and Livestock Production Systems, integrated by the Universidad Austral in Chile (Carmen Gallo), the Universidad de la República in Uruguay (Stella Huertas), and the Universidad Nacional Autónoma de México. The role of the OIE collaborating centre is discussed in more detail below.

So far, research on applied ethology in Latin America has enriched academic programmes and has helped to provide better private and official veterinary services. Research, specifically on farm animals, has helped to develop good handling practices for producers in the private sector, policy development, improvements in animal welfare in production systems and knowledge transfer through capacity building for farmers, transport drivers and slaughter house workers (Paranhos da Costa *et al.*, 2012). These activities have also allowed development and improvement of legislation on animal welfare in many countries.

10.6 ISAE conferences in Latin America

There has been an active, although fluctuating, involvement of the region in ISAE meetings. Usually ISAE membership from Latin America has changed according to the number of graduate students doing PhD's in countries and research groups highly involved in the society. Membership and participation has increased in years when the annual ISAE conference has been hosted in Latin America (Figure 10.1). As of 2016, three ISAE conferences have been held in Latin America: the 34th conference in Florianopolis, Brazil in 2000, the 41st congress in Mérida, México in 2007, and the 47th meeting again in Florianopolis in 2013 (Figure 10.2 and 10.3).

When some of us first joined the society in the late 1980s and early 1990s it was still called the Society for Veterinary Ethology, and the Latin American region of the society had not

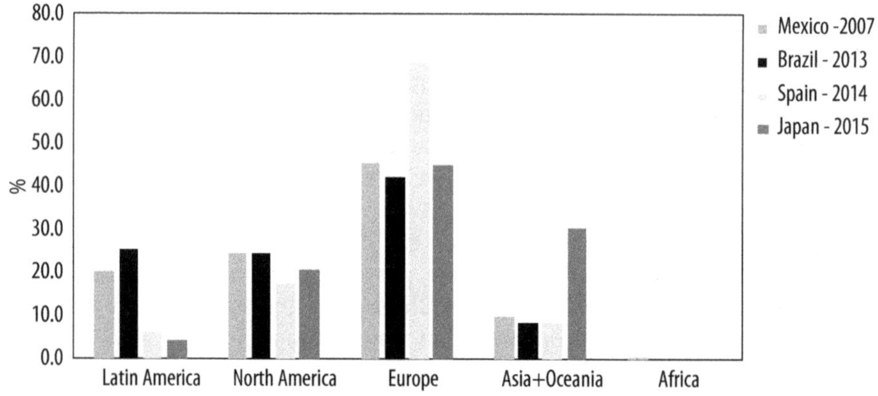

Figure 10.1. Total presentations from Latin America and other regions at ISAE conferences held in 2007 (Mexico), 2013 (Brazil), 2014 (Spain) and 2015 (Japan).

Figure 10.2. Mike Mendl, Don Broom, Adroaldo Zanella, Jeremy Marchant-Forde, and Francisco Galindo in the aftermath of the ISAE conference in Merida in 2007. The beers were hidden!

been formed. The first members from the region were connected to the Mediterranean region (probably as a sort of fraternity of the former colonies!) and it was precisely in 1991, after the meeting in Edinburgh when the society changed its name to International Society for Applied Ethology, that the Latin American region was formed. Soon after that it was decided that the conference would be hosted in Florianopolis in 2000 and the first secretary of the region was elected. This position was filled by Luiz Carlos Pinheiro Machado Filho (Universidade Federal de Santa Catarina, Brazil). Since then four other regional secretaries have been elected: Francisco Galindo (Universidad Nacional Autónoma de México), Rodolfo Ungerfeld (Universidad de la República, Uruguay),

Figure 10.3. Part of the organising committee on the last day of the conference in Florianopolis 2013.

Maria José Hötzel (Universidade Federal de Santa Catarina, Brazil) and Jessica Gimpel (Pontificia Universidad Catolica de Chile).

Several attempts to organise regional meetings have been made, but it has not been easy especially because of travel expenses in such a large region. However, in 2008 and in 2011 two ISAE regional meetings were held, the former organised by Rodolfo Ungerfeld in Uruguay with active participation from several countries and the latter in Ilhéus, Brazil, organised by Selene and Sergio Nogueira. As there are now so many different meetings on topics related to applied ethology and animal welfare in the region and globally, the possibility of having joint regional ISAE meetings in conjunction with other events in the region, such as the OIE researchers meeting, has been contemplated. Certainly, any efforts to reduce costs for students and minimise travel would help towards having more regional meetings.

10.7 World Organisation for Animal Health (OIE) actions in the Americas: opportunities for applied ethology

10.7.1 OIE Collaborating Centre for Animal Welfare and Livestock Production Systems

In 2009 the OIE Collaborating Centre for Animal Welfare Research in the Americas was established though a bipartite initiative between Chile and Uruguay (Faculty of Veterinary Sciences of the Austral University of Chile, and the Faculty of Veterinary

Medicine of Uruguay's University of the Republic). In 2013, the Faculty of Veterinary Medicine and Husbandry of the National Autonomous University of Mexico (UNAM) was incorporated into the existing Collaborating Centre, in order to address the needs of all three Americas, at which point it was renamed the OIE Collaborating Centre for animal welfare and livestock production systems. This is one of four current OIE collaborating centres worldwide on animal welfare.

The Collaborating Centre's main aim is to promote research on animal welfare, with emphasis on farm animals and livestock in production systems specific to the Americas, and to serve as a reference group to promote OIE standards and guidelines within the region. The tasks assigned to this centre are to carry out research, provide advice on animal welfare issues, disseminate management methods and techniques that contribute to animal welfare, propose and develop methods to facilitate the harmonisation of international standards on animal welfare, provide scientific and technical training on animal welfare to other countries as required, and to organise scientific meetings and coordinate studies in collaboration with other institutions or organisations.

The Collaborating Centre works closely with Official OIE delegates and in particular with national Focal Points for animal welfare in the Veterinary Services of countries in the Americas. It also works with other public and private institutions in the food and agriculture sector and NGOs in countries of the region, promoting improvements in animal welfare and the sector's economy, with special emphasis on training for stakeholders.

10.7.2 OIE Regional Animal Welfare Strategy for the Americas

In 2012, the 'Regional Animal Welfare Strategy for the Americas' was adopted. Among other things, this strategy sets a framework for achieving results related to animal welfare, based on scientific evidence and knowledge. Thus, it recognises the importance of education, training and research on animal behaviour and welfare. It also promotes the implementation of OIE standards, promoting regional research as a way to bring about improvement in animal welfare and to build partnerships.

This strategy recognises the relationship between animal welfare, animal health, environment, production and food safety and security, and acknowledges the need to take into account the region's distinctive characteristics. The strategy is not limited to farm animals, as it includes the welfare of animals not destined for trade and food production, such as wild animals, companion, work, and laboratory animals, and those used for recreation, among others.

The strategy is aimed at a broad spectrum of individuals and organisations in the region, including those responsible for animals, animal users, veterinary professionals and those working in related disciplines, such as agricultural sciences and animal science, research institutions, livestock farmers, processors, industry and transport operators, conservation and animal protection organisations, national veterinary associations, academics, researchers, consumers and other governmental and nongovernmental organisations.

10.8 Conclusions

During the past 25 years, and especially during the past decade, there has been a rapid expansion of academic programmes on applied ethology in many Latin American countries. There has been much involvement of some NGO's promoting teaching materials and training courses, especially in institutions where academic staff are not trained in this area. There is much more to be done to improve how courses in behaviour are taught and integrated, and also to change how behaviour and welfare are perceived within some sectors of the academic community. Strengthening graduate programmes in the region and having academics specifically trained in applied ethology will help in this regard.

Research has also expanded, moving from a Eurocentric perspective to a more integral vision of Latin America, considering local perceptions and needs. Although most research has been carried out in farm animals the number of projects and research groups working with other groups of animals (companion, working equines, wildlife, laboratory animals) has increased considerably. More efforts are needed to support research and the development of policies in Latin America with a sound scientific basis. In this sense, the OIE regional strategy on animal welfare will be a useful tool to strengthen networking and promote collaborative research on applied ethology. It is important to raise the profile of the ISAE globally and at the same time to raise awareness among decision makers and intergovernmental organisations, such as the OIE or FAO, of the work that the ISAE is doing and of the role of ethology in science-based policies. Animal behaviour studies will be a necessary tool for the development of sustainable livestock production systems, for better legislation on animal ownership and schemes for wildlife conservation, and also for efficient social programmes that are based on human-animal interactions.

References

Borges, T.D., Sans, E.C.O., Braga, J.S., Machado, M.F. and Molento, C.F.M., 2013. Teaching pain and animal welfare in Veterinary Medicine courses in Brazil. Arquivo Brasileiro de Medicina Veterinaria e Zootecnia 65: 29-36.

Broom, D.M. and Fraser, A.F., 2010. Comportamento e bem-estar de animais domésticos. Manole, Barueri – SP, Brazil.

Broom, D.M., Galindo, F.A. Galindo and Murgueitio, E., 2013. Sustainable, efficient livestock production with high biodiversity and good welfare for animals. Proceedings of the Royal Society B 280: 20132025.

Coimbra, P.A.D., Pinheiro Machado Filho, L.C. and Hötzel, M.J., 2012. Effects of social dominance, water trough location and shade availability on drinking behaviour of cows on pasture. Applied Animal Behavior Science 139: 175-182.

Corti, P., Wittmer, H.U. and Festa-Bianchet, M., 2010. Dynamics of a small population of endangered huemul deer (*Hippocamelus bisulcus*) in Chilean Patagonia. Journal of Mammology 91: 690-692.

Edwards, C., Heiblum, M., Tejeda, A. and Galindo, F., 2007. Experimental evaluation of attachment behaviors in owned cats. Journal of Veterinary Behavior Clinical Applications and Research 2(4): 119-125.

Food and Agriculture Organisation (FAO), 2012. Livestock production in Latin America and the Caribbean. Available at: http://tinyurl.com/hzlrg8z.

Franchi, V., Aleuy, O.A. and Tadich, T., 2016. Fur chewing and other abnormal repetitive behaviors in chinchillas (*Chinchilla lanígera*), under commercial fur-farming conditions. Journal of Veterinary Behavior: Clinical Applications and Research 11: 60-64.

Galindo, F., Williams, D., Zarza, H., Avila, R. and Suzán, G., 2015. Sustainable livestock production as a tool for conservation in the neotropics: a case study in Yucatán, México. Oxford University Press, Oxford, UK.

Gallo, C., Lizondo, G. and Knowles, T.G., 2003. Effect of journey and lairage time on steers transported to slaughter in Chile. Veterinary Record 152: 361-364.

Gallo, C., Tadich, T., Huertas, S., César, D., Paranhos Da Costa, M. and Broom, D.M., 2010. Animal welfare education in Latin America. Proceedings of the International Conference on Animal Welfare Education: Everyone is responsible. October 1-2, 2010. Brussels, Belgium.

González-Rebeles, C. and Galindo, F., 2003. Departamento de etología, fauna silvestre y animales de laboratorio. In: Quiroz, R.H. and Cervantes, J.M. (eds.) Historia de la facultad de medicina veterinaria y zootecnia de la UNAM, 1853-2003. Facultad de Medicina Veterinaria y Zootecnia, Universidad Nacional Autónoma de México, Mexico, Mexico.

Heiblum, M., Labastida, R., Chavez-Gris, G. and Tejeda, A., 2007. 'Didy' a clinical case of cognitive dysfunction syndrome. Journal of Veterinary Behavior 2: 68-72.

Hewson, C., Baranyiová, E., Broom, D.M., Cockram, M.S., Galindo, F., Hanlon, A.J., Hänninen, L., Lexer, D., Mellor, D.J., Stookey, J., Waldau, P., Ödberg, F. and Serpell, J.A., 2005. Approaches to teaching animal welfare at thirteen veterinary schools in Europe, North America and South America. Journal of Veterinary Medical Education 32: 422-437.

Hötzel, M.J., Machado, L.C.P., Yunes, M.C. and Da Silveira, M., 2005. An aversive milker causes fear, but does not influence milk yield of Holstein cows. Brazilian Journal of Animal Science 34: 1278-1284.

Huertas, S., Gallo, C. and Galindo, F., 2014. Drivers of animal welfare policy in the Americas. Scientific and Technical Review of the Office International des Epizooties 33: 67-76.

Macfadyen, S., Cunningham, S.A., Costamagna, A.C. and Schellhorn, N.A., 2012. Managing ecosystem services and biodiversity conservation in agricultural landscapes: are the solutions the same ? Journal of Applied Ecology 49: 690-694.

Machado Filho, L.C.P. and Hötzel, M.J., 2003. Etologia aplicada. In: Del-Claro, K. and Prezoto, F. (eds.) As distintas faces do comportamento animal. Livraria Conceito, São Paulo, Brazil, pp. 246-253.

Molento, C.F.M., 2008. Ensino de bem-estar animal nos cursos de Medicina Veterinária e Zootecnia. Ciência Veterinária nos Trópicos 11: 6-12.

Molento, C.F.M. and Calderón, N., 2009. Essential directions for teaching animal welfare in South America. Scientific and Technical Review of the Office International des Epizooties 28: 617-625.

Navas, A., 2010. Importancia de los sistemas silvopastoriles en la producción del esters caloric en sistemas de producción ganadera tropical. Revue de Médecine Vétérinaire 19: 113-122.

Nogueira, S. and Nogueira Filho, S.L.G., 2011. Wildlife farming: an alternative to unsustainable hunting and deforestation in neotropical forests? Biodiversity and Conservation 20: 1385-1397.

Organización Mundial de Sanidad Animal (OIE), 2012. Recomendaciones de la OIE sobre las competencias mínimas que se esperan de los veterinaries recién licenciados para garantizar Servicios Veterinarios Nacionales de Calidad. Proceso PVS. Organización Mundial para la Sanidad Animal.

Organización Mundial de Sanidad Animal (OIE), 2013. Veterinary education core curriculum. OIE Guidelines. PVS Process. World Organisation for Animal Health.

Paranhos da Costa, M.J.R., Huertas, S., Gallo, C. and Dalla Costa, O., 2012. Strategies to promote farm animal welfare in Latin America and their effects on carcass and meat quality traits. Meat Science 92: 221-226.

Pifarré, M., Valdés, R., González-Rebeles, C., Romano, M. and Galindo, F., 2012. The effect of zoo visitors on the behaviour and faecal cortisol of the Mexican wolf (*Canis lupus baileyi*). Applied Animal Behaviour Science 136: 57-62.

Poletto, R. and Hötzel, M.J., 2012. The five freedoms in the global animal agriculture market: challenges and achievements as opportunities. Animal Frontiers 2: 22-30.

San Martín, F., 2003. Homologación de planes de studio de la carrera de medicina veterinaria en latinoamérica. Revista de Investigaciones Veterinarias del Perú 14: 178-180.

Tadich, N., Gallo, C., Brito, M.L. and Broom, D.M., 2009. Effect of weaning and 48h transport by road and ferry on some blood indicators of welfare in lambs. Livestock Science 121: 132-136.

Tadich, N., Molento, C.F.M. and Gallo, C., 2010. Teaching animal welfare in some veterinary schools in Latin America. Journal of Veterinary Medical Education 37: 69-73.

Taylor, J.J., Ruvalcaba, S., Michel, H., Taylor, A., Angueano, R. and Estrada, G.S., 2014. Bienestar animal y educación veterinaria en latinoamerica. Diagnóstico, pp. 75-96.

Vargas-Terán, M., Vidal, M. and Del Barrio, L., 2012. Notas para un medico veterinario del 2020 en el continente Americano. In: Taylor, J.J. (ed.) Competencias profesionales en medicina veterinaria. Asociación Panamericana de Ciencias Veterinarias (PANVET).

11. Applied animal behaviour science in Japan and the culture of '*aigo*'

S. Sato
Faculty of Life and Environmental Sciences, Teikyo University of Science, 2-2-1 Senju-Sakuragi, Adachi, Tokyo 120-0045, Japan; shusato@ntu.ac.jp

Abstract

Under the rice-eating culture, Japanese people practised the concept of '*aigo*' which means 'loving, not killing, and protecting animals'. Japanese culture has a unique perspective on animal welfare based on the influence of religious philosophies such as the indigenous *Shinto* religion, Buddhism, and Confucianism and the concept of *aigo* ethics. This characteristic cultural orientation towards animals has resulted in the applied behaviour sciences developing a little differently in Japan than in Western countries. Initially, applied animal behaviour science developed to study free-ranging horses and cattle with the objective of establishing guidelines for free-ranging animal production. After the establishment of the Agricultural Basic Law of 1961 having selectively facilitated animal production in Japan, behavioural problems in farm animals in indoor rearing systems became prominent, and during this time studies on social behaviour also developed with the introduction of applied ethology. Applied animal behaviour research became more structured after 1984, and the development of textbooks in Japanese provided the foundation for the prosperous development of applied animal behaviour science. In 2002, the Japanese Society for Applied Animal Behaviour Science was established, and with the international congress of ISAE held in Japan in 2005, the third stage of behaviour research began. The Japanese Society for Applied Animal Behaviour Science will continue to play an important role in '*aigo*' ethics and the adoption of animal welfare concepts, including greater respect for wild animals and fishes.

Keywords: *aigo*, religion, animal welfare, behaviour, Japan

11.1 Introduction

In Japan, the use of livestock for food became popular only after establishment of the Agricultural Basic Law in 1961. Previously, only one or two cattle or horses were reared for draft and transportation on each farm, and farmers held strong psychological attachments to their animals. The Japanese did not generally kill farm animals, but killed only wild animals and fish for eating. Farm animals were reared in barns, while breeding stock were raised extensively in the communal native grassland and forest.

Under the rice-eating culture, Japanese people practised the concept of '*aigo*' which means 'loving, not killing, and protecting animals'. This is a different concept from that of animal welfare. In Japan, ethics regarding animals has developed to emphasise the

DOI 10.3920/978-90-8686-828-5_11, © Wageningen Academic Publishers 2016

motivation to respect animals and not kill them, rather than from a belief in the need to protect the welfare of animals. Whereas the animal is the subject in animal welfare ethics, the motivation of attachment to animals in humans is the focal point of *aigo* ethics. The animal is regarded as 'sentient' in animal welfare ethics, and 'living' in *aigo* ethics. While the purpose of animal welfare ethics is to foster good conditions for animals; in *aigo* ethics, it involves encouraging sympathy and respect for the animals (The Act on *Aigo* and Management of Animals in Japan, enforced in 1973; revised 2012). Therefore, in the context of *aigo*, animals must not be killed before they naturally die.

This results in confusion for consumers regarding the welfare of farmed animals. The public have difficulty considering the welfare of farmed animals when they are ultimately killed for food, as this goes against the ethics of *aigo*.

This characteristic cultural orientation towards animals has resulted in the applied behaviour sciences developing a little differently in Japan than in Western countries. This can be understood by looking at differences in the development of applied animal behaviour science from the views of culture, science history, and research subjects. These themes will be discussed within this chapter, and it concludes with some predictions on how applied animal behaviour science will develop in the future in Japan.

11.2 Characteristics of respect towards animals (*aigo* ethics) in Japan

Some key distinctions between animal welfare and *aigo* ethics mentioned above are discussed here. It is believed that the Japanese respect for animals, *aigo*, was formed by combining the ethics of the indigenous *Shinto* religion, Buddhism, and Confucianism; each of these belief systems encourages people to take care of and value living things. The basis of these beliefs is described briefly here.

11.2.1 The indigenous culture (*Shinto* religion)

There are several thousand shrines in Japan, which we regularly visit on New Year's Day, during marriage ceremonies, festivals, and so on. The 'Eternals' in the *Shinto* shrines are not absolute deities, such as in the Christian and Islamic religions, but various abstractions, such as ancestors, big trees or rocks, mountainous areas, and so on. Kubota (1997) argued that the Eternals in *Shinto* religion are those in the atmosphere that represent some components of nature; these are derived from the polytheism of the rich natural areas of Japan's warm and wet climate. The myth says that three gods arose from chaos. The first one is the god governing the whole world, while the rest are gods who produce everything in the world. '*Musu*' and '*Hi*' contained in the names of the second and the third gods mean 'to connect' and 'to produce', respectively. Moreover, gods in Japan are seen as auras having the power of connecting one life with another and produce new life. There is no conflict between humans and nature in the *Shinto* religion, which has a worldview of harmonious coexistence derived from the concept that humans and nature are the same as the other component parts of the ecosystem.

Nakamura (1984) investigated how the Japanese regard animals by comparing the transfigurations in the Grimm fairy tales and Japanese folk stories. Of the transfigurations in the Grimm fairy tales, 92% were cases of transforming humans into animals as a means of alienation by devils or witches. On the other hand, these accounted for only 13% of the transfigurations within Japanese folk stories. Of the transfigurations in Japanese folk stories, 69% were cases of transforming animals to humans, who had an entirely human nature. He concluded that the Japanese do not have a tendency to regard animals as lower-class form of life, and that Japanese have an affinity towards animals without any conflict of interest.

Japanese culture has an ideological denial of speciesism, without distinction between the lives of animals and plants, rather than a shallow but scientific denial of speciesism, as advocated by Singer (1976), where the distinction is made between sentient and insentient beings.

11.2.2 Buddhism

Buddhism was introduced into Japan in AD 538 and became the established religion in Japan in AD 604. Buddhism was of central importance to the old imperial dynasty and in the warrior age. Buddhism was the first religion prohibiting the killing of living things, and prohibiting not only humans but also animals (and sacrificial animals) from being killed, including those such as deer, boars, and rabbits, which were popular to sacrifice in the indigenous culture of Japan (Harada, 2014). Buddhism was readily adopted by Japan's agricultural cultivators; as horticultural products, fish, and wild animals were abundant, and husbandry of domestic animals was not needed. As there is a rank of animals eaten, from fishes to bipeds, and tetrapods, in that order, the Japanese still experience a conflict between the prohibition on killing animals and eating animals, especially for tetrapods. Fishes are still not included within the Act on *Aigo* and Management of Animals in Japan, including whales and dolphins (Nisbett (2003) says that whereas Westerners emphasise differences, Asians emphasise similarity within their epistemic logic).

Cattle were first imported from the continent and used for work rather than food production during the Yayoi-style stage of culture, around the first and second centuries BC. At this time, cattle were reared mainly on the rice paddy areas of western Japan, and the Yamato administration required that farmers keep one cow for every two hectares for draft use from the fourth to the sixth century. At this time, cattle and horses were important for farming, transportation, and military affairs. There was one cow for every several farmers, even into the Edo era, around the nineteenth century. There were no fundamental changes in the management of animals for several centuries, with each farmer raising one or two cows for the purposes of farming labour and manure production, until the ratification of the Agricultural Basic Law of 1961. This law was responsible for establishing government policy on agriculture, in which animal production had been selectively facilitated in Japan.

The earlier traditional food producing system, prohibiting killing of farmed animals and recommending the release of captive wild animals into nature, may have been part of the cultural foundation for accepting Buddhism. However, prior to the development of animal production systems which began in 1961, people still ate meat, but this came from wild animals rather than farmed animals. It was considered necessary to resolve the conflict between the prohibition on killing living beings, while accepting the killing of wild animals. The Suwa shrine resolved this conflict by giving the indulgence '*kajikimen*' (permission to eat meat of wild animals) to the public, and advocating the doctrine of '*suwa-no-kanmon*' (the report on religious aspects of eating animals); the belief that animals could only attain Buddhahood as components of humans ingesting their meats. Thus killing and eating animals while giving thankfulness to the animals was considered supportive to animals, and ensured compatibility between killing and compassion. Wild animals have also been left out of the Act on *Aigo* and Management of Animals in Japan.

As Japanese are still suffering conflicts about killing farm animals under *aigo* ethics, concepts such as '*suwa-no-kanmon*' are still used in the fundamental education of school children in Japan when teaching the compatibility of killing and *aigo*. As a result, the public generally do not spend much time thinking about farm animal welfare.

11.2.3 Confucianism

Kellert (1996) suggested that the Japanese prefer nature to be managed and modified by humans rather than nature retaining its wild character. The animals and the landscape protected by the Japanese are restricted to artificial naturalness. Kawai (1995) described the species of animals appearing in the nursery tales by a still popular Japanese writer, Kenji Miyazawa (1896-1933), to discuss the image of animals in Japanese culture. Mammals were very popular in his tales, and especially domestic animals, such as horses, cats, cattle, dogs, goats, and sheep. Chickens were also popular, as well as wild animals, including foxes, mice and rats, raccoons, deer, and rabbits. Kawai concluded that the Japanese consider the domestic and wild animals living around their domestic dwellings as peers. Though nature itself is protected, the managed and sophisticated form of nature that is especially revered by the Japanese may be derived from Confucianism. Kaji (1990) explained that nature itself has a rude character and the artificial world made sophisticated by ethics is 'right' in Confucianism. Thus, the Japanese prefer to see an enriched system for farm animals within a controlled environment (and with little in the way of behavioural restrictions), rather than extensive outdoor grazing in harsh weather.

Death is the condition separating the soul and the body in Confucianism. The Japanese do rites for ancestors to invite their souls from the next world into this world, to incorporate the soul and the body of the deceased in this world, and to acquire immortality. '*Kou*' is the ethical sense in Confucianism of respecting ancestors and elderly peoples, and leaving offspring, which is extended to pet animals, raccoons living near houses, and one or two farm animals being reared by a farmer, as well as the act of building memorials for dead animals. Yoda (2007) reported that 71% of animal experimenters in Japan feel a sense of guilt and resistance after experiments and that 41% of them perform a memorial service to avoid the sense of guilt and resistance. This tradition of '*Irei*' (conciliating souls of dead humans and animals) seems to be derived from a mixture of ethics from animism, the *Shinto* religion, Buddhism, Confucianism, and Taoism. Not to hurt, not to cause suffering, and not to kill living things, to conciliate the mystic force of the soul, and to invite the soul from the next world when living things die is the attitude of the Japanese. There are many memorial monuments across the country (Figure 11.1) for experimental animals, hunted animals, fur and draft animals, pets, food animals, and even weed plants.

The word '*aigo*' was firstly used in 1908 to describe an integrated ethics regarding animals based on the religions and philosophies cited above. Under *aigo* ethics, animal welfare has been studied mainly in zoo animals and pets, rather than farm animals, wild animals and fishes (including whales and dolphins). Behavioural studies on farm animals have been biased towards animal management and grazing from the perspective of production instead of animal welfare. In addition, behavioural studies have been done on the widespread phenomenon of wild animals such as deer, wild boar, monkeys and crows invading dwellings and cultivated lands. This is due to the dramatic increase in wild animals which resulted from them having not been regarded as food or culled after the establishment of the Agricultural Basic Law in 1961. As animal production has developed rapidly since the establishment of the Agricultural Basic Law, sources of protein foods in Japan have changed from wild animals to domestic animals. As well, the cultural view on wild animals has changed from 'culling things' to 'preserving species'.

Figure 11.1. Various memorial monuments such as for horses, fishes, cats, foodstuff, and weed plants in Japan.

11.3 The emergence of farm animal behaviour research in Japan

The first applied studies in animal behaviour looked at the home range and tracks of horses in the forests of Kemanai and Hanawa. This was published by the Akita regional forestry office in 1939, during the China-Japan war to investigate a minimum free-ranging area required for rearing of horses. As the national forestry office disliked using the national forests for grazing, it tried to restrict the use of these lands for grazing. Up until the 1970s, these early applied studies in farm animal behaviour were done by experts in forestry. The behaviour of two groups of horses, raised for the army, that were turned out into two separated fields was studied. The fields, located in northern Japan, consisted of about 800 and 200 ha, with each containing forest and grassland. The behaviours of lead horses in this semi-natural environment were focally sampled continuously in each of two groups of 10 and 14 heads. In addition to behaviour observations, the weather conditions, migration pathways, topography, vegetation, plant species, and parts eaten were recorded for 21 days from 02:00-23:00. Although this was the first report on applied animal behaviour in Japan, it was not investigated systematically during the war. After World War II, one of the most important national policies was the expansion of food production, in which animal production was promoted in the natural grasslands. Research in this area was conducted by Kurosaki *et al.* (1956), who studied cattle behaviour in semi-native pasture consisting of Chinese silvergrass (*Miscanthus sinensis*), Japanese lawngrass (*Zoysia japonica*), and the broadleaf tree forests, also in northern Japan. Kurosaki was based in the laboratory of animal reproduction in Tohoku University and was one of the first to study egg transfer techniques in farm animals in Japan, after this he was transferred to the university farm and began doing cattle behaviour research by himself. He is the founder of farm animal behaviour science in Japan from the view of animal production. At the same time, Imanishi and Kawai (1955) started to investigate a group of about 100 feral horses living on a peninsula of 473 ha consisting of Japanese lawngrass and evergreen broadleaf forest in southern Japan. Though Imanishi is the founder of primatology in Japan (Matsuzawa and McGrew, 2008), he firstly studied feral horses in Mongolia and Japan. His observation style was based on individual identification and recording of social interactions through prolonged and continuous observations. These observational styles of Kurosaki and Imanishi have strongly influenced the subsequent study of behaviour in Japan.

In addition, books on ethology, such as '*Social Behaviour in Animals*' (Tinbergen, 1953), '*King Solomon's Ring*' (Lorenz, 1960) and '*The Study of Instinct*' (Tinbergen, 1951) were first translated into Japanese by Watanabe *et al.* (1955), Hidaka (1963), and Nagano (1975), respectively. I still remember the initial shock I experienced from reading these books on ethology as a university student in 1969, because of this very new discipline of biology. It confirmed that behaviour can also be analysed by the methods of natural science. After this I began to study free-ranging cattle behaviour with self-taught knowledge in 1973, based on individual identification, and recording of social interactions through prolonged observations. Observations based on individual identification caused me to develop an empathy for cattle which then grew into an interest in the study of animal welfare.

Farm animal behaviour science in Japan, led by Mimura (1988) and Kurosaki, developed initially without the influence of ethology. Mimura studied especially the thermal environment of farm animals, and became the founder of animal management science in Japan. He recognised the importance of behaviour management in addition to managing the thermal environment of animals in indoor housing systems. In 1984, the research group on farm animal behaviour was established, in which Sato (1985) and Kondo (1985) advocated the importance of applying ethological methods to studies on farm animal behaviour. This included establishing ethograms for each farm animal species, investigating mechanisms and functions of behaviour, and utilising specific behaviours for animal management, such as control of the rearing environment or selection of useful behavioural characteristics in farm animals. Farm animal behaviour has developed as one of the disciplines in animal management science for promoting productivity in Japan. Examples of this work include gentle handling of calves for improving tractability (Sato *et al.*, 1984), studies on the size and design of paddocks and of group size for promoting exercise in race horses (Kusunose *et al.*, 1985), and methods for suppressing agonistic behaviour in cattle (Kondo *et al.*, 1989). Sato *et al.* (1991) presented the first paper from Japan at the Society for Veterinary Ethology at the 25th anniversary in Edinburgh, in which the society changed its name to the International Society for Applied Ethology. Sato (1991), soon after coming back to Japan, excitedly introduced this international congress, developing horizontally from farm animals to pets, zoo animals, experimental animals, and wild animals, and vertically from pure ethology towards greater cooperation with physiology and psychology.

Up until 1984, applied animal behaviour science developed based on the observational styles of Kurosaki and Imanishi. More typical methods of ethology were adopted after that; in particular, Tinbergen's four questions related to the causation, function, development, and evolution of behaviour. The research group on farm animal behaviour held 24 annual meetings from 1984 to 2001, during which time it produced books on the behavioural ethograms of farm animals and a textbook of farm animal behaviour science. Three criteria were used to identify key behaviours in these books: (1) the behaviour must be identified as functional; (2) function is divided into two aspects: maintenance and reproduction; and (3) there is a hierarchical structure for naming. The results were detailed in two books, '*Ethogram of Farm Animals*' (ed. by Sato *et al.*, 1995) and the text book, '*Farm Animal Behaviour*' (revised edition, ed. by Mimura, 1997) published in Japanese.

11.4 Main themes of applied animal behaviour research in Japan

In 2002, the research group on farm animal behaviour became the Japanese Society for Applied Animal Behaviour Science (the first president: Shusuke Sato) and the 39th meeting of the ISAE was held in Japan in 2005. The society dealt not only with farm animals but also with pets, zoo animals, and wild pest animals as the ISAE enlarged the study field in 1991. Now, there are over 200 members in this society, who are well educated in animal science and biology in universities in Japan. The papers presented at the annual meeting of the Japanese Society for Applied Animal Behaviour Science from 2005 to 2015, categorised by research species are listed in Figure 11.2. The most common theme involved studies on housing facilities in cattle, such as stall designs, feeding stations, and feeders to promote resting and feeding behaviours, and to investigate their accessibility to cattle. Another important research theme was behavioural control in cattle and pig production, such as physiological control of temperament in cattle and control of tail-biting in pigs. Other studies looked at the control of defecation sites in cattle (Saito *et al.*, 2009) and improving learning in cattle to control breakout from an electric fence during grazing (Fukasawa *et al.*, 2008). The automatic control of management in farm animal production and the psychological relationship between humans and farm and pet animals, including animal-assisted education, was the third most common theme. Until 2014, the study of cattle behaviour was prominent in applied animal behaviour research in Japan. However, the focus on farm animal behaviour has declined gradually, and in 2015 zoo animals replaced cattle as the main subject of research. This decline may be due to the fact that animal production is predicted to decrease by half following implementation of the Trans-Pacific Partnership (TPP) agreement of 2015. Past studies on cattle have mainly consisted of production oriented studies on grazing, free-stalls, and automated management, which have been drastically decreased.

More recently, animal welfare studies have focused on enrichment in confined rearing systems (Ninomiya and Sato, 2009; Shimmura *et al.*, 2010a), temperament (Tachi *et al.*, 2014), sleep as a welfare indicator (Ninomiya *et al.*, 2006), stress control during birth

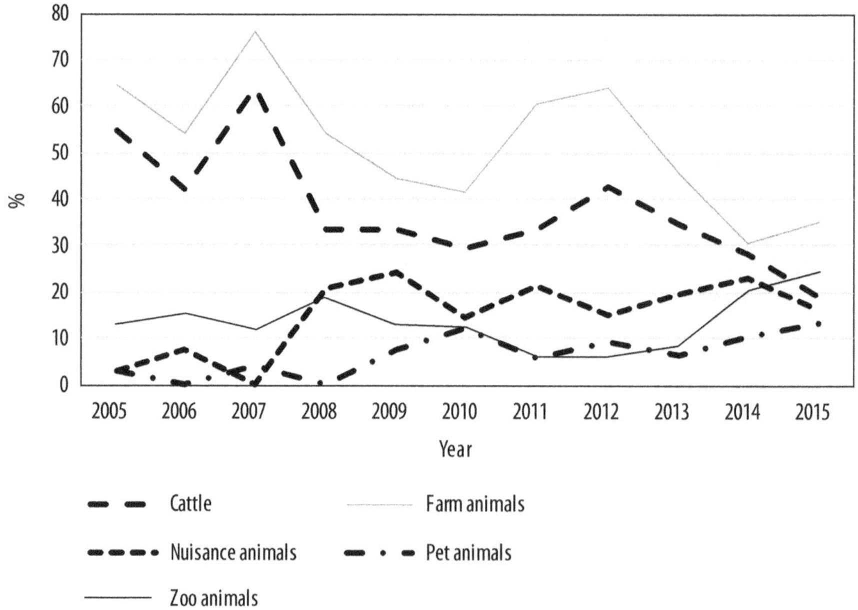

Figure 11.2. Percentages of the papers categorised by research object animals presented at the annual meeting of the Japanese Society for Applied Animal Behaviour Science from 2005 to 2015.

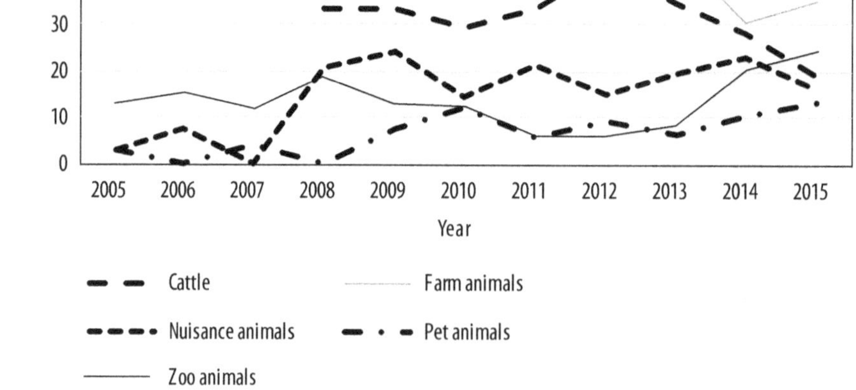

and weaning (Chen *et al.*, 2015; Kohari *et al.*, 2014; Shimmura *et al.*, 2010b; Yayou *et al.*, 2015), and stress during transport and slaughter (Uetake *et al.*, 2011). In addition, there are studies on sexual behaviour and social network analysis. Most of the zoo animal studies are related to enrichment, and pet studies have focused on human-animal relationships, stress, training, and animal-assisted education. Nuisance studies are on indigenous wild boars, deer, and crows and exotic animals, such as the masked palm civet (*Paguma larvata*).

Farm animal studies will develop further only if new rearing systems are developed. In conjunction with this, it will be necessary to develop new aspects of farm animal research, such as epigenetics and behavioural genetics. There has been increasing interest in this area, and it is more attractive and relevant to young scientists compared to older methodologies. As zoo and pet animals diversify and nuisance problems increase, studies on these animals will develop further. As the role of pet animals is qualitatively changed to companion animals, with greater emphasis on the human-animal bond, this line of study will also develop further.

11.5 The future of applied ethology in Japan

In the coming years there will be an increasing need for behaviour research in Japan due to the globalisation of trade in animal products and the declining birth-rate and growing proportion of elderly people. Farm animal behaviour science will also be needed to establish animal welfare-friendly production systems in Japan. As Japan experiences high temperature and humidity in summer and very low temperature and dry conditions in winter, welfare-friendly production systems must be designed according to these weather conditions. Pet and wild animal behaviour science will be needed to develop a more symbiotic relationship between humans, their pets, and wild animals.

11.5.1 Leading animal welfare research

While little animal welfare research has been conducted with farm animals in Japan, problems related to farm animal welfare may become increasingly important in the coming years due to global trade in animal products. Japanese consumers may be influenced by welfare-friendly animal products from abroad. In addition, the Japanese government is encouraging the export of animal products, such as the Kobe beef. The International Organization for Standardization (ISO) is working to establish Technical Specifications for animals in the food supply chain, which meets the standards of the Terrestrial Animal Health Code of World Organisation for Animal Health (OIE). This OIE code indicates that the welfare of animals should be assessed using outcome-based measures, and that these indicators and appropriate thresholds should be adapted to the different situations where animals are managed (http://tinyurl.com/hft4k8w). These requirements will necessitate a greater focus on farm animal welfare research in Japan, to determine and ensure that production systems can meet the required standards. These standards will be adopted in response to pressure from foreign trading partners, not in response to pressure from domestic consumers. Ensuring compliance would be basically the responsibility of the Japanese agricultural industries which will need the support of government and scientists to determine appropriate thresholds for animal-based measures, because these must be established based on scientific evidence and peer reviewed articles.

The Japanese are required to love animals under the Act on *Aigo* and Management of Animals, established in 1973 and revised in 1999 and every five years after. Article 1 of the act says, 'The purpose of this act is to engender a spirit for *aigo* among citizens and contribute to the development of a respect for life and sentiments of amity and peace by providing for the prevention of cruelty to animals'. Thus, there is an incentive to maintain good animal welfare not only in pets, but also in farm animals, wild animals, and fishes. For Japan to advance with respect to animal welfare, it must be understood that improving the welfare of animals involves the practice of *aigo*, and that not only pet and zoo animals, but also farmed animals, wild animals and fishes must be loved. Even so, the Japanese give importance to life even in suffering and do not defend freedom to perform normal behaviour. Although applied animal behaviour scientists in Japan consider it very important to allow animals to perform normal behaviour, the government document, 'The Care and Handling Guideline for Farm Animal Welfare in Japan', established in 2011, recognises that it is difficult to maintain the freedom to perform normal behaviour in animal production systems, especially considering the lack of support among consumers. Japanese scientists must therefore continue to investigate the motivation of animals to perform normal behaviour for the development of future animal production systems in Japan.

11.5.2 Establishing new technologies in animal production

After the huge earthquake and the following nuclear accident in Japan in 2011, traditional animal production systems were largely destroyed. Two development options emerged: large scale animal production systems for global supply, and ecological and sustainable ones for local supply. The development of both these systems in Japan has created a need for increased study of applied animal behaviour. Automated systems for the assessment of animal health and management are needed for the former, while studies on the impact of ecological production systems on animal welfare are needed for the latter (Sato, 2013). In 2013, in response to labour shortages and an aging population in the animal production sector, the Ministry of Agriculture, Forestry, and Fisheries set a goal for animal management by information and communication technology (ICT) to be achieved within 10 years. Plans for ICT include automated monitoring of body temperature, body condition score (BCS) and oestrus, especially in cattle and pigs. Research on the development of small scale farming systems is also ongoing, and includes the use of deep litter systems for pigs (Kohari, 2009), a movable housing system for grazing laying hens (Hosokawa and Saito, 2008), a very short term grazing of one hour for confined cattle (Ariga *et al.*, 2015), and a welfare-oriented production system for broilers by feeding unhulled rice (Ohara *et al.*, 2015). Traditionally, local consumers and organic farmers have connected on mutual trust, known as *Teikei* in Japan. If the systems described above are adopted in organic animal production, these systems may be supported by more selective consumers.

11.5.3 Multidisciplinary research

Animal behaviour has recently been studied in collaboration with various other disciplines and this work has resulted in some very interesting findings. For example, behavioural genetics has been used as a tool to study relationships among gene polymorphism in dopamine and serotonin receptors and temperament in dogs and horses (Inoue-Murayama, 2009), and the relationship between gene polymorphism in growth hormone and the stress response in cattle (Tachi *et al.*, 2014). These studies may contribute to improved relationships between humans and animals, enhanced social stability in group rearing, and greater productivity. Concepts related to experimental psychology, such as the application of consumer demand theory (Dawkins, 1983), the study of cognition bias (Harding *et al.*, 2004), and the study of convergent evolution between humans and dogs (Miklosi, 2007), can also be applied to farm animals. As well, epigenetic studies, for example looking at stress in neonates and behaviour modification (Jensen, 2014) from the perspective of animal welfare can also be explored in domestic animals.

Domestic animals, including not only dogs and cats but also farm animals, have been selected from among over 6,000 species of mammals and 9,000 species of birds as being suited for mutual dependence with humans. Therefore, farm animals also have utility and value beyond food production. Farm animals have been used for animal-assisted therapy (AAT) and for animal-assisted activity (AAA) for elderly people and for school children. As these activities are financially supported by the Japanese government, behavioural research in this area will develop by further coordination with education and medical facilities in Japan.

11.5.4 Behaviour and management of companion animals

In 2014, the number of dogs and cats in Japan was estimated to be 10.3 and 10.0 million, respectively (Japan Pet Food Association, 2015), while estimates for pigs and cattle were 9.5 and 4.0 million, respectively (Ministry of Agriculture, Forestry and Fisheries in Japan, 2015). Interest in the behaviour of companion animals is growing rapidly compared to that of farmed animals. There are many problems in companion animals, including abnormal behaviour, obesity, aggression, inappropriate elimination, demented elderly dogs and cats confined to houses, and inadequate socialisation in young dogs and cats usually sold in pet shops in Japan; these will be investigated by utilising the accumulated knowledge and methods of animal behaviour science.

11.5.5 Control of wild animals invading cultivated lands

Crop damage by wild and introduced animals has been a serious problem since 1990. Damage to property accounts for about 20 billion yen per year in Japan. Deer, wild boars, and monkeys are the main pest animals. Feeding behaviour around crop fields, daily and seasonal home ranges, and effectiveness of guard fence designs have been intensively investigated, especially in these three species. In addition, effective culling methods must be developed when considering animal welfare because it is popular to use live traps in

Japan. A recent study in this area by Takeda *et al.* (2014) reported on the effects of decoy, courtship call, and artificial turf on trapping wild deer.

11.6 Conclusions

Japanese culture has a unique perspective on animal welfare based on the influence of religious philosophies and the concept of *aigo* ethics. This cultural uniqueness has influenced the development of applied animal behaviour science in Japan. Initially, applied animal behaviour science developed to study free-ranging horses and cattle with the objective of establishing guidelines for free-ranging animal production. After the establishment of the Agricultural Basic Law of 1961, behavioural problems in farm animals in indoor rearing systems became prominent, and during this time studies on social behaviour also developed with the introduction of applied ethology. Applied animal behaviour research became more structured after 1984, and the development of textbooks in Japanese. The translation of important books and creation of textbooks in mother languages by domestic scientists are very important activities for developing relevant science around the world. This work provided the foundation for the prosperous development of applied animal behaviour science in Japan. In 2002, the Japanese Society for Applied Animal Behaviour Science was established, and with the international congress of ISAE held in Japan in 2005, the third stage of behaviour research began. The Japanese Society for Applied Animal Behaviour Science will continue to play an important role in '*aigo*' ethics and the adoption of animal welfare concepts, including greater respect for wild animals and fishes. Applied animal behaviour science will also develop in new directions, including not only farm animals but also pets, zoo, and wild animals and with a focus not only on animal welfare but also on precision management by ITC, animal-assisted activity and therapy, and pest control.

References

Ariga, S., Tanaka, S., Chiba, T., Shibuya, K., Kajiwara, I., Chen, S. and Sato, S., 2015. Long-term exercise on soil floor improves the health and welfare of Japanese Black steers. Proceedings of the 49th Congress of the International Society for Applied Ethology, Sapporo, Japan.

Chen, S., Tanaka, S., Ogura, S., Roh, S.G. and Sato, S., 2015. Effect of suckling systems on serum oxytocin and cortisol concentrations and behavior to a novel object in beef calves. Asian-Australasian Journal of Animal Sciences 28: 1662-1668.

Dawkins, M.S., 1983. Battery hens name their price: consumer demand theory and the measurement of ethological 'needs'. Animal Behaviour 31: 1195-1205.

Fukasawa, M., Kohari, D., Kosako, T., Oikawa, K. and Tsukada, H., 2008. The behavioral patterns of breakout from an electric fence in Japanese Black cows. (In Japanese) Nihon Chikusan Gakkaiho 79: 535-541.

Harada, N., 2014. God and meat: animal sacrifice in Japan. (in Japanese) Heibonnsha, Tokyo, Japan, pp. 97-177.

Harding, E.J., Paul, E.S. and Mendl, M., 2004. Animal behaviour: cognitive bias and affective state. Nature 427: 312.

Hidaka, T., (translator) 1963. King Solomon's ring, by K. Lorenz. (in Japanese) Hayakawa Shoten, Tokyo, Japan.
Hosokawa, Y. and Saito, K., 2008. The development of movable chicken house for organic broiler production by local breeds. (in Japanese) Animal Behaviour and Management 44: 154-155.
Imanishi, K. and Kawai, M., 1955. Horses on the Toi peninsula and domestic rabbits. (in Japanese) Kobunsha, Tokyo, Japan.
Inoue-Murayama, M. 2009. Genetic polymorphism as a background of animal behavior. Animal Science Journal 80: 113-120.
Japan Pet Food Association, 2015. The actual condition survey of rearing dogs and cats in 2015. (in Japanese) Available at: http://www.petfood.or.jp/data/chart2015/index.html.
Jensen, P., 2014. Behaviour epigenetics – the connection between environment, stress and welfare. Applied Animal Behaviour Science 157: 1-7.
Kaji, N., 1990. What is Confucianism? (in Japanese) Chukoshinsho, Tokyo, Japan.
Kawai, M., 1995. Animal in Miyazawa Kenji's world. (in Japanese) In: Kawai, M. and Haniwara, K. (eds.) Animals and culture. Asakurashoten, Tokyo, Japan, pp. 8-25.
Kellert, S.R., 1996. Culture. In: The value of life. Island Press, Washington, DC, USA, pp. 131-152.
Kohari, D., 2009. Improved pig-keeping. (in Japanese) Proceedings of the 1st Kobe International Conference on the Care for All Creatures 2009. Available at: http://tinyurl.com/zybsuj2.
Kohari, D., Namikawa, S. and Tanaka, M., 2014. How do visual and auditory cues influence calf stress reactions at weaning? Animal Science Journal 85: 336-341.
Kondo, S., 1985. Future applied animal behavior science. (in Japanese) Japanese Journal of Livestock Management 20: 138-141.
Kondo, S., Sekine, J., Okubo, M. and Asahida, Y., 1989. The effect of group size and space allowance on the agonistic and spacing behavior of cattle. Applied Animal Behaviour Science 24: 127-135.
Kubota, N., 1997. Atmosphere of polytheism in Japan. (in Japanese) PHP Institute, Tokyo, Japan, pp. 23-32.
Kurosaki, J., Iizumi, S. and Sugawara, K., 1956. On the relationship between grazing habits and vegetation of grassland, In: Grazing habits and feeding. (in Japanese) Bulletin of the Agricultural Research Institute, Tohoku University 8: 125-137.
Kusunose, R., Hatakeyama, H., Kubo, K., Kiguchi, A., Asai, Y., Fujii, Y. and Ito, K., 1985. Behavioral studies on yearling horses in field environments, 1. Effects of the field size on the behavior of horses. (in Japanese) Bulletin of Equine Research Institute 22: 1-7.
Matsuzawa, T. and McGrew, W.C., 2008. Kinji Imanishi and 60 years of Japanese primatology. Current Biology 18: R587-R591.
Miklosi, A., 2007. Dog behaviour, evolution and cognition. Oxford University Press, Oxford, UK, 276 pp.
Mimura, K., 1988. Farm animal behavior science. (in Japanese) Yokendo, Tokyo, Japan.
Mimura, K., 1997. Farm animal behavior (revised edition). (in Japanese) Yokendo, Tokyo, Japan.
Ministry of Agriculture, Forestry and Fisheries in Japan, 2015. The statistical data on animal husbandry. (in Japanese) Available at: http://tinyurl.com/hxybby4.
Muramoto, Y., (translator) 2004. The geography of thought, by R.E. Nisbett. (in Japanese) Diamond, Tokyo, Japan.
Nagano, T., (translator) 1975. The study of instinct, by N. Tinbergen. (in Japanese) Sankyo-Shuppan, Tokyo, Japan.
Nakamura, T., 1984. Idea to animal in Japanese. (in Japanese) Kaimeisha, Tokyo, Japan, pp. 9-31.

Ninomiya, S., Sato, S., Kusunose, R., Mitumasu, T. and Obara, Y., 2006. A note on a behavioural indicator of satisfaction in stabled horses. Applied Animal Behaviour Science 106: 184-189.

Ninomiya, S. and Sato, S., 2009. Effects of 'Five Freedoms' environmental enrichment on the welfare of calves reared indoors. Animal Science Journal 80: 347-351.

Ohara, A., Oyakawa, C., Yoshihara, Y., Ninomiya, S. and Sato, S., 2015. Effect of environmental enrichment on the behavior and welfare of Japanese broilers at a commercial farm. Journal of Poultry Science 52: 323-330.

Saito, T., Seo, T., Furumura, K. and Kashiwamura, F., 2009. The effects of limited time to enter a feeding station after standing up on defecation place of heifers. (in Japanese) Japanese Journal of Zootechnical Science 80: 71-76.

Sato, S., 1985. Emergence of ethology and applied animal ethology. (in Japanese) Japanese Journal of Livestock Management 20: 133-137.

Sato, S., 1991. Developing applied animal behaviour science. (in Japanese) Anima 228: 68-69.

Sato, S., 2013. Current status and the future of livestock farming after the great eastern Japan earthquake. (in Japanese) Tohoku Journal of Animal Science and Technology 62: 23-27.

Sato, S., Kondo, S., Tanaka, T. and Kusunose, R., 1995. Ethogram of farm animals. (in Japanese) Asakura-Shoten, Tokyo, Japan, 128 pp.

Sato, S., Siki, H. and Yamasaki, F., 1984. The effects of early caressing on later tractability of calves. Japanese Journal of Zootechnical Science 55: 332-338.

Sato, S., Tarumizu, K. and Sonoda, T., 1991. Social, behavioural and physiological functions of allo-grooming in cattle. In: Appleby, M.C., Horrell, R.I., Petherick, J.C. and Rutter, S.M. (eds.) Applied animal behaviour: past, present and future. UFAW, Hertfordshire, UK, pp. 77-78.

Shimmura, T., Hirahara, S., Azuma, T., Suzuki, T., Eguchi, Y., Uetake, K. and Tanaka, T., 2010a. Multi-factorial investigation of various housing systems for laying hens. British Poultry Science 51: 31-42.

Shimmura, T., Kamimura, E., Azuma, T., Kansaku, N., Uetake, K. and Tanaka, T., 2010b. Effect of broody hens on behaviour of chicks. Applied Animal Behaviour Science 126: 125-133.

Singer, P., 1976. Animal liberation. Thorsons, Wallingborough, UK, pp. 1-27.

Tachi, N., Tanaka, T., Ardiyanti, A., Katoh, K. and Sato, S., 2014. Bovine growth hormone gene polymorphism affects stress response in Japanese Black cattle. Animal Science Journal 86: 722-728.

Takeda, K., Danjo, R., Yamazaki, T., Kamiya, K. and Ihara, K., 2014. Investigation of trapping methods using behaviour of female deer to visual and phonetical cues. (in Japanese) Journal of Veterinary Medicine (Japan) 67: 656-660.

The act on *Aigo* and management of animals, 2012. Available at: http://tinyurl.com/hsraahg.

Uetake, K., Tanaka, T. and Sato, S., 2011. Effects of haul distance and stocking density on young suckling calves transported in Japan. Animal Science Journal 82: 587-590.

Watanabe, M., Hidaka, T. and Uno, H., (translators) 1955. Social behaviour in animals, by N. Tinbergen. (in Japanese) Misuzu, Tokyo, Japan.

Yayou, K., Ito, S. and Yamamoto, N., 2015. Relationships between postnatal plasma oxytocin concentrations and social behaviors in cattle. Animal Science Journal 86: 806-813.

Yoda, K., 2007. Why do we build graves for animals? (in Japanese) Shakaihyouronsha, Tokyo, Japan, pp. 12-37.

12. China perspective: emerging interest in animal behaviour and welfare science

J. Bao[1] *and Y. Li*[2*]
[1]*Northeast Agricultural University, School of Animal Sciences, Harbin 150030, Heilongjiang, China P.R.*
[2]*West Central Research and Outreach Center, University of Minnesota, 46352 State Hwy 329, Morris, MN 56267, USA; yuzhili@umn.edu*

Abstract

The development of interest in animal behaviour and welfare science in China mirrors the political and economic changes in the country over the last three decades. Traditionally, animal farming in China was small-scale and family-based with the main purpose of feeding the family rather than supplying food to society. At that time, the health status, productivity and efficiency of farm animals were usually overlooked. As large-scale confinement production systems were introduced from western countries in 1980s, problems in animal health and animal behaviour emerged. In seeking solutions to the problems, Chinese scientists started to teach and research 'Applied Ethology' at universities. After joining the WTO in 2001, China has been challenged on animal welfare issues in international markets. In domestic markets, Chinese consumers generally believe that improving animal welfare would improve the food safety of animal products, such as meat and eggs. As a result, research in applied ethology and animal welfare is better funded now as compared to 20 years ago. The presence of international organisations for animal welfare, animal protection and animal rights in China has encouraged Chinese citizens to get involved in animal welfare and protection. The Chinese government has constituted laws for animal protection. This chapter describes historical, political and economic impacts on the development of animal behaviour and welfare science in China.

Keywords: animal welfare, applied ethology, China

12.1 Introduction

The development of interest in animal behaviour and welfare science in China mirrors the political and economic changes in the country over the last three decades. Traditionally, animal farming in China was small-scaled, family-based with the main purpose of feeding the family rather than supplying food to society. At that time, most farmers did not utilise animal and veterinary sciences to improve productivity, efficiency or health status of farm animals. As large-scale confinement production systems were introduced from western countries in 1980s, China started to experience the challenge in animal health that was associated with the productivity- and efficiency-driven systems. In seeking solutions to the problem, animal scientists and producers realised the importance of applied ethology. Since then, people's interest in applied ethology has grown in China. After the concept of

DOI 10.3920/978-90-8686-828-5_12,

'animal welfare' was introduced into China in 1990s, research and education in applied ethology and animal welfare have been stimulated. A Chinese survey (You *et al.*, 2014) done in 2011 revealed that about one third of the general public had heard about animal welfare. Among these individuals, 73% believe that improving rearing conditions for swine and poultry would improve food safety of meat and eggs, and 54% expressed willingness to pay more for products produced from welfare-friendly operations.

One of the reasons for the rapid development in applied ethology and animal welfare awareness in China is that consumers are more sensitive and concerned about food safety than 20 years ago, especially the quality and safety of food animal products, such as meat, milk and eggs. It is widely believed that there is a connection between animal welfare and food safety. Recent incidences of food poisoning have attracted the attention of the general public and the central government. In response, the government has started to promote research on good production practices in order to produce healthy animals for food. This has encouraged educational and research activities in applied ethology and animal welfare. As a result, research in these areas is better funded now as compared to 10 years ago. In addition, the presence of international organisations for animal welfare, animal protection and animal rights in China has encouraged Chinese citizens to get involved in animal welfare and protection. Consequently, the Chinese government has constituted laws for animal protection, with the primary focus of this legislation being the protection of wildlife. In this chapter, we will discuss the development of animal behaviour and welfare science in China and our projections for future research in this area.

12.2 Historical impacts

Chinese culture maintains very different views about the treatment of animals compared to western countries and cultures. Traditionally, Chinese are heavily influenced by Confucianism and Taoism philosophies which advocate the idea of 'harmony between nature and humans' (Bao, 2013). Both philosophies promote the respect of nature and natural principles to achieve harmony. They believe that humans are an integral part of nature, and that man and nature are connected in an organic circle. The principle of 'harmony between nature and humans' requests that humans respect, and not dominate nature. Although the ancient Chinese philosophy advocates that humans should respect nature, it does not oppose human use of natural resources, including animals. Therefore, ancient Chinese philosophy does not emphasise a particular view in regards to animal welfare. In addition, people's attitude towards treating animals depends on not only their moral values but also on their personal welfare and quality of life. In countries, including China, which struggle to improve human welfare, animal welfare is often overlooked.

The First World Congress of Ethology and Applied Zootechnology held in Madrid, Spain in 1978 drew the attention of Chinese scientists to ethology for the first time. Three prestigious Chinese professors: Zhen-Ying Xu, Yi Rong, and Zheng-Kang Han, attended the conference. After returning from the conference, they disseminated information in China through a newsletter of the Chinese Animal and Veterinary Science Association. In the newsletter, they made sure to emphasise and describe the emerging discipline of

ethology. At that time, China had just resumed a high education system after 10 years of a so called 'cultural revolution'. All agricultural universities were looking for new courses for undergraduate students, and a course in 'Ethology' was very appealing to both professors and administrators of universities. Meanwhile, animal production was shifting from small-scale, family-based farming to large-scale, intensive production. It was believed that applied ethology would help large-scale animal production. Some Chinese scientists were excited about information from the conference and started to explore the link between ethology and animal production. In 1980, an article (Han, 1979) entitled 'Application of Ethology in Livestock Production' was published in the Journal of Animal Husbandry and Veterinary Medicine, the most popular journal for animal and veterinary sciences in the country. This article introduced the concept of applied ethology, outlined potential research areas, and described specific behaviours such as normal vs abnormal behaviours, social behaviour, sexual behaviour, parental behaviour, learning, exploration, territorial behaviour, and preening behaviour. This was the first time that applied ethology was systematically described in the Chinese language.

It seems that it is easier for Chinese scientists to accept the concept of applied ethology than accepting the concept of 'animal welfare'. In early 1990s when the idea of 'animal welfare' was introduced, there were intensive debates in Chinese academic communities. The initial debates among animal scientists were about how to define animal welfare, how to implement animal welfare on the farm level, and how important animal welfare was to animal production and animal keeping in China. Soon after, the debates were extended to pets kept and wildlife. Then the focus of the debates was whether or not it was necessary to promote research activities in animal welfare in China. At that time, neither the Chinese government nor the academic community was convinced that research in animal welfare was critical to animal production in China.

12.3 Key pioneers

Two influential Chinese professors in animal science during the last three decades are considered key pioneers in the development of applied ethology. They were Professor Zhen-Ying Xu from the Northeast Agricultural University in Harbin, Heilongjiang Province, and Professor Zhen-Kang Han from Nanjing Agricultural University in Nanjing, Jiangsu Province.

Professor Zhen-Ying Xu (1907-1993) was the most prestigious animal scientist in China (Figure 12.1). In 1956, he was invited by the Chinese Premier En-Lai Zhou to participate in development of The First Five-Year Development of Science and Technology Plan for the country. Professor Xu was an animal nutritionist by training, however, he did not limit his work to this field only. Instead, he devoted his life to the animal sciences at large. He especially emphasised the application of animal science to advancing animal production. Realising the importance of animal behavioural science to modern animal agriculture, Professor Xu became the first advocate of applied ethology in China. He encouraged his former student, Professor Shi-An Li, to publish the first textbook of '*Applied Animal Behaviour Science*' in China (Li, 1985).

On returning from the First World Congress of Ethology and Applied Zootechnology, Professor Zhen-Kang Han (1924-) led his research team to translate scientific papers from international journals related to applied ethology. He published the first article (Han, 1979) about the application of ethology, which was considered a cornerstone in the development of applied ethology in China. Professor Han is an animal physiologist by training. He is well known for his pioneering research in digestive physiology of swine and ruminants, and lactation physiology of Chinese Yellow cattle, buffalo, and yak. He has not conducted any research project in applied ethology himself. Instead, he has trained graduate students who now are actively involved in applied ethology and animal welfare research (Nielsen and Zhao, 2012; You *et al.*, 2014) in China.

Another key pioneer in advocating applied ethology in China was Professor Shi-An Li (1931-2015) from the Northeast Agricultural University (Figure 12.2). Professor Li was the first to offer a course in Applied Animal Behaviour to undergraduate students in 1983, and recruited the first graduate student in applied ethology in 1984 in China. Professor Li devoted his life to studying the behaviour of different species, including pigeons, dogs, cats, rabbits, and pigs. He built pigeon and dog houses himself on his own property so that he could observe and study their behaviours. His research mainly focused on social behaviour, consumptive behaviour and fixed action patterns in dogs and pigeons. An interesting experiment that Professor Li conducted in early 1980s was that he put a mirror on a side of a room to enlarge the image of the room and observed the behavioural reactions of the hens in the room. His passion about animals, and animal behaviour in particular, made him publish a series of articles in both Chinese and Japanese. Professor Li was not only a cornerstone of applied ethology in China, but also an international advocate of ethology.

Figure 12.1 Professor Zhen-Ying Xu, agricultural scientist and the first advocate of animal behaviour in China.

Figure 12.2. Professor Shi-An Li, behavioural researcher and educator.

Besides these key pioneers, many scientists across the country have contributed to the development of applied ethology and animal welfare science in China. For instance, Professor He Zhou from Henan Agricultural University explored the application of applied ethology to solve problems in animal production and explained the relationship between applied ethology and animal welfare in 1978, which helped the academic community find directions in ethology research. Professor Chen-Ping Lu from the Nanjing Agricultural University authored the first textbook on animal welfare, titled '*An Introduction to Animal Protection*' in 1998. The third edition of the book was published in 2009, indicating a high demand for information on animal protection and welfare in China. In addition, Professor Xu-Xian Zhu from the An-Hui Medical Science University translated Peter Singer's '*Animal Liberation*' into the Chinese in 2004. This landmark publication was the first book on animal rights translated into Chinese. The book triggered great reactions from the general public about animal rights. After reading the book, many Chinese citizens exchanged their opinions on social media, expressing that they never thought about that the liberty of animals was just as same as the liberty of African Americans or women. Some readers pointed out that eating the meat of cats and dogs is violating animal rights.

12.4 Economic influences

The rapid development of the economy has influenced animal production, and consequently, the development of applied ethology in China. Before the political reform of the 1980s, livestock production in China relied mainly on small-scale, family-based or collective farming. Productivity was low and production efficiency was very poor. As a result, animal products were scarce. In 1978, the national per capita share of meat, eggs and milk was only 9.1 kg, 2.4 kg, and 1.0 kg, respectively (CAAC, 2009), which was far below the world average. Meanwhile, the annual output value of animal production was only 20.93 billion Chinese Yuan ($3.18 billion USD). Three decades later, China has become the largest meat and egg producer, and is ranked 3rd in milk production in the world. As of 2010, the per capita share of meat, eggs and milk has increased to 45.8, 26.7 kg and 20.7 kg, respectively (CAAC, 2009). The annual output value of animal production has increased to over 2 trillion Yuan, accounting for more than 30% of total agricultural output value. This rapid development of animal production was not only driven by changes in production scale, but also by advances in science and technology, including genetic selection, animal nutrition, and improved management on the farm level. Meanwhile, applied ethology emerged and developed rapidly, largely because of the problems that emerged with the application and popularisation of large-scale production systems. These problems include reduced resistance of animals to diseases which results in more sick animals and high mortality, and increased abnormal behaviours. It seems that application of modern facilities and advances in veterinary medicine have increased the occurrence of these problems (Li, 2009). Also, people began to realise that these problems are largely symptomatic, and that the root of the problem was related to the animals' living environment, which failed to meet their biological requirements, resulting in 'unadapted-environment syndrome'. As people began exploring solutions to this syndrome, applied ethology and animal welfare found its nursing soil in China. Producers, governmental officials and researchers all realised that understanding animal behaviour and animal welfare is key to improving modern animal production (Qi *et al.*, 2009; Wan, 2014).

In addition, food safety issues have raised animal welfare awareness among Chinese citizens. Incidences of food poisoning over the last few years have caused great concerns for consumers about food safety of animal products. In response to consumers' concerns, the government has encouraged academic communities to find solutions. This has provided applied ethologists with opportunities to educate producers and consumers about animal welfare. For instance, Professor Peishi Yan at the Nanjing Agricultural University offered a TV series on hot topics about animal welfare with animal producers and the general public. Through these educational events, along with national and international academic conferences, both producers and government officials realised that the animal welfare does not only mean to provide comfortable environment for animals, but these types of standards are also essential for producing safe food animal products.

12.5 Political influences

Since 2001 when China became a member of the WTO, Chinese animal products have been challenged on international markets for their lower animal welfare standards, which became a trade barrier for western markets. The barrier resulted in losses to Chinese exporters. For instance, poultry meat from Heilongjiang Province exported to European markets was once rejected because the housing conditions for the birds were too crowded, and failed to meet the European standards (Li, 2005). This incidence has triggered greater vigilance by the Chinese government on animal welfare regulations. In 2005, the Ministry of Commerce of China held the International Forum of Farm Animal Welfare and Meat Safety. This was the first international conference officially held by the State Ministry on the issue of animal welfare. More than 10 world-leading scientists (including Drs. D.M. Broom and D. Fraser), Chinese government officials, scientists, and journalists attended the conference. The conference inspired the participants to engage in conducting and dissemination of applied ethology and animal welfare research.

Currently in China, there is no national legislation to protect farm animal or companion animal welfare. Most regulations are for protection of either wildlife or laboratory animals, such as the 'Regulation on Management of Laboratory Animals' promulgated in 1988, and the 'Wildlife Protection Law' of 1989. At the local level, there are some regulations pertaining to companion animal welfare. For instance, Beijing Municipal Government implemented 'Provisions of Park Management according to the Regulation on Management of Dogs' in 2003, which laid out fines for mistreating animals. With the continuous development of China's constitution, an 'Animal Protection Law of People's Republic of China' is currently under discussion. This law will cover protection of all kinds of animals, including wildlife, companion animals, laboratory animals, farm animals, and other animals during captivity and transportation. This will be a non-mandatory regulation with recommendations for animal welfare standards. The animal welfare standard in this law is based on the Five Freedoms. Although the regulations are not mandatory, it is a clear signal that the government has taken the first step towards creating legislation to protect animal welfare. It is believed that this law will help speed up the development of formal animal welfare legislation in China.

12.6 Research in animal behaviour and welfare

In the early 1990s, research in applied ethology was mainly part of graduate programs in agricultural universities. Since these projects focused on graduate education, most of the projects were funded internally and with limited money. Researchers could not implement the results to solve problems in animal production. Instead, they focused on understanding animal behaviour, such as abnormal behaviours of farm animals, aggression among pigs and chickens, tail biting in pigs, and effects of the environment on animal behaviour. At that time, companion animals were not common and were rarely researched in China.

For the last two decades, the central government has increased financial support for animal behaviour and welfare research. Since 1996, the Natural Science Foundation of China (NSFC) has allocated 14.8 million Chinese Yuan ($2.5 million USD) for competitive grants to support 44 projects related to farm animal behaviour and welfare. Among these projects, 40% were poultry-related, 25% were swine-related (Table 12.1), and the rest included other species, such as sheep, cattle and horses. All these projects emphasised the understanding of animal behaviour in relation to genetics, stress, environment, management and nutrition. While the NSFC grant is very competitive and considered the most prestigious research

Table 12.1. Selected research projects on farm animal behaviour and welfare supported by the National Science Foundation of China (1996-2015).

Species	Project title
Poultry	Effect of photoperiod-regulated melanopsin expression on behavioural rhythm of laying hens
	Analysis of MAQ and COMT of genes related to aggression in Turpan game cocks
	Effect of cold stress on welfare of Lindian chicks during brooding period
	Behavioural and physiological response and molecular mechanism of feather-pecking in Sanhui ducks that are raised without access to pond
	Design of animal welfare friendly cages for laying hens
	Identifying vocal characteristics of laying hens by applying digital models
	Effect of light-induced prolactin secretion and regulation on behavioural pattern of nesting and egg-laying in Beijing Oil Chicken
	Effect of gene expression of prolactin and prolactin-receptors on nutritional regulation of pigeon hens during nesting and egg-laying period
	Investigation of inversion mechanism of sexual behaviour in laying hens
	Relationship between feeding behaviour and performance of laying hens
	Genetic selection and management of high efficiency laying hens
	The role of hypothalamus on regulation of feeding behaviour in poultry
	Effect of gonadotropin-suppressing hormones on regulation of breeding activities and prolactin secretion in Magang geese
	Application of high-volume sequencing technology to study genetic mechanism for egg quality in green shell egg-laying hens
	Study on reproductive performance of Taihe hens
	Relationship between feather pecking and MHC type of social recognition
	Controlling feather pecking and cannibalism: Strategies based on social recognition through odour clues
Swine	Association of short-term/long-term depression with oral/nasal behaviours in sows
	Effect of farrowing crates on behavioural and hormonal responses of sows to novel environments
	Effect of groundwater cooling on heat stress in sows during summer
	Application of high-volume sequencing technology to study genetic mechanisms of abnormal behaviour in swine
	The role of fatty flavour and fatty acid receptors on regulation of feed intake in swine
	Effect of music on growth performance and hormone concentrations in newly weaned piglets
	Video recognition of respiration rate and gait score in swine
	Tonic pupil reflection and its association with depression in swine
	Identifying genes that related to savaging behaviour of sows using QTL mapping
	Development of abnormal behaviour in sows
	Assessment of animal welfare through behavioural indicators
	Brain morphology and physiological mechanisms of pupillary reflex latency in sows

award in the country, the amount of funds for individual projects is quite limited. In fact, NSFC grants usually act as seed funds, which mainly support basic research and do not require commercial application or demonstration of the results.

In contrast to the NSFC, the Ministry of Agriculture of China (MAC) recently released some Special Funds for Agro-scientific Research in the Public Interest. One of the projects under this initiative was entitled 'Investigation and Demonstration of Key Technical

Measures to Improve Animal Welfare in Livestock and Poultry Production' (Nielsen and Zhao, 2012). A major goal of this project was to establish Chinese welfare standards for swine and poultry. The aim is to further enhance the competitiveness of Chinese food animal products in international markets, to assure consumers of food safety and quality, to promote human health, and to safeguard sustainable agriculture in China. Three universities, three agricultural research institutions and one producer across the country participated in this 18-million Yuan ($3 million USD) project. International symposiums on animal welfare were organised by the project leaders. Experts on animal behaviour and welfare across the world were invited to speak at the symposia to exchange information on animal welfare research. European experts were invited to train Chinese scientists and producers in on-farm animal welfare assessment. Meanwhile, a national survey on swine and poultry welfare was conducted to identify areas for improvement. The areas identified for swine were: thermal environment and air quality in the barn, gestation and lactation housing design, piglet processing protocols (teeth clipping, tail docking, and castration), rearing environment, bacteria, toxins and antibiotics in feed. The areas identified for broilers included stocking density, quality of bedding, perch design, and low resistance to stress. For laying hens, the main areas were housing systems, stress and metabolic diseases. Results and technologies developed from the project were demonstrated on 14 swine and poultry farms across the country. This is a good example of how research results in animal welfare have been implemented on-farm in China.

12.7 Higher education

The first Applied Ethology course in China was offered at the Northeast Agricultural University, Harbin, in 1983. Soon after, a few key agricultural universities, including Beijing Agricultural University, Nanjing Agricultural University, and Shandong Agricultural University, followed the path of the Northeast Agricultural University and started offering applied ethology as an elective course to undergraduate students. Over the next decade, most agricultural universities developed curricula in applied ethology for undergraduate students. Meanwhile, a number of universities established graduate programs at both the MS and PhD level in applied ethology. Due to the emerging interest in animal welfare, this has also become an important part of courses in applied ethology. Many of the courses offered are broad and cover the welfare of wildlife, laboratory animals, aquatic animals, as well as animal welfare related legal issues and international trade regulations. Currently, courses in animal welfare are offered not only at agricultural universities, but also at liberal arts universities, business schools, and law schools. For instance, Fuzhou University offered an animal welfare course for undergraduate students of the International Trade Major, which covers animal welfare issues from the perspective of non-tariff trade barriers. The Northeast Forestry University offers the course 'An Introduction to Animal Conservation' to undergraduate students, and the course 'Wildlife Conservation and Research Methodology' to graduate students. In addition, universities such as the China Agricultural University and Northeast Agricultural University offer seminars to students and faculty on a regular basis to address animal welfare issues and to discuss on how to properly care for farm and companion animals. In general, these courses are getting popular among students across the country. In addition to students from animal science major, postgraduates from non-animal science majors, such as life science and environmental science major, are also attracted to enrol in these courses.

12.8 National academic organisations and international collaboration

Researchers from three national academic associations are highly trained to conduct research related to applied ethology and animal welfare. Two of the associations (Livestock Ecology Society, and Livestock Environment and Hygiene Society) are branches of the Chinese Association of Animal Husbandry and Veterinary Medicine Science, and another (Livestock Engineering Society) is a branch of the Agricultural Engineering Association. Most members of these associations are university faculty and scientists from research institutes. These researchers along with graduate students have formed the major work force in applied ethology and animal welfare research. Currently, there are neither ISAE branches nor members of the ISAE in China. However, Chinese scholars do publish their research in the journal of Applied Animal Behaviour Science and attend the ISAE conferences.

An international academic body, the International Research Center for Animal Environment and Welfare (IRCAEW; www.ircaew.org/), was established in 2011 in China. The IRCAEW facilitates international scientific and technological exchanges and promotes collaborative research on animal environment and welfare issues. Since its establishment, the IRCAEW has organised annual international symposia to promote communication and sharing of research findings in animal behaviour and welfare, environmental control, biosecurity, manure management, and farm facility design. Currently, scientists from 8 countries, including the United States, Australia, Canada, the Netherlands, Belgium, Denmark, Brazil and China, sit on the Board of Directors, which organises the academic activities of this centre.

12.9 Presence of animal welfare organisations

The emergence of both international and national organisations of animal welfare, animal protection and animal rights escalated the animal welfare movement in China. These organisations include World Animal Protection (WAP), Royal Society for the Prevention of Cruelty to Animals (RSPCA), Compassion in World Farming (CIWF), and Chinese Animal Protection Network (CAPN). In 2004, CAPN, the first animal rights group, launched its campaign 'Chinese Companion Animal Protection Network' in China. Meanwhile, the international organisations for animal welfare co-hosted forums with Chinese research institutions to encourage Chinese researchers to devote more time and funding to applied ethology and animal welfare research. For instance, the RSPCA and the Chinese Economists' Society for Forestry, Livestock and Fisheries co-hosted the 'International Forum of Importance of Farm Animal Welfare Science to Sustainable Agriculture' in Beijing in 2008. In 2013, the WAP and the International Cooperation Committee of Animal Welfare (ICCAW – China) co-hosted the Animal Welfare Forum in Beijing. The aim of these forums was to promote animal welfare in China, especially in areas where large-scale confinement production systems prevail.

12.10 Looking to the future

In recent years, food safety issues have attracted great attention from consumers. In 2008, a significant incidence reported was that a chain-dairy processor added melamine in infant milk powder to increase the nitrogen content (a measurement of protein concentration of milk), which caused many infants to become sick and hospitalised. Soon after in 2011, a large-scale pork producer in Henan Province was found to supplement an excessive amount of beta-agonists in feed to promote lean growth of pigs, resulting in some consumers hospitalised after eating the pork. After the turmoil, Chinese citizens started to pay attention to the way that farm animals were raised and food was processed. They believe that ultimately, improved farming systems will be a solution to the problem of food safety. We expect that future research in applied ethology in China will focus on developing sustainable farming systems. This is because sustainable farming systems are required to meet the biological and ethological needs of farm animals. It is the applied ethologists' job to identify the needs of farm animals by conducting research. Currently, the Chinese government has enhanced financial support to develop sustainable farming systems. For example, one of the high priorities of the 13th Five-Year Plan (2015-2020) of National Science and Technology is to investigate key technical components for sustainable farming systems (Xinhua News, 2016). Research priorities will focus on development of techniques to improve farming systems for dairy, beef, swine, poultry and sheep production. All proposed research projects should emphasise performance, health, and animal comfort, with the ultimate goal of improving food safety and sustainability of animal production. We believe that this will require animal scientists to master ethological knowledge in order to accomplish their research projects. In addition, researchers will need to explore in-depth research and multi-disciplinary collaborations among applied ethology, animal welfare, livestock-environmental engineering, housing systems, and facility design. We expect that in contrast to other traditional disciplines, applied ethology will become a very active research area in coming years in China. Thus we truly believe that with strong financial support from the Chinese government, Chinese ethologists may be able to collaborate in a greater capacity with scientists from other countries in the future.

On the farm level, there is evidence that some farmers began to replace intensive production systems with alternative systems, such as free-range, raising animals on pasture or in forests, or using animal welfare-friendly housing systems such as loose housing for dairy cattle or enriched cages for hens. Consumers believe that these farming systems can improve the living environment for animals, and ultimately will safeguard animal health and food safety of animal products. With an increase in average income level, we expect that many Chinese consumers will be willing to pay a premium for high quality and safe animal products (You *et al.*, 2014). We predict that products from sustainable or alternative farming systems will be more welcomed by consumers than those from intensive farming systems. Consequently, alternative farming will be encouraged by market demand, and animal welfare will grow its roots deep in China.

In summary, China has come a long way in advancing applied animal behaviour and welfare science. Over the 20 years between 1980 and 2000, Chinese pioneer scientists worked hard to identify applications for behavioural science in animal production and to build teaching and research teams in applied ethology. Compared to the fact that no courses in animal behaviour or applied ethology were offered at any agricultural university in China until the early 1980s, many courses related to applied ethology have now been offered at most agricultural universities in China since the beginning of 21st century. Subsequently, increasing numbers of graduate students are being trained in applied ethology (MSc and PhD level), and are building a core of expertise in applied ethology in China. As the Chinese economy started booming and large-scale confinement animal production systems emerged, applied ethology research developed rapidly in China, supported by increased funds from the central government. We expect that as China works to meet consumers' demand for high quality and safe food animal products, applied ethology will play an important role in developing sustainable animal production systems in China.

References

Bao, J., 2013. Current status of farm animal welfare in China. In: Gates, R., Xin, H., Li, B. and Zhao, R. (eds.) Proceedings of International Symposium on Animal Environment and Welfare. October 19-22, 2013. Rongchang, Chongqing, China.

Chinese Association of Animal Science and Veterinary Medicine (CAAC), 2009. Progress in livestock industry during 30 years of reform and opening up in China. (in Chinese) The Guide to Chinese Poultry 26: 22-25.

Han, Z.K., 1979. Application of ethology in livestock production. Chinese Journal of Animal Production 1: 40-50.

Li, K.N., 2005. The broiler welfare status in EU and in China. China Broiler Guide 22: 7-9.

Li, S.A., 1985. Applied ethology. People's Publishing of Heilongjiang, Harbin, China.

Li, X., 2009. The role of animal behavior in 'good' animal farming practice. Journal of Heilongjiang Livestock Veterinary Medicine 10: 19-20.

Neilsen, B.L. and Zhao, R., 2012. Farm animal welfare across borders: a vision of the future. Animal Frontiers 2(3): 46-50.

Qi, L., Bao, J. and Li, J.H., 2009. The importance of animal ethological research in improving animal production. Animal Quarantine of China 26(9): 68-69.

Wan, Q., 2014. Animal ethological theory applied to animal production practice. Contemporary Livestock 20: 29-30.

Xinhua News, 2016. National medium and long term science and technology development outline 2006-2010. (in Chinese) Xinhua News Agency, September 2, 2016. Available at: http://tinyurl.com/gnomoq6.

You, X., Li, Y., Zhang, M., Yan, H. and Zhao, R., 2014. A survey of Chinese citizens' perceptions on farm animal welfare. PLoS ONE 9: e109177.

13. Variety is the spice of life: developing applied ethology

M.C. Appleby
World Animal Protection, 5th floor, 222 Gray's Inn Road, London WC1X 8HB, United Kingdom; michaelappleby@worldanimalprotection.org

Abstract

There is substantial variation between people, between countries and between cultures in attitudes towards and use of animals. This chapter considers this variation, and the influences of economic circumstances and political processes, with particular reference to developing countries and emerging economies, and with particular relevance to animal welfare. Interaction between ideas and people from different backgrounds and countries was important from the foundation of the Society for Veterinary Ethology in 1966, and recognised by our change of name to the International Society for Applied Ethology in 1991. International links are vital in applied ethology, because few questions about animals are confined to individual countries and so both the science and solutions to problems benefit from international communication and collaboration. Variety will not disappear, and our understanding of animals and their behaviour will continue to be enhanced by the rich variety of environments, societies and animals around the world. In the development of applied ethology, in developed and developing countries and emerging economies, variety is the spice of life.

Keywords: developing countries, emerging economies, internationalism, religion, economics, political processes

13.1 Introduction

When proposals to change the name of the SVE were discussed in 1991, addition of the word International was generally agreed upon (whereas addition and deletion of other words caused considerably more debate). In a way that is surprising, because there was nothing in the name Society for Veterinary Ethology to limit international collaboration and communication. Yet it was clear that members wanted to encourage worldwide participation, and indeed it was in 1992 that the first congress was held outside Europe, in Pittsburgh, USA. So International we explicitly became. For the record, it was Norwegian ethologist Bjarne Braastad who proposed the name International Society for Applied Ethology, adopted at the Annual General Meeting in Edinburgh, UK, in 1991. And Mark Rutter, working in Edinburgh, who adapted the SVE logo into one appropriate for ISAE by placing the horse (emblazoned with ΗΘΟΣ or Ethos) in the globe.

We are all – as scientists and citizens concerned with animals – international now. The globalisation of transport, trade and governance, and human impacts on climate,

DOI 10.3920/978-90-8686-828-5_13, © Wageningen Academic Publishers 2016

habitats and societies, affect every person and animal in the world. Some of those effects are negative, some are positive, but one aspect which is almost universally accepted as advantageous is increased communication. While there is substantial variation between people, between countries and between cultures in attitudes towards and use of animals, increased communication (about facts, science and ethics) can help us to comprehend, allow for or address such variation. In the long term, the impact of this shared knowledge could be a general, more broadly-based understanding of animals themselves. This can often benefit people, animals and also the environment.

Here's an example that made me think. I have been fortunate to travel to many places in the course of my work, but I can only think of two named after animals. According to local legend, the founder of Bern, now capital of Switzerland, decided to name the city after the first animal he met on a hunting trip, and this turned out to be a bear. Switzerland is known for strong protection of animals, but for centuries bears have been kept on public view in Bern. Until recently they were in an old-fashioned, barren bear pit (Figure 13.1), which contrasted strangely with that reputation. I am glad to say that they are now in much better conditions. By contrast, the name of Surabaya in Indonesia derives from the words 'suro' (shark) and 'boyo' (crocodile). Visitors can see a fine sculpture of the legendary fight between these creatures (Figure 13.2) that ended in an agreement that the former would rule the sea and the latter the land. The sculpture celebrates both the importance of animals in our culture and their beauty in the natural world. While these instances are not representative, it is interesting that the country with a strong reputation for animal protection nevertheless confined animals in poor conditions for human amusement, whereas the country with no such reputation did not do so in this case.

This chapter will consider variation in attitudes to and use of animals, with particular reference to developing countries and emerging economies, and with particular relevance to animal welfare. Applied ethology is already helping people, animals and the environment around the world, and there are many other opportunities for it to do so in developing countries. In particular (FAO, 2009, p. 1):

Figure 13.1. Bern in Switzerland is believed to be named after a bear hunted by its founder. Until recently bears were on view in this old-fashioned bear pit. A cub can be seen in this photograph, which I took on a Kodak Instamatic camera in 1967.

Figure 13.2. The foundation story of Surabaya in Indonesia includes a fight between the 'suro' (shark) and the 'boyo' (crocodile) for mastery, demonstrating the importance of animals, their environments and their behaviour in our culture. I am very grateful to Jen Brown for this lovely drawing of the sculpture in Surabaya.

> Roughly one billion people, including many of the world's poor, depend directly on animals for income, social status and security as well as food and clothing, and the welfare of their animals is essential for their livelihood. Moreover, positive relations with animals are an important source of comfort, social contact and cultural identification for many people.

Applied ethology helps to understand and improve those relationships, to the benefit of both people and animals.

13.2 Internationalism – a personal perspective

Interaction between ideas and people from different backgrounds and countries was part of my life from an early age. My father was a veterinary pathologist (and a council member of the Universities Federation for Animal Welfare). He worked with colleagues across Europe, including eastern countries where travel was difficult, to share information about disease and mortality – hitherto largely secret – between zoos. He also taught many foreign students. My parents welcomed many of these people to our house, a tradition I have continued. When I was 12 we lived in Switzerland for three months. Much later, I worked for the Humane Society of the United States (HSUS) for four years, and learned for myself the truth that to get to know a country's culture and ways of thinking, you actually have to live there, not just visit.

I have always sought opportunities to travel for work. For example, for my degree in Zoology at the University of Bristol I had to do a vacation project, and got the chance to work on elephants with an ecologist in Kenya. Those months living with wild animals in east Africa gave me direct experience of species and behaviour I learned about in lectures and subsequent reading, and have therefore been a valuable grounding for my science ever since. After a PhD in animal behaviour I went to work with Ian Duncan at the Poultry Research Centre near Edinburgh (now Roslin Institute, famous for Dolly the sheep), on the nesting behaviour of hens, and then to a lectureship at the University of Edinburgh as a colleague of David Wood-Gush.

As indicated above, I believe that international links are vital in applied ethology, because few questions about animals are confined to individual countries and so both the science and solutions to problems benefit from international collaboration. Throughout my career I have continued to pursue and encourage international links wherever possible. Here are three examples. Of course, many others in the field have also recognised the importance of international participation.

When we started the Masters course in Applied Animal Behaviour and Animal Welfare at Edinburgh in 1990 (led by David Wood-Gush, with Alistair Lawrence another key organiser), we sought students from as far afield as possible – and funding for as many of them as possible – from the very beginning. We wanted to enable a diverse range of people from different countries to become educated in animal behaviour and welfare. Many of these graduates now work in the field, all over the world.

When we organised the SVE conference in Edinburgh in 1991 (the 25th Anniversary meeting at which, perhaps surprisingly, the name change to ISAE was decided), we tried to support attendance by colleagues who might find it difficult to attend. We were successful at getting support from Ruth Harrison's Farm Animal Care Trust for Marek Špinka and Ľubor Košťál to come from Czechoslovakia. Ruth asked whether it was worthwhile to bring two people from the same country, but I persuaded her that the Czech and Slovak regions had differences as well as things in common – as became clear when they later became separate countries. Marek (later President of ISAE) and

Ľubor participated enthusiastically, both in the science and in the Scottish dancing. The tradition of dancing at congresses has been maintained ever since.

And lastly, in editing the book '*Animal Welfare*', my colleagues and I strove to make the authorship as international as possible, for both coverage and credibility. In the second edition (Appleby *et al.*, 2011) we were particularly successful: the book had 46 authors from 15 countries. Bas Rodenburg (2012) reviewed the book for the journal of the same name. He told me that he was pleased to be asked ... until he realised that he was 'the only applied ethologist in the world who wasn't an author'. It was an exaggeration but, I am pleased to say, one that rang true.

13.3 National variation

While the ISAE tries to be as 'I' as possible, the uneven distribution of members to date is striking. In 2015 we have:

- Western Europe, USA, Canada, Australasia and Japan: 533 members in 20 countries;
- Eastern Europe, Africa, Middle East, rest of Asia and Latin America: 32 members in 14 countries.

A glance at the map (Figure 13.3) reminds us of how many countries are unrepresented in the latter list, without even a single member. There are, of course, people working in applied ethology and related fields who are not members of our society. However, these gaps are an indication of the variation between countries in attitudes towards animals, combined with variation in awareness of scientific developments and ability to participate in such activity. We tend to forget, for example, that access to the internet is heavily restricted in some countries.

Remember then, that Africa, Asia and Latin America – countries for which ISAE representation is extremely scant – contain by far the majority of both the people and the animals of the world, with many major issues (including welfare) that could be addressed by applied ethology.

It is well recognised that concern about animal welfare varies between countries. Within Europe, it has historically been stronger in the north and west and weaker in the south and east. The most persuasive explanation is that such concern largely developed in urban people whose involvement with animals differed from that in rural areas – who kept pets more often than farm animals. The UK and the Netherlands, for example, were more industrialised than many other countries, and pressure for animal protection came mostly from city dwellers rather than from those involved in farming. A revealing snapshot was provided in 1981 by a review of which countries had ratified the Council of Europe's 1976 Convention on the Protection of Animals kept for Farming Purposes (the foundation for much of Europe's legislation on farm animal welfare). Of the 21 countries that were then members (Table 13.1), the 11 that ratified first were mostly from the north and had an average of only 6% of the population involved in agriculture. Countries that ratified later were mostly southern, with an average of 21% involved in agriculture. As well as

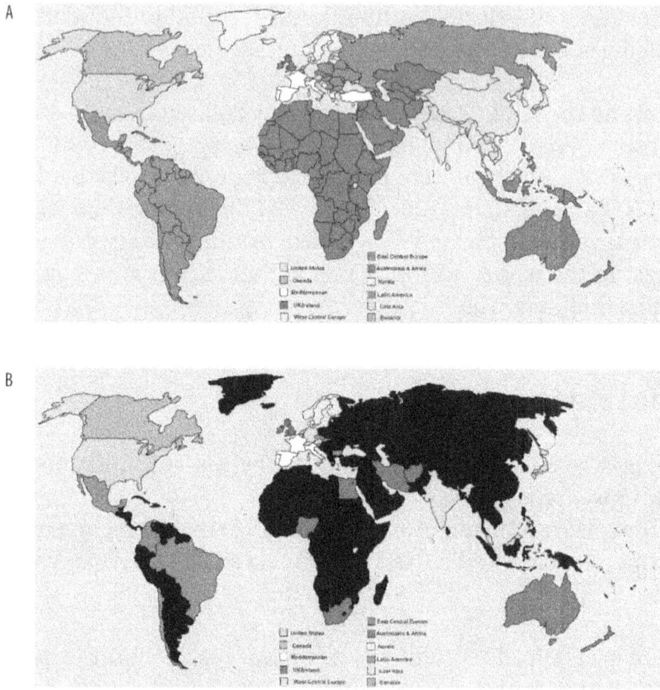

Figure 13.3. (A) The ISAE Regions, (B) Countries with ISAE members. There are no members in many countries, particularly in Eastern Europe, Africa, Asia and Latin America.

Table 13.1. Ratification of the Council of Europe's 1976 Convention on the Protection of Animals kept for Farming Purposes by 1981, and the percentage of each country's population involved in agriculture (Ludvigsen et al., *1982).*

Ratified	Agricultural labour (%)	Not yet ratified	Agricultural labour (%)
Belgium/Luxembourg	4	Austria	9
Cyprus	-	Greece	30
Denmark	8	Iceland	9
France	9	Ireland	23
Netherlands	5	Italy	12
Norway	8	Liechtenstein	-
Sweden	5	Malta	5
Switzerland	5	Portugal	26
UK	2	Spain	17
West Germany	4	Turkey	54
Average	**6**		**21**

differences in attitudes, this dichotomy may have reflected the fact that, where many people are engaged in agriculture, governments are less willing to impose restrictions that affect their livelihood. However, in recent years attention to animal welfare has also grown in southern Europe.

Attitudes towards animal welfare also tend to be positive in areas such as North America (particularly Canada) and Australasia. This is partly because emigration from Europe led to cultural similarities, and partly because in these developed countries, increasing affluence has allowed people to diversify their interests beyond the question of whether they can afford their next meal. In addition, intensification of agriculture in these countries has provoked concern for farm animal welfare and for other issues, such as food safety and effects on the environment.

Evidence of concern for animal welfare in developing countries and emerging economies has tended to be less. Some of the explanation is cultural while much of it is economic, as discussed in the next sections. However, these countries are increasingly adopting intensive systems of agriculture similar to those in developed countries, and at least part of their populations have both rising incomes and access to international communication. Correspondingly, interest in animal welfare, food safety and environmental protection is rising. The needs and opportunities for applied ethology in these countries are therefore increasing too.

13.4 Cultural variation

Some differences in the use of animals within and between countries are cultural, including religious. While these issues are sensitive, it is possible to address them sensitively. For example, one issue discussed in this context is slaughter, including questions about how animals should be handled beforehand and whether or not they should be stunned (made insensible) first. It is important to note that many methods of religious slaughter have their origins in the avoidance of previous unsanitary or inhumane practices. For example, one of the factors in the objection to stunning prior to slaughter for Halal food is the principle that animals killed for food should be uninjured. Indeed, most religions have traditions or laws requiring humane treatment of animals (Raj, 2004). If these are followed there may be many benefits to welfare, including prior to and during slaughter. So while some religious beliefs may not assist in improving animal welfare, others may be positive for welfare, and scientific evidence may assist in favouring the latter.

Furthermore, there has often been less attention given to animal treatment earlier in life, in both small-scale and commercial production systems. Islamic texts, including sayings of the Prophet Mohammad (peace be unto him), provide considerable support for the importance of animal welfare during handling and husbandry: for example, one such text says 'The worst of shepherds is the ungentle, who causes the beasts to crush or bruise one another'. It can therefore be argued that many commercial practices of animal rearing, feeding, handling and transport do not follow strict Islamic guidelines, and that meat produced by such practices (for example from animals housed intensively) cannot properly be described as Halal. There is considerable potential for applied ethologists to work with authorities, including religious authorities, to address such issues with mutual benefit. As an example, the World Society for the Protection of Animals (WSPA, now World Animal Protection) developed a draft standard for welfare of livestock in collaboration with the International Halal Integrity Alliance (Rahman *et al.*, 2009) that requires such work for its implementation. For instance, the standard for housing and penning requires that 'Animal accommodation should allow all animals space to stand, turn round, stretch, sit and lie down comfortably, and to escape from each other'. If this standard is to be implemented, it will need specification by applied ethologists for the particular animals and conditions (including climates) for which it is to be used.

13.5 Economic variation

Economic issues clearly affect the use of, and attitudes towards animals, and are also sensitive. Countries that have difficulty feeding their people understandably place more emphasis on food availability and raising the standard of living for people than on the welfare of animals. This is one reason why concern for animals is often less active in developing than in developed countries. However, this is not always so: as noted in the Introduction, many of the world's poorest people depend on animals for income, social status and security as well as food and clothing. The survival, performance and welfare of their animals are important to them, and many measures to improve animal welfare can also benefit their owners (McCrindle, 1998). This section outlines the principles involved, which suggest that future development of animal science, including applied ethology, in developing countries will identify many such opportunities.

Sometimes there is a negative relationship between animal welfare and production: increasing production reduces welfare, or conversely increasing animal welfare costs money, at least in the short term. But sometimes the relationship is positive: both welfare and production can be improved to benefit both animals and people. This has been well expressed by John McInerney, an agricultural economist at the University of Exeter (Figure 13.4). Here it is important to emphasise that there are many situations, perhaps even more so in developing than in developed countries, where better understanding, treatment and care of animals can benefit owners as well as animals. Sometimes this requires help with information and perhaps financial outlay. For example, problems of animal disease and poor health are widespread. If owners can be helped with advice on veterinary care or husbandry, and perhaps also with basic costs for medicines and housing, both they and their livestock will gain. Indeed, the conventional approach to farming, emphasising financial efficiency, has not always identified the best methods to achieve these aims. It took an alternative approach, looking at individual animals and aimed at reducing problems for the animals concerned, to identify the fact that understanding animal behaviour can improve design of handling systems and hence efficient use of labour in handling livestock. By considering the impact of the facilities on the animal, US behavioural scientist Temple Grandin (2007) devised systems such as curved races (with no sharp corners) for cattle, that decreased animal stress and injuries, improved working conditions for operators, and increased meat quality, yield and profit.

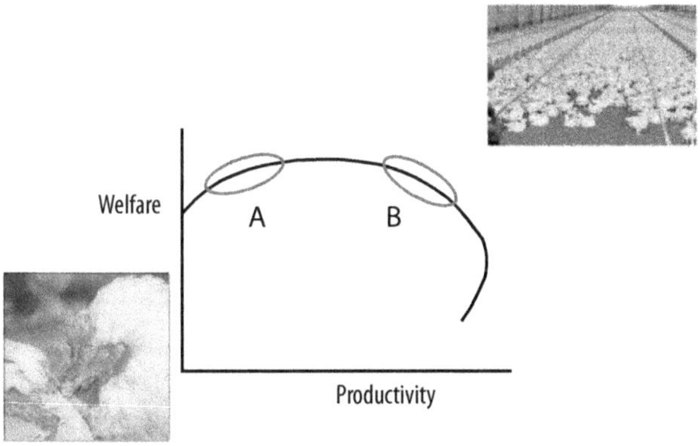

Figure 13.4. McInerney's (2004) hypothetical relationship between animal welfare and productivity. Up to a point, increases in productivity arising from better nutrition, health and housing should also enhance animal welfare. For example, improvement of health of the chicken in the left picture could be represented by a move up the curve within envelope A. Beyond the point of maximum welfare, though, increases in productivity may be possible from greater intensification, genetic selection and pharmaceutical intervention, at a cost to the welfare of animals. Thus increasing stocking density in the house shown in the right picture might involve a move down the curve within envelope B, at least in the short term. However, it is also possible that higher density might increase the long-term risk of disease, in which case reducing density will benefit both welfare and long-term productivity (a move within envelope A). With long-term, broad perspectives, there are more positive relationships between animal welfare and production than generally supposed, useful in developing as well as in developed countries.

This approach has been implemented in developing countries such as Uruguay, where slaughterhouse design and practice has been modified as suggested by Grandin (Appleby and Huertas, 2011), as well as in developed countries.

Businesses at all stages of the food chain – producers, processors and retailers, including global companies active in developing countries – are now rapidly recognising the importance of livestock care, both because of the direct advantages just outlined and because of interest and concern among their consumers. One of the leaders in this field is McDonald's, which started to implement minimum requirements for care of the animals supplying their products in about 2000. Foremost in designing and leading their welfare assurance programme has been Temple Grandin. Another prominent example is Nestlé, one of the biggest food companies globally. My colleagues and I at World Animal Protection are working with Nestlé to assure appropriate animal treatment on farms providing products to their many thousands of supply companies around the world.

Economic opportunity also arises through trade. Many companies and countries are keen to export animal products, and while that may sometimes lead to competition on prices (with negative impacts on animals) it can also be a mechanism for the application of standards. Thus many importing countries, for example in Europe, have standards for animal welfare as well as food safety. These standards have to be met by suppliers, including in their slaughterhouses. World Animal Protection has been involved in humane slaughter training in several countries, including China and Brazil, for this reason. These programmes are developed and led by veterinarians and behavioural scientists. To date, changes that they have introduced to slaughterhouse facilities and practices have improved the welfare of over 8 billion animals.

In addition, as previously said, in most or all countries of the world, the incomes of at least some of the population are rising. This is increasing both the attention given to issues other than food security, including animal welfare and environmental protection, and the opportunities (and sometimes funding) to address those issues. One development from this has been food labelling programmes, for quality, for environmental protection (especially organic) and for animal welfare assurance – such as Freedom Food, Certified Humane and products assured by the Global Animal Partnership. The standards for all of these are drafted and kept up to date by committees in which most members are scientists, and ethologists are prominent in all of these programmes.

This is not to say that priority should be given to the preferences of affluent people over those who are poor or malnourished. On the contrary, applied ethology is helping to address the animal-related problems in all sectors of the population – and can do so more in future. It has already been pointed out that there is considerable scope for expansion of applied ethology in developing countries, and we should now recognise that such expansion is important. This science can help developing countries make many improvements, in areas such as the economics of livestock production, food security and environmental protection. Sometimes it can do so by helping an existing situation, such as pre-slaughter handling, with minimal cost interventions. But achieving even that in many countries requires investment in education and infrastructure. Furthermore, some of the problems in developing countries will require a different approach, not just taking knowledge from abroad and applying it locally. This suggests a particular need for local ethologists within those countries. The ISAE should give more consideration to helping in the identification, training and support of such people.

13.6 Political variation

All the aspects of variation considered so far – social, cultural and economic – are affected by political processes. These set the context for science, for economics and for the use of animals, not just through legislation but in areas such as education. For example, in the European Union (EU) there is extensive legislation protecting animals at national and EU level. By contrast the USA has very little national (federal) legislation on animal protection: some legislative measures are taken by individual states, but much more is left to the free market approach than in Europe. Thus there is variation in the degree to which governance is controlled by politicians or by other stakeholders: the involvement of other stakeholders is a reminder that politics does not just involve the actions of professional politicians but includes developments in policy and public affairs at multiple levels. Some applied ethologists work in government, while others are involved in trade associations, scientific societies (notably ISAE) and non-governmental organisations concerned with animal protection and other matters such as conservation. To give two examples of ethologists working for NGOs, after many years of teaching and research in applied ethology at the University of Edinburgh, I moved to The HSUS and then WSPA (now World Animal Protection), because I was keen to try a different approach to getting change for animals actually implemented. As a second example, the current Director of the Universities Federation for Animal Welfare, Robert Hubrecht, has a background in primate and canine ethology.

To return to the earlier theme of internationalism, while the ISAE has long worked internationally, other layers of politics are also increasingly international. Don Broom (Chapter 14) discusses the contributions of ISAE members to deliberations and decisions of the Council of Europe and the EU. Other groups of countries, such as the Association of Southeast Asian Nations (ASEAN), are also active on trade and other issues that strongly affect animals; no such group invites as much advice on applied ethology or animal welfare as the EU, but avenues for such advice and implementation are beginning to open. For example, when the EU negotiates Free Trade Agreements (FTAs) with other countries, it sometimes requires consideration of animal welfare. This was particularly successful in the EU's FTA with Chile, which included provisions for animal protection (particularly during transport and slaughter) and veterinary training and support. This led to increased involvement by government personnel, by livestock producers in relation to global trading, and by academics, with technical collaboration not only between the EU and Chile, but also within the region. As a result, Chile established a Collaborating Centre on Animal Welfare with Uruguay with the support of the World Organisation for Animal Health (OIE).

Don Broom (Chapter 14) also mentions contributions of applied ethologists to the OIE, and the importance of these – including by Don himself – cannot be overstated. In 2005 the first global standards on transport and slaughter were agreed unanimously by the member countries (which at the time of writing in 2015 number 180). The OIE has since produced standards for the rearing of broiler chickens and cattle, and the treatment of street dogs and laboratory animals, and is working on standards for other farm animals and equids. All of these standards have had input from applied ethologists in drafting, and will require much more input for implementation. Their implementation (which is starting in a number of countries, for example Japan) will lead to improved animal management and welfare in the majority of OIE member countries. This issue is also gaining ground with another major intergovernmental decision-maker, the World Trade Organisation (WTO). In 2014 the WTO upheld an EU ban on importing seal products from Canada and Norway, established because EU citizens believe the methods used to kill seals are inhumane. The judgment was made on the basis of a clause in the WTO's founding agreement that countries can take measures 'necessary to protect public morals'. This was the first international decision that animal welfare can be protected in trade, and again it was informed by input from applied ethologists. Andy Butterworth of the University of Bristol and colleagues provided scientific evidence of the welfare problems in the seal hunt (Butterworth and Richardson, 2013), and contributed to the hearings that produced the WTO's judgment (Butterworth, 2014).

While intergovernmental organisations have an increasing impact, it remains true that political systems within countries have huge influence on animal use and treatment, including the scientific study of animal use, and the participation of scientists in decision making about how animals are to be managed and treated. There are considerable differences in political systems between countries, some of them receptive to practice and advice by applied ethologists but others more resistant. It is clear, though, that both the need and the opportunities for applied ethologists in these processes are increasing. As one example, in response to the OIE process of establishing animal welfare standards, Japan decided to produce its own rules for farm animal management. That process was led by Japanese ethologist, Shusuke Sato (Chapter 11), who also gave the Wood-Gush Memorial Lecture at the ISAE Congress in 2015.

Here is one final, major, international demonstration of the vital roles of applied ethology. As discussed, animal management has impacts on many critical issues such as hunger, poverty, disease control and environmental protection. This was recognised by the 2012 UN Rio+20 Conference on Sustainable Development, which included the following wording in the conference outcome document 'The Future We Want' (emphasis added):

> 111. We reaffirm the necessity to promote, enhance and support **more sustainable agriculture, including crops, livestock, forestry, fisheries and aquaculture**, that improves food security, eradicates hunger, and is economically viable, while conserving land, water, plant and animal genetic resources, biodiversity and ecosystems, and enhancing resilience to climate change and natural disasters. We also recognise the need to maintain **natural ecological processes that support food production systems**.
>
> 112. We stress the need to enhance **sustainable livestock production systems, including through improving pasture land** and irrigation schemes in line with national policies, legislation, rules and regulations, enhanced sustainable water management systems, and efforts to eradicate and prevent the spread of animal diseases, recognising that the **livelihoods of farmers including pastoralists and the health of livestock are intertwined**.

Those bold phrases (bold in both senses of the word) could hardly be a clearer call to arms for applied ethologists. On this subject, I called the talk I gave at the ISAE Japan Congress in 2015 'Applied ethology for ever: animal management and welfare are integral to sustainability'.

13.7 Internationalism – working together

Discussion of the variation between countries in cultural, economic and political respects emphasises the fact that some have made more progress than others in addressing problems concerning human-animal interactions. However, a theme of this chapter is that global approaches including communication are enabling advantages – such as the opportunities for applied ethology to help both people and animals – to be shared.

Intergovernmental organisations such as the OIE are playing a major part in this, and in addition in most parts of the world there is increasing involvement of stakeholders (including scientists and civil society) in national and international decision making.

It is worth emphasising this point, because until recently the relationship between international decision making and public opinion was generally tenuous. When trade representatives of the member countries of the WTO or national veterinary officers who make up the constituency of the OIE met, they represented the citizens of their countries to some extent but represented government, industries and companies to a much greater extent. So the influence of public opinion about animal welfare and other matters on these discussions was in the past weak at most. The influence of public opinion is now increasingly strong – or to put it another way the international involvement of all relevant stakeholders including scientists is now more complete, along similar lines to that within countries – because of the growth of internationalism.

I sometimes think that the word 'stakeholder' has had more impact on recent public affairs than almost any other word in the dictionary, because once it was established in public discourse it could not be ignored. As applied ethologists, we need to make it clear that in discussions of animal use, we are among the stakeholders who also cannot be ignored. The ISAE has been effective in doing so in those countries in which it is active. But as commented above, this suggests an urgent need to increase ISAE representation and activity in countries where it is almost or completely absent, particularly in Eastern Europe, Africa, Asia and Latin America.

13.8 Conclusions

It is often said that communication in general is irreversible, in that once you have learned something you cannot unlearn it. That is why the international work and communication of the ISAE is essential. Gaining and publicising a scientific understanding of animals and their behaviour can benefit people, animals and environment, and is therefore central to the sustainability of our future. That understanding counters the negative aspects of globalisation such as the competition for ever-cheaper animal production caused by increasing trade. In addition, trade can be positive for animal care, as discussed in Section 13.5 on economics, because it is often involves the application of food safety and animal welfare standards. In general, there is strong reason to believe that increased attention to animal management and animal welfare worldwide is leading not to a 'levelling down' effect, as was feared by some people, but to 'levelling up'.

Variety will not disappear, and our understanding of animals and their behaviour will continue to be enhanced by the rich variety of environments, societies and animals around the world. In the development of applied ethology, in developed and developing countries and emerging economies, variety is the spice of life.

References

Appleby, M.C. and Huertas, S.M., 2011. International issues. In: Appleby, M.C., Mench, J.A., Olsson, I.A.S. and Hughes, B.O. (eds.) Animal welfare. CAB International, Wallingford, UK, pp. 304-316.

Appleby, M.C., Mench, J.A., Olsson, I.A.S. and Hughes, B.O., 2011. Animal welfare. CAB International, Wallingford, UK.

Broom, D.M., 2016. World impact of ISAE: past and future. In: Brown, J.A., Seddon, Y.M. and Appleby, M.C. (eds.) Animals and us – 50 years and more of applied ethology. Wageningen Academic Publishers, Wageningen, the Netherlands, pp. 269-278.

Butterworth, A., 2014. The moral problem with commercial seal hunting. Nature 509: 9.

Butterworth, A. and Richardson, M., 2013. A review of animal welfare implications of the Canadian commercial seal hunt. Marine Policy 38: 457-469.

Food and Agriculture Organisation (FAO), 2009. Capacity building to implement good animal welfare practices. Available at: http://tinyurl.com/h8fue7z.

Grandin, T., 2007. Livestock handling and transport. CAB International, Wallingford, UK.

Ludvigsen, J.B., Empel, J., Kovacs, F., Manfredini, M., Unshelm, J. and Viso, M., 1982. Animal health and welfare. Livestock Production Science 9: 65-87.

McCrindle, C., 1998. The community development approach to animal welfare: an African perspective. Applied Animal Behaviour Science Journal 59: 227-233.

McInerney, J.P., 2004. Animal welfare, economics and policy. Report to department of environment, food and rural affairs. Available at: http://tinyurl.com/hmkau9y.

Rahman, S.A., Appleby, M.C., Kolesar, R. and Parente, S., 2009. Draft international halal integrity alliance standard for animal welfare. World Halal Forum, Kuala Lumpur, Malaysia.

Raj, A.B.M., 2004. Cultural, religious and ethical issues associated with animal welfare. In: Global conference on animal welfare: an OIE initiative. OIE, Paris, France, pp. 235-247.

Rodenburg, B., 2012. Animal welfare. Animal Welfare Journal 21: 301-302.

Sato, S., 2016. Applied animal behaviour science in Japan and the culture of '*aigo*'. In: Brown, J.A., Seddon, Y.M. and Appleby, M.C. (eds.) Animals and us – 50 years and more of applied ethology. Wageningen Academic Publishers, Wageningen, the Netherlands, pp. 227-240.

Part 4. Future directions

14. World impact of ISAE: past and future

D.M. Broom
Centre for Animal Welfare and Anthrozoology, Department of Veterinary Medicine, University of Cambridge, Madingley Road, Cambridge CB3 0ES, United Kingdom; dmb16@cam.ac.uk

Abstract

When the ISAE was formed in 1966, studies of farm, companion or other animals kept by humans were thought of as a side issue for real science, and those who did such work were considered to be second-rate scientists. Since that time, applied ethology has moved on to become an established and respected scientific discipline. Many of the major contributions to our understanding of motivation in recent years have come from work on domestic animals. The causation of many behaviours that are used as welfare indicators has been explored, and much work on the mechanisms controlling behaviour development has been done on farm animals. On a global scale, the work of ISAE members, through their collective knowledge and expertise, has played a significant role in the development of animal welfare legislation and global policies. This chapter explores how far reaching the impact of the ISAE has been. Since the early 1980s the work of ISAE members has been widely used by the EU in determining the current state of the scientific knowledge prior to drafting new animal welfare legislation. Consistently ISAE members have been part of the EU scientific working groups as well as serving on committees for the World Organisation for Animal Health and the Food and Agriculture Organisation of the United Nations. The work of ISAE members continues to spread through publications in Applied Animal Behaviour Science, the official journal of the ISAE, and other related journals. Annual ISAE conferences and regional conferences have focused on broadening the scope of and influence of this science, with meetings aimed at increasing the utilisation of research in particular areas, such as encouraged collaboration between applied ethologists and agricultural economists. In recent years, the ISAE has become involved with education and information provision on animal welfare. Looking forward, it is anticipated that the ISAE will continue to have strong impact in education as well as in research related to genetics, environment and management of captive and domestic species.

Keywords: applied ethology, OIE, animal welfare, legislation, education

14.1 Introduction: developing the science

When the Society for Veterinary Ethology (SVE) was founded in the UK in 1966, animal behaviour was a young and expanding science and animal welfare was not thought of as a scientific concept. I was working for my PhD, on the effects of experience on behaviour development in domestic chicks, in the Department of Zoology, Sub-Department of

DOI 10.3920/978-90-8686-828-5_14,

Animal Behaviour in Cambridge University at that time. My supervisor was Bill Thorpe, who introduced the concept of animals having needs with a biological basis, including needs to show certain behaviours, when he was a member of the Brambell Committee (Broom, 2011, 2014; Thorpe, 1965). My principal advisor was Robert Hinde whose publications on motivation made most ethologists realise that understanding how the brain controls what animals do and when they do it is a central issue in animal behaviour, including human behaviour (Hinde, 1970). Thorpe had clearly been influenced by Hinde when he was writing about needs. Thorpe and Hinde were trying to persuade psychologists and psychiatrists that ethology had valuable lessons for these subject areas, Thorpe by organising regular seminars with Oliver Zangwill, then Professor of Psychology in Cambridge (Thorpe and Zangwill, 1961), and Hinde by a long series of publications on the behaviour of non-human primates and children (Hinde, 1974). However, although applying ethology to humans, the Sub-Department of Animal Behaviour in Cambridge was not then applying ethology to domestic animals. The emphasis of Niko Tinbergen's ethology group in Oxford was the evolution of behaviour and domestic animals were thought of as changed by humans, perhaps as degenerate (Hemmer, 1983) and therefore inappropriate for the study of adaptive behaviour. There was not much thought of applied ethology in the other main ethology centres in the world, for example those of Baerends and van Iersel in the Netherlands, Lorenz in Germany, Fabricius in Sweden, Hediger in Switzerland, or Lehrman and Rosenblatt in the USA. Lorenz had previously written about dogs and Hediger about zoo animals but none of the mainstream ethologists were working principally on applied ethology. Scientists who moved their research area to ethology in the 1960s, and contributed to the development of applied ethology, included Wood-Gush in Scotland, Signoret in France and McBride in Australia.

Where the main aim of an ethologist was to understand how the mechanisms controlling behaviour had evolved, studies of farm, companion, working, laboratory or other animals kept by humans were thought of as a side issue for the real science. Those who did such work were considered to be second-rate scientists. However, if the aim of the ethologist, or other behavioural scientist, was to understand the causation of behaviour, it was acceptable to study laboratory animals. Some of those who worked on motivation also started to use farm animals as subjects but this was also thought of by many mainstream ethologists as questionable, perhaps because the researchers might have tainted, applied objectives. If the principal question for the researcher was how to increase animal productivity, or make animal management easier and more efficient, or how to improve animal welfare, it was obvious that the species used commercially should be the subject of the research. Many of those who carried out research with applied objectives were also investigating fundamental questions of ethology in the course of their work.

Since the early days of the SVE, ethology has moved on to become an established discipline in science and its application to animals used by humans has developed rapidly (Broom, 2011) and has had a major impact on the public. The aspects of behaviour best known to many of the general public are now abnormal behaviours, considered as an indicator of poor welfare of animals, and studies of learning and sentience in domestic animals (Broom, 2014). Many of the major contributions to our understanding of motivation in recent years have come from work on domestic animals. The causation of many behaviours that are used as welfare indicators has been explored, and much work on the mechanisms controlling behaviour development has been done on farm animals. There has also been rapid progress in studies of cognitive abilities and feelings of farm animals and companion animals, as well as on wild animals in captivity. Some of the researchers have had objectives to show that the animals have complex abilities and should not therefore be treated in ways that result in poor welfare. At the same time they could find out about the cognitive and emotional mechanisms themselves. Studies of interactions between farm or companion animals and humans have also been of great benefit to those who use these animals, and many studies of animal behaviour have helped our understanding of human non-verbal behaviour.

The world impact of the Society for Veterinary Ethology (SVE), called International Society for Applied Ethology (ISAE) after 1991, is described below with reference to a range of activities of its members. The biggest impact has come from the actual research results. Scientific understanding of cognition, motivation and the important fundamental issue of the methods of animals coping with negative impacts of their environment, has been greatly aided by the research of applied ethologists. Many of the research results are reviewed by Broom and Fraser (2015) and discussed in other chapters of this book so will not be detailed here. Scientists, government officers, veterinarians, farmers and other animal users now use the information about animal welfare and behaviour produced by the many thousands of publications of ISAE members. Some future areas where there is likely to be further impact are discussed in the last section of this chapter.

14.2 Actions on behalf of ISAE in international committees

The constitution of SVE/ISAE, stated in 1990, included under Aims of the Society: 'To provide a pool of expertise to government, industry and animal welfare organisations which deal with problems involving animal behaviour'. As a consequence, if ISAE has been formally asked to send experts to meetings, this has been done where possible. There have also been many occasions where ISAE members have been appointed to committees because of their scientific knowledge and have informally reported back to ISAE about their activities, as discussed in the next section.

The Council of Europe has produced a series of Conventions that have had very important influence in the world. In the years before and soon after the European Union (EU) started, the Council of Europe, which had representation from all European countries except communist countries, was the principal forum for discussion of ethical issues affecting all countries and its meetings on many subjects were often attended by observers from outside Europe. Council of Europe Recommendations, once adopted by the member countries, were legally binding. The Recommendations of the Standing Committee of the European Convention for the Protection of Animals Kept for Farming Purposes (T-AP) were generally adopted and put into legislation within one year. Hence they formed the basis for much EU and national legislation. This continued until the expansion of the EU to over 20 countries, after which the EU committees became relatively more important and the influence of the Council of Europe Standing Committees waned.

The main stimulus for the discussions that led to the European Convention for the Protection of Animals kept for Farming Purposes (Council of Europe, 1976) was Ruth Harrison's book '*Animal Machines*' (Harrison 1964) and the ensuing Brambell Report (Command paper 2386, 1965) and legislation in the UK. Ruth Harrison later became a member of SVE/ISAE and used the scientific information available at its meetings to good effect (Broom, 2013). This same stimulus had led to the formation of SVE. In the early 1980s, after the T-AP Committee was set up, SVE was asked to send a scientist to provide evidence for the Committee. In order to pay the costs of travel and accommodation for ISAE scientific experts, the Society had to set up a fund and asked various organisations for donations to this fund.

When the welfare of pigs was being discussed, Gerrit van Putten attended meetings in Strasbourg and when laying hens were discussed, Henrik Simonsen attended. In 1987 when the welfare of cattle was to be discussed, Donald Broom was appointed by SVE Council to attend and I continued to do so until 2000, during which time there were discussions of Recommendations on the welfare of cattle, goats, sheep, domestic fowl, ducks, Muscovy ducks and mulards (Muscovy × mallard hybrids used for foie gras production), geese, ratites, turkeys, animals kept for fur production, as well as amendments to the Convention that dealt with several issues including genetics and farm animal welfare. Other ISAE members who acted as scientific advisors have included Neville Gregory on the European Convention on the Protection of Animals for Slaughter, Code of Conduct on Stunning and Slaughter; Frank Ödberg on the European Convention for the Protection of Vertebrate Animals used for Experimental and other Scientific Purposes; and Markus Stauffacher on the European Convention for the Protection of Companion Animals. Other ISAE members who contributed scientific information to the T-AP Committee included Ingvar Ekesbo who was the Swedish representative and Andreas Steiger who was the Swiss representative. Both of these also served as Chairman of the T-AP Committee. The Council of Europe Recommendations included many that have subsequently been incorporated in legislation, not just in Europe but in many other countries around the world.

Another organisation with international impact that has received scientific advice from ISAE is the Association for Assessment and Accreditation of Laboratory Animal Care (AAALAC), which sets standards for laboratory animal management, treatment and care. The current ISAE advisor is Brianna Gaskill.

14.3 Other actions of ISAE members on international committees

The research of scientists, such as ISAE members, becomes known in the world to some extent if it is presented at scientific meetings but it must be published in respected scientific journals and books if it is to have any real impact. The information can then be found by those who might be able to use it. However, mere presence within the scientific literature does not necessarily lead to useful influence. The information has to be put in a form that is intelligible to the user and in a place that is accessed. Work on domestic animal behaviour and welfare can be communicated to students and users by summarising the results in books, in magazines and journals read by animal users, on internet sites, in newspapers and in radio and television programmes. Some ISAE members have done this and there is much public interest in such material. Governments and major animal user and animal protection groups need scientific information in high quality reviews in order to utilise it. As explained above, the Council of Europe, whose committees are made up of national representatives, sought scientific advice backed up by reference to actual publications when producing its influential Recommendations. Other national and international bodies are now adopting this 'science based' approach in relation to animal welfare.

The European Economic Community (later EU) started to organise small scientific meetings, publish their proceedings and use scientific working groups to summarise the state of scientific knowledge prior to drafting new legislation relevant to animal welfare in the 1980s. Publications included Bessei (1982), Baxter (1983), Wiepkema *et al.* (1983) and Smidt (1983). SVE members who were members of scientific working groups included Donald Broom, Ian Duncan, Gerrit van Putten, and Piet Wiepkema. The working group meetings were organised by staff of the relevant Directorate of the European Commission and reports were submitted directly to the European Commission. Such reports preceded each of the Directives on animal welfare matters, e.g. on welfare during transport, welfare of calves and welfare of pigs. In 1990, this was formalised when the Scientific Veterinary Committee was formed with a section on animal welfare whose chairman was Donald Broom. This was followed in 1997 by the Scientific Committee on Animal Health and Welfare run by the Public Affairs Directorate of the European Commission and then in 2003 by the Scientific Panel on Animal Health and Welfare of the European Food Safety Authority (EFSA: see also Chapter 7; Algers, 2016). The members were solely scientists selected for scientific merit and no one was a representative of a member state, industry or animal protection society. ISAE members who were Chair or Vice Chair of these committees were Pierre Le Neindre, Donald Broom and Jörg Hartung. Others who have been members of the committees, as well as having major influence on scientific development in their own countries, include: Piet Wiepkema, Jean-Pierre Signoret, Robert Dantzer, Per Jensen, Harry Blokhuis, Marina Verga, David Morton, Xavier Manteca, Bo Algers, Linda Keeling, Mohan Raj, Lotte Berg and Antonio Velarde. Figure 14.1 shows retiring members of the EFSA Panel in 2012. Many other ISAE members served on the working groups of these committees.

The reports, later called opinions, of these committees concerned a wide range of animal welfare issues and were used in subsequent EU legislation discussions. They are available on the internet and have been used by animal producer and animal protection organisations as well as by governments. The sequence of events that led to EU legislation on veal calves is described by Broom (2009, 2014).

ISAE members have also served on OIE, FAO, and other international and national committees. For example, David Fraser is on the OIE (World Organization for Animal Health) committee, he and Mateus Paranhos da Costa have spent time in Rome advising FAO (Food and Agriculture Organization of the United Nations). Francisco Galindo,

Figure 14.1. Retiring members of the EFSA Scientific Panel on Animal Health and Welfare in 2012: (from left) Don Broom, Mike Sharp, Philippe Vannier, Martin Wierup, Oriol Ribo (EFSA staff), Mo Salman and David Morton.

Stella Huertas, Paolo Dalla Villa, Lindsay Matthews, Clive Phillips and others have been involved with OIE regional activities.

The consequences for animal welfare of ISAE contributions to Council of Europe, EU, OIE, FAO and other committees have been substantial. It is now recognised in many countries of the world that evidence from animal welfare science is valuable, and should be used in the formulation of laws or codes of practice. The scientific information has been used in the laws and codes that require stunning before painful killing methods are used to slaughter food animals. Animal transport in over-stocked vehicles, with journey lengths, handling procedures and driving methods that cause poor welfare have become less common. Housing systems that do not meet the needs of the animals, such as close-confinement of calves and pregnant sows in pens where they cannot turn around, keeping of hens in battery cages, and systems that lack manipulable materials for pigs or bedding for several species, are gradually being prohibited or made economically unviable because consumers are forcing food retail companies to ban practices. Some of the changes have still occurred only in pioneer countries or states, whilst others are now supposed to be changed in all 180 countries that are signatories to the World Organization for Animal Health (OIE).

14.4 ISAE and scientific publishing

When Elsevier set up the journal Applied Animal Ethology, later to become Applied Animal Behaviour Science, the first Editor was a founding member of SVE, Andrew

Fraser. Andrew continues to be active in writing about the behaviour and welfare of domestic animals (e.g. Broom and Fraser, 2015; Fraser, 2010). In 1993 there was an agreement made between ISAE and Elsevier that the society would liaise over the appointment of the Editor of Applied Animal Behaviour Science and that members would receive a reduced subscription rate. As a result, Per Jensen, Carol Petherick and Sylvie Cloutier have been editors. The academic standard of this journal has clearly been helped by ISAE involvement. Members of ISAE have had editorial involvement with several other journals such as Animal Behaviour, British Poultry Science, Animal Welfare and Animals.

14.5 Collaborative activities with other organisations

After becoming an international organisation in the 1970s, SVE/ISAE has held international conferences in many countries around the world. There have also been a series of regional meetings. Some of these conferences have been aimed at helping in the utilisation of research. For example, as early as 1998 an ISAE meeting was held at the Royal Veterinary College in London entitled: 'Can ethological research contribute to standards in marketing schemes?' This encouraged collaboration between applied ethologists and agricultural economists.

In addition, ISAE has had links to animal science and veterinary societies and joint meetings. Examples include the ISAE joint meeting with the European Association for Animal Production and the Deutsche Veterinärmedizinische Gesellschaft in Germany in 1993, the joint meeting with the Association for the Study of Animal Behaviour on 'Motivation, cognition and applied aspects' in 1994 and a series of meetings from 2002 with the British Society for Animal Science and the World Poultry Science Association in the UK. In Australia and New Zealand, and also in Brazil, there have been many collaborative meetings involving ISAE and veterinary, animal science and animal protection organisations.

Many of the most important consequences of the original scientific work and spread of information about the behaviour and welfare of animals resulting from ISAE activities have influenced the education of veterinary, animal science, biology, psychology and philosophy students. Whilst most of those who took courses 40 years ago heard little or nothing about the subject, many more students today take courses in this area (Broom, 2005; Fraser, 2008; Gallo *et al.*, 2010; Hewson *et al.*, 2005). The relationship between ethics and this subject is discussed in the context of many religions and philosophies (Szücs *et al.*, 2012; Waldau and Patton, 2009). The knowledge about how to manage farm, companion and other animals that is available via books, on-line sources and instructors includes much more good quality information about behaviour currently than it did in the past.

14.6 Future international influence and impact of ISAE

Looking forward, it is anticipated that the ISAE will continue to have impact in the many ways described above. In particular, there is a continuing need for scientific studies of animal welfare, and precise reviews of these, so that good laws and codes of practice can be formulated. There are also exciting new possibilities for education and information provision. The ISAE is now providing scientific evaluation for the Animal Welfare Science Hub. This internet source provides information about animal welfare, how to teach aspects of it, and what teaching material is available. The hub was initiated by the Animal Welfare Indicators science research project, which used some of the valuable information produced by the EU Welfare Quality project. Those who visit www.animalwelfarehub.com can find learning tools, details about welfare courses, protocols for welfare assessment and welfare research information. ISAE validates the quality of the information before it is put on the website. There are also ISAE contributions to the FAO animal welfare gateway and other internet sites, although these do not have the content quality checking that the animal welfare science hub has.

In future, the need for good quality information, such as that on the Animal Welfare Science Hub, will continue and with this comes an ongoing role for scientific societies, such as ISAE, to expand knowledge and check quality. Behaviour and welfare subjects requiring ongoing study include research related to genetics, environment and management of captive and domestic species. Whilst the genetic selection of domestic animals has promoted some adaptations that improve animal welfare and ease of animal management, selection for high productivity has resulted in major animal welfare problems. These negative effects of selection can be carefully investigated, and means of penalising and controlling the harmful traits found. Controlling the early experience of domestic animals can help them to adapt more rapidly to their captive environment. This may be particularly useful in extensive farming situations where the animals have to adapt to humans (Hemsworth and Coleman, 2010; Le Neindre *et al.*, 1996).

It is also likely that there will be developments in how to utilise the new information about the cognitive and emotional functioning of animals. Developments in welfare assessment, appreciation of system sustainability and ethical procedures will affect human actions in relation to keeping animals (Broom, 2014). The ISAE can contribute to such developments and, in doing so, can further benefit the animals that we use and the people who use them.

Acknowledgements

I thank many colleagues, in particular Pierre Le Neindre, for helpful suggestions.

References

Algers, B., 2016. Applied ethology in the EU: development of animal welfare standards and actions. In: Brown, J.A., Seddon, Y.M. and Appleby, M.C. (eds.) Animals and us – 50 years and more of applied ethology. Wageningen Academic Publishers, Wageningen, the Netherlands, pp. 155-173.

Baxter, S.H., 1983. Farm animal housing and welfare. Martinus Nijhoff, Dordrecht, the Netherlands.

Bessei, W., 1982. Disturbed behaviour in farm animals. Hohenheimer Arbeiten 12. Eugen Ulmer, Stuttgart, Germany.

Broom, D.M. and Fraser, A.F., 2015. Domestic animal behaviour and welfare, 5th edition. CABI, Wallingford, UK.

Broom, D.M., 2009. Animal welfare and legislation. In: Smulders, F.J.M. and Algers, B. (eds.) Welfare of production animals: assessment and management of risks. Wageningen Academic Publishers, Wageningen, the Netherlands, pp. 339-352.

Broom, D.M., 2011. A history of animal welfare science. Acta Biotheoretica 59: 121-137.

Broom, D.M., 2013. Ruth Harrison's later writings and animal welfare work. In: Dawkins, M. (ed.) Animal machines. Reprint with new commentaries. CABI, Wallingford, UK, pp. 21-25.

Broom, D.M., 2014. Sentience and animal welfare. CABI, Wallingford, UK.

Broom, D.M., 2005. Animal welfare education: development and prospects. Journal of Veterinary Medical Education 32: 438-441.

Command Paper 2836, 1965. Report of the technical committee to enquire into the welfare of animals kept under intensive husbandry conditions. Her Majesty's Stationary Office, London, UK.

Council of Europe, 1976. European convention for the protection of animals kept for farming purposes. Available at: http://tinyurl.com/h2brumw.

Fraser, A.F., 2010. The behaviour and welfare of the horse, 2nd edition. CABI, Wallingford, UK.

Fraser, D., 2008. Toward a global perspective on farm animal welfare. Applied Animal Behaviour Science 113: 330-339.

Gallo, C., Tadich, N., Huertas, S., César, D., Paranhos da Costa, M. and Broom, D.M., 2010. Animal welfare education in Latin America. In: Proceedings of the International Conference on Animal Welfare Education: Everyone is responsible. October 1-2, 2010. Brussels, Belgium, pp. 90-97.

Harrison, R., 1964. Animal machines. In: Dawkins, M. (ed.) Animal machines, CABI, Wallingford, UK, pp. 31-207.

Hemmer, H., 1983. Domestikation: verarmung der merkwelt. Vieweg, Braunschweig, Germany.

Hemsworth, P.H. and Coleman, G.J., 2010. Human-livestock interaction: the stockperson and the productivity and welfare of intensively farmed animals. CABI, Wallingford, UK.

Hewson, C.J., Baranyiova, E., Broom, D.M., Cockram, M.S., Galindo, F.A., Hanlon, A.J., Hanninen, L., Lexer, D., Mellor, D.J., Molento C.F.M., Odberg, F.O., Serpell, J.A., Sisto, A.M., Stafford, K.J., Stookey, J.M. and Waldau, P., 2005. Approaches to teaching animal welfare at 13 veterinary schools worldwide. Journal of Veterinary Medical Education 32: 422-437.

Hinde, R.A., 1970. Animal behaviour: a synthesis of ethology and comparative psychology, 2nd edition. McGraw Hill, New York, NY, USA.

Hinde, R.A., 1974. Biological bases of human social behaviour. McGraw Hill, New York, NY, USA.

Le Neindre, P., Boivin, X. and Boissy, A., 1996. Handling of extensively kept animals. Applied Animal Behaviour Science 49: 73-81.

Smidt, D., 1983. Indicators relevant to farm animal welfare. Current topics in veterinary medicine and animal science. Martinus Nijhoff, Dordrecht, the Netherlands.
Szücs, E., Geers, R., Jezierski, T., Sossidou, E.N. and Broom, D.M., 2012. Animal welfare in different human cultures, traditions and religious faiths. Asian-Australasian Journal of Animal Science 11: 1499-1506.
Thorpe, W.H. and Zangwill, O.L., 1961. Current problems in animal behaviour. Cambridge University Press, Cambridge, UK.
Thorpe, W.H., 1965. The assessment of pain and distress in animals. Appendix III in Command Paper 2836. Her Majesty's Stationary Office, London, UK.
Waldau, P. and Patton, K., 2009. A communion of subjects: animals in religion, science, and ethics. Columbia University Press, New York, NY, USA.
Wiepkema, P.R., Broom, D.M., Duncan, I.J.H. and Van Putten, G., 1983. Abnormal behaviours in farm animals. Commission of the European Communities, Brussels, Belgium.

15. New answers – old questions; new questions – old answers: how applied ethology is cross-fertilised by other disciplines

P. Jensen
IFM Biology, Avian Behaviour Genomics and Physiology Group, Linköping University, 58183 Linköping, Sweden; perje@ifm.liu.se

Abstract

Applied ethology is mainly concerned with aspects of animal behaviour, which are relevant from the perspective of human utilisation. During the last 50 years, it has moved from being a highly practical and solution-driven branch of science, to becoming a vital part of science, affecting and being affected by fundamental ethology. This chapter gives some examples of fields where cross-fertilisation has been most prominent. Physiology and psychology influenced applied ethology heavily in developing ideas about behavioural needs, and have in turn been enriched by the development of new motivational models, developed to understand farm animal behaviour in restrictive environments. The original gap between evolutionary biology and applied ethology has largely disappeared. Thinking in behavioural ecology has been important in understanding many aspects of domesticated animal behaviour, and new evolutionary theories have been developed by applied ethologists. The increased interest in understanding domestication has been fruitful in both fundamental and applied ethology and sparked a growing research in genetics and epigenetics of behaviour. Personality theory is another example of a vibrant field of fundamental evolutionary biology of today, highly influenced by original work in applied ethology. The chapter finishes with a look towards an exciting future, where applied ethology will benefit from a range of novel technologies and methods developed in various fields of biology.

Keywords: genetics, evolution, personality, motivation, ethology

15.1 Applied ethology – a dynamic science

Applied ethology does not exist in a vacuum – it is consistently fuelled by new questions posed by reality, but also by fundamental science. Farming inventions such as robot milking machines, transponder-feeding devices for pigs and skip-a-day feeding of broiler breeders bring new issues into research. How are the animals coping with technology like this, and how can it be improved to fit their natural behaviour better? But basic ethology (looking at wild animals) is also dynamic, and insights and progress in that field in turn feed our applied science with fuel for thought. New insights with respect to genetics, evolution and neurobiological mechanisms in behaviour inevitably make the

DOI 10.3920/978-90-8686-828-5_15, © Wageningen Academic Publishers 2016

minds of applied ethologists start to tick. What do these novel findings mean with respect to animal breeding, domestication and cognition, for example?

Of course, the results of research in applied ethology also affect fundamental science, even if this is a largely overlooked aspect of the scientific dynamic. Applied ethology does indeed contribute to the development of basic science. In this chapter, I will give some examples of how interactions between basic research, applied ethology and practical husbandry have fertilised all aspects of our understanding of animal behaviour. To me, this is the most rewarding aspect of being a scientist in this field, and I hope others will share this view.

As an undergraduate, I attended a lecture by the Swedish ethologist Lars Fält, who spoke about the behaviour of pigs, and I became fascinated. It had never occurred to me that domesticated animals possessed such a wide range of fascinating behaviour, and I was surprised by the fact that so few pure ethologists actually studied them. How could anyone not be thrilled by a large mammal, which builds nests, cares extensively for large litters and shows a complex social organisation? This lecture was a turning point for me and since then, I have devoted my career to scientific studies of domesticated animals.

15.2 Applied ethology emerged from fundamental questions

I will confess that the concept of 'applied ethology' never appealed to me as a student and young researcher, since I associated the term with very down-to-earth, practical studies of solutions to the various problems of farmers. Questions such as how wide a farrowing crate should be to allow a sow to rise and lie down properly, or how many square centimetres are needed in a battery cage to allow a chicken to flap its wings, simply never caught my scientific curiosity (although I fully appreciated the practical importance of answering such questions). I was and am interested in how animals function and why. So I invented my own term for the field I decided to pursue: 'applicable ethology'.

To me, the best ethological science is applicable. This means that the scientific questions can be of a fundamental nature, focusing on mechanisms and function, but the knowledge produced is also essential for understanding and solving practical problems – at least in a remote perspective. Over the years, I have found that when you address domesticated animals with basic research questions, the applicability comes more or less automatically.

In fact, applied ethology arose when scientists first started to ask fundamental questions on motivation in relation to farm animals. Jerry Hogan, Klaus Vestergaard and Fred Toates are three pioneers who all contributed extensively to this area. For example, would a pig, which has been raised under confined stall conditions for generations and never experienced open air and natural surroundings, still be able to express its natural maternal behaviour if allowed to do so? If so, what does that mean for the welfare of the sow confined in a farrowing stall? And does it matter to a chicken if, after generations spent in battery cages, they have access to a perch or a dust bath?

Hogan and Toates developed theoretical models explaining how the motivation to perform instinctive behaviour develops during ontogeny (Hogan, 2014; Toates, 2004) and came to the conclusion that the behavioural control systems, as developed during evolution, are often preserved in captivity and during domestication. Preventing their proper expression may cause poor welfare and even suffering. Vestergaard, inspired by these ideas, showed in a series of ingenious experiments that dust-bathing motivation in chickens develops largely independently of external stimuli, and concluded that their welfare in battery cages is severely affected by the inability to perform this behaviour properly (Vestergaard, 1982).

Klaus Vestergaard was devoted to the application of fundamental ethology in the world of farm animals. He became a good friend, and together we carried out several experiments on free-ranging pigs. To properly study the motivation of pigs to express their ancestral behaviour, we released groups of pigs into large, naturalistic enclosures and observed them over their different life phases. In a memorable study, we spent 24 h a day for weeks in the forests, recording the onset and progress of nest building in a group of sows. The sows always started with the same sequence of behaviour, including pawing and rooting in the ground, and went on with a more flexible part, which consisted of collecting and arranging nest material (Figure 15.1). Whereas the first part was highly similar in all nest building events, the latter could vary widely in duration and intensity depending on external conditions, such as temperature and precipitation. From this we concluded that internal factors appeared to trigger the behaviour, whereas its detailed structure depended largely on external factors (Jensen *et al.*, 1993).

15.3 How physiology and psychology influence applied ethology

The central and most influential motivational theories have generally lacked proper physiological substrates. Famously, Konrad Lorenz's so called 'psychohydraulic model' (originally a joke term, invented by Desmond Morris) pictured neural processes as water

Figure 15.1. A free-ranging sow building a farrowing nest. The control systems underlying this behaviour have enabled a wide understanding of motivation and behavioural needs, adding fruitful insights to applied as well as fundamental ethology (photo P. Jensen).

flowing through a container, and most other models have used a similar 'black-box' approach.

Of course, black-box models can allow far-reaching insights into how behaviour is controlled and structured, without necessarily considering the detailed physiological mechanisms. For example, Lorenz's model triggered the previously mentioned experiments by Vestergaard on dust-bathing in chickens, showing that depriving birds of the opportunity to perform the behaviour will cause a gradual build-up of motivation, even in birds from genetic strains that lack feathers (Vestergaard *et al.*, 1999).

Even when obviously wrong, black-box models have been helpful in advancing our understanding of behaviour. For example, Baxter (1983) imagined a control-systems model suggesting that nest-building in sows would be controlled by feedback from a smooth nest surface to the skin of the udder. Based on this, he predicted that offering a sow soft flooring would provide the correct feedback, and would reduce her urge to perform nesting before farrowing. Subsequent research showed that this was fundamentally wrong. Even though feed-back as well as feed-forward appeared to be involved in the behaviour, attempts at experimentally short-cutting the system actually led to increased motivation to perform nest-building (Jensen, 1993; Jensen *et al.*, 1993).

However, many scientists are not content with describing behavioural control in terms of flow diagrams and control systems; they want to know the hard facts of the physiological mechanisms underlying the motivation. With respect to nest building in sows, pure behaviour data suggested two distinct control systems working in concert. The mechanisms causing the initial, rather inflexible, rise in nest building appeared to be separate from those regulating the more flexible motivation to collect nest material. My good friend and colleague Bo Algers and his PhD student Heli Castrén explored the endocrine basis for this, and found that a sudden increase in prolactin is the most likely physiological cause of the onset of nest building, whereas the collection of nest material is unrelated to endocrine events (Castrén *et al.*, 1994).

The study of nest building in sows is a nice example of how 'applied' (or rather, 'applicable') ethology can fertilise and advance fundamental science. Using rigorous models for motivational control allowed for precise experiments, with proper, falsifiable hypotheses. It also drove the science towards more fundamental studies of physiological mechanisms. As a result, nest building in domestic sows is today one of the best understood behaviour systems in any species, and this approach can serve as a model for any scientist wishing to understand the basic mechanisms of behaviour. Furthermore, the studies have really advanced our understanding of the behavioural needs of sows. It seems almost impossible to circumvent the drive to nest build before farrowing by means of alterations in the environmental conditions of the sow, since the behaviour is triggered by a hormone which is central to the onset of lactation and maternal care. A sow that will nurse her litter is also a sow that is motivated to build a nest before farrowing!

15.4 Evolutionary biology changes our views

In fundamental ethology, studies of motivation and mechanisms largely went out of fashion during the 1970s in favour of a growing interest in evolutionary and ecological questions about behaviour after seminal contributions from scientists such as William Hamilton, Robert Trivers, John Krebs and Nick Davis. This change of focus was so dramatic that leading scholars even predicted an emerging end of classical ethology, to be replaced by behavioural ecology.

Since applied ethologists have traditionally been quite occupied with animal welfare, which includes a focus on motivation and mechanisms, it was not completely obvious how the functional views of evolutionary biology could easily be considered in a more practical framework. A fundamental problem was the concept of fitness, which does not appear immediately applicable in the context of farm and companion animals. How could one study fitness aspects of behaviour, when reproduction is completely controlled by humans? What does the concept of fitness mean to millions of castrated male fattening pigs?

However, a number of researchers realised that objections such as these are less important when you ask questions on the evolutionary background of different behaviours. Most aspects of the behaviour we can observe in modern farm animals are adaptations inherited from their ancestors, so understanding functional aspects of the behaviour of a pig or a chicken requires proper considerations of the ecological context in which their ancestors, such as wild boars and Red Junglefowl, evolved prior to domestication. This resulted in important experiments, where pigs and chickens were studied in semi-natural settings in order to allow the full expression of behaviour in conditions similar to those their ancestors experienced (Duncan *et al.*, 1978; Jensen *et al.*, 1993; Stolba and Wood-Gush, 2010).

From my own perspective, the apparent evolutionary strangeness of pig behaviour was what originally caught my interest in domestic animal behaviour, so it was a natural thing to pursue this line of work. For example, pigs are exceptional mammals in combining large litters with precocial offspring. Could that be related to the extraordinary suckling behaviour, characterised by intense communication between mother and offspring, rigid teat order, and an unusually long non-nutritional post-massage period at every nursing? Bo Algers and I tried to view nursing from an evolutionary perspective, applying theories of mother-offspring conflict and honest signalling, to explain these complex interactions.

Pursuing this idea, we found that post-massage (the snout rubbing performed towards the udder after milk has been obtained) actually fits the predictions of evolutionary theories quite well. It appears to act as an honest signal to the mother, determining later milk production in the teat occupied by the signalling piglet (Jensen *et al.*, 1998). Using similar evolutionary approaches, David Fraser and co-workers showed that the eye-teeth of new-born piglets, eight needle-sharp, forward-pointing teeth, serve their function mainly as weapons against siblings during the first days of life, corroborating theories of sib-sib competition (Fraser and Thompson, 1991).

I have met people questioning whether this kind of research actually serves applied ethology and animal welfare at all, or whether it 'merely' adds to the literature in behavioural ecology. Personally, I am convinced that most knowledge about animals has potential applicability, although the context may not always seem clear at the time of publication. For the studies mentioned above, let me mention two important practical implications.

With respect to the communication between the sow and her piglets during suckling, Bo Algers and I found that the regular grunting emitted by the sow serves to communicate to the piglets the release of oxytocin (which triggers the milk ejection) and thereby synchronises suckling behaviour within the litter. At that time, there was a debate in Sweden with respect to acceptable levels of fan noise in pig houses, and those discussions basically concerned possible hearing damage and stress caused by too high noise levels. We realised that the vocal communication could also potentially be affected, and ran an experiment where we studied piglet synchronisation during suckling when they were exposed to fan noise at a level according to the current legislation. The effects were rather dramatic – the synchronisation of litters in the noisy pens was severely disrupted, with potentially detrimental effects on milk transmission and consequently piglet survival (Algers and Jensen, 1985).

Regarding the function of the eye-teeth, David Fraser and his colleagues noted that farmers usually regard these as a nuisance. Therefore, they are often removed through clipping or grinding, to avoid injury to littermates or damage to the sow's udder. This removal has a number of potential welfare problems for the piglets. Fraser and Thompson (1991) found that the larger the litter, the bigger the risk of damage to siblings and mothers when eye-teeth are present. Hence, the current practice of breeding for increased litter size will increase the demand of farmers to handle the problems caused by sibling rivalry, potentially compromising piglet welfare.

The paper by Fraser and Thompson (1991) also has another story to it. It may have been the first 'applied ethology' paper to be published in a pure evolution journal, the rather high profile 'Behavioural Ecology'. Moreover, it was the first (and only, as far as I am aware) paper ever to have the entire abstract published as poetry, with perfect rhyme and meter. The starting lines of the abstract are etched into my memory forever: 'A piglet's most precious possession / Is the teat which he fattens his flesh on' (p. 9).

Other scientists have pursued the evolutionary and functional approach in fruitful manners as well, and widened our view on the behaviour of domestic animals. In Norway, Inger-Lise Andersen has been an enthusiastic promoter of this. She used maternal investment theory to study the relationship between litter size and age of the mother on one hand, and the rates of growth and survival among piglets on the other. Generally, she found that larger litters and older mothers are associated with greater mortality, much as predicted since large litters lead to less maternal investment per offspring, and older mothers are generally less fit due to previous investment. This may have important consequences for the breeding goals in current pig breeding (Andersen *et al.*, 2011).

Another scientist who has explored the functional aspects of domestic animal behaviour in depth is Marek Špinka. Together with Susan Held, he has made a number of theoretical contributions on the topic of animal play, perhaps one of the most poorly understood behaviour phenomena in ethology in general. Why do young animals of so many different species devote so much time and energy to seemingly meaningless activities (from a fitness point of view)?

Held and Špinka made major advances in understanding this when they took a functional approach to the question of whether play can serve as a welfare indicator. Considering different possible ultimate explanations of play and combining them with theories of animal welfare, they produced a novel and intriguing view of animal play. They argued, for example, that play is performed in situations where there is absence of fitness threats, that it is associated with positive emotional states and contagious, so playing animals can also generate good welfare for others. Hence, this theory offered a potential new tool for assessing good as well as poor welfare in a variety of species (Held and Špinka, 2011).

15.5 Domestication

When Charles Darwin finally, after decades of thought and hesitation, decided to publish his theory of the evolution of species by natural selection, he needed to set the scene to convince his readers. In the absence of any proper understanding of genetics (yet to be discovered by Gregor Mendel and others), it was not at all obvious how one species could gradually change and be transformed into another. Darwin chose to start his book by considering the domestication of animals and plants. This became the proof-of-principle, demonstrating that with strict selection pressure, populations can change substantially in appearance and behaviour over a relatively limited number of generations.

To this day, domestication remains a central proof-of-principle for evolution. Nowadays, we are more interested in exactly how species change, in how they do not change (resilient traits), and in the mechanisms that cause the specific changes we observe. Domestication is the largest evolutionary experiment on this planet. To those of us who are interested in behaviour of domesticated animals, it is a gold mine. Unlike evolutionary biologists, applied ethologists generally know which ancestral forms have given rise to the variants we see today, and often these ancestral breeds still exist in their original form and context. We also know the time course of the process and can define the specific selection pressures that have acted upon populations. As I will come back to later, contemporary molecular genetics offers us unprecedented tools to investigate the genetic mechanisms underlying changes associated with domestication.

The most striking result from domestication research is how little animal behaviour has changed compared to ancestral breeds, despite rather intense selection and large differences in appearance and reproductive ability. In many species, fundamental behaviour is more or less indistinguishable from that of the ancestors. For example, chickens urge to rest on perches high over the ground and to lay their eggs in secluded areas remains strong, even when kept in battery cages for generations (Schrader and

Müller, 2009), and similarly, nest building in domestic sows is performed in an almost identical manner to that of their ancestor, the wild boar (Gustafsson *et al.*, 1999).

However, in some cases the differences between domestics and their ancestors are profound. This primarily concerns their propensity for being tame and general temperament. In some cases, the ability to communicate with humans has also changed dramatically.

The Russian geneticist, Dmitry Belyaev, envisaged that tameness may be a central trait to domestication, and that many other aspects of the domesticated phenotype are in fact secondary changes to being tame. In order to investigate this, he selected farmed silver foxes, a species never fully domesticated, based on their fear response towards humans. Within a few generations of selecting for 'tameness', many phenotypic aspects typical of domestication appeared, such as loss of pigmentation, shorter legs and skulls, and increased reproduction (Trut *et al.*, 2009). Similarly, my colleagues and I selected ancestral Red Junglefowl for reduced fear of humans, and within a few generations observed increased reproduction and food conversion (Agnvall *et al.*, 2012, 2014). It appears that when animals are selected for the ability to function well in close proximity to humans, this creates a cascade of correlated responses, which may help to explain how domesticated phenotypes have arisen.

But the most striking aspect of domestication is the development of increased ability to communicate and relate to humans. This represents an exciting example of how new behavioural traits evolve in a historically short period, as an adaptation to modified selection pressure. The most exciting examples of this have come from recent studies in dogs.

Ethological studies of dogs have an intriguing history. For a long period of time, this work did not produce much in the way of interesting results. With several other scientists, I committed the fundamental error of trying to control for human interactions, since these were considered as noise. We wanted to study 'pure dog behaviour', and approached the subject largely as we would study a rat or a pig. We attempted to arrange study situations where humans were ruled out as affecting factors, for example by studying dogs in isolation or with other dogs only. The results produced were mostly rather uninteresting.

The error we committed was that trying to exclude humans from the affecting factors actually removed one of the central features influencing what it is like to be a dog. More than any other domestic animal, dogs have evolved in close symbiosis with humans, having been under close contact and selection for almost twice as long as any other species. Therefore humans cannot be treated as irrelevant noise, but perhaps should rather be seen as their most important environmental factor. Without its human companions, a dog is not really a complete dog.

Once scientists realised this, research exploded. The ability of dogs to communicate with humans is astonishing, and studies have since shown how they outperform ancestral wolves and also primates in reading and responding to human body language, understanding facial expressions and even understanding words and simple grammar

(Miklosi, 2008). The Border collie 'Chaser', studied in detail during a five-year period, mastered over 1000 different names of objects, was able to categorise them into larger categories, and could differentiate verbs from nouns!

Dog studies over the past fifteen years have accumulated a large body of evidence on adaptations and breed differences under domestication. The dog has become one of the most important evolutionary models, and opened up novel and innovative research approaches, as we will also see later.

15.6 Genetics at centre stage

When Darwin suggested that behaviour develops under evolution just like any other trait, he was implicitly assuming a genetic basis for behavioural variation. This was not an obvious assumption at the time, and it took a long time for the scientific community to accept that behaviour has a strong innate component that is coded in the genome (there are in fact still fields outside biology where this is questioned, at least for humans). However, fundamental research over many years and in many different species has shown beyond any doubt that even very complex species-specific behaviours can develop largely independent of environmental input, and must therefore be encoded in the genes.

These are fundamental text book facts today, and our main ideas of animal welfare rely on the notion that domesticated animals are born with genetically encoded behavioural needs, which should be taken into consideration when housing them. Given that the genetic influence on behaviour has been a basic insight for more than 150 years (since Darwin's publications), and the DNA code has been understood for 50 years, it is a bit embarrassing that we have so little knowledge of how this happens. How can a piece of DNA, containing the code for a protein, cause its bearer to build a nest before giving birth (if the DNA happens to reside in a pig) or to jump up on perches when night approaches (if it is in the nuclei of a chicken cell)?

The reason for our lack of knowledge here is of course that the process is immensely complex. Genes interact with each other and not only the actual codes matter, but perhaps more importantly the timing of their expression; and other gene networks in turn control this. And of course, the genes do not cause behaviour in a direct manner. They code for the building of a nervous system with sensory organs, and for all the molecules needed to make the system work in the right way: hormones, enzymes, transmitter substances, etc.

Hence, it is not surprising that scientists have generally relied on relatively simple or standardised organisms in order to find the connections between genes and behaviour. Much of the literature here consists of studies of *Drosophila*, mice and the nematode model organism *Caenorhabditis elegans*. This has produced many beautiful examples of genetic systems controlling various types of behaviour, such as foraging in *C. elegans*, social behaviour in mice and mating in fruit flies (Robinson *et al.*, 2008). However, in most cases, detailed mechanisms linking particular gene variants to a specific behaviour remain elusive.

When it comes to studying genes and behaviour, applied ethology has a major advantage. We have access to breeds of domesticated animals, often with well-defined selection histories, and in many cases their wild ancestors. Since domestication is still comparatively young as an evolutionary phenomenon, reproductive isolation has not occurred, and so modern farm and companion animals can be easily mated, both with other breeds and with their ancestors, to create fertile offspring. Domestic animals therefore offer unprecedented possibilities to breed genetically interesting mapping populations, where inheritance patterns and molecular mechanisms can be examined in detail.

One of the first large-scale behaviour genetic studies was in fact performed on dogs. The American geneticists John Paul Scott and John L. Fuller crossed Basenjis and Cocker Spaniels to create a segregating F2-generation, and back-crossed to the parental generation, allowing a broad Mendelian genetic mapping of inheritance patterns of the major behaviour differences between the breeds. They found, for example, that the famous absence of barking in Basenjis can be attributed to two genes where the Basenji-trait is dominant (Scott and Fuller, 1965). At the time when this study was performed, the genetic code was not well understood, so it was not possible to elucidate which genes were actually involved, or how they might exert their effects.

Contemporary genetics of course comes with a completely different toolbox. My co-workers and I crossed ancestral Red Junglefowl with a modern, domesticated White Leghorn egg-laying chicken (Figure 15.2) and behaviourally characterised more than 700 F2, and about 600 F8 birds. In addition, we analysed hundreds of DNA-markers in every individual, and were able to link many cases of specific behaviour variation to certain genes and mutations. For example, together with Linda Keeling, we demonstrated that the risk of being exposed to feather-pecking, a dramatic behavioural disorder detrimental for chicken welfare, is closely linked to a pigmentation related gene called *PMEL17*. A domestication related mutation, consisting of the deletion of three amino-acids from the transmembrane region of the protein, not only stops melanin expression – and thus pigmentation – but also exerts various pleiotropic effects on the behaviour of the carrier, including protecting it against feather-pecking (Karlsson *et al.*, 2010; Keeling *et al.*, 2004).

However, the ultimate dream of any behaviour geneticist may actually be the dog. Different breeds have been reared in controlled, closed populations for 150 years, and divergently selected for various behaviour traits. Some breeds excel in herding, others in retrieving. Some have been bred for independent, high-energy work (hunting dogs), and others for intense cooperation with humans (herding breeds). These behaviour differences can be mapped with modern molecular tools, and many researchers have already started pursuing this. We have recently shown that the propensity for Beagles to seek human cooperation has a strong genetic component, which is presently being mapped down to the level of actual genes (Persson *et al.*, 2015). When the causal genes are identified, together with the proteins for which they code and their regulation, it will be possible to understand how behaviour is affected on a molecular level. This will indeed represent a true advancement for behaviour science.

Figure 15.2. Chicks of Red Junglefowl (in the foreground) and a domesticated White Leghorn layer. Domestication offers a model for evolutionary biologists and behaviour geneticists alike, and provides invaluable knowledge for the applied ethologist (photo P. Jensen).

15.7 Personality

One area where applied ethology has really paved the way for an avalanche of research in fundamental science is in personality theory. For decades, individual differences in populations studied by behavioural biologists were treated as noise. They were considered to be variation around the mean value (representing the 'true' behaviour of the population), which made it more difficult to obtain statistical significance, but were otherwise of little importance.

Major breakthroughs were achieved by Dutch biologists in the 1990s, who discovered that the response to various types of stress differed among wild caught house mice in a consistent manner. Rather than dismissing this as noise, the scientists, led by Jaap Koolhaas, pursued the differences and found that individuals are actually consistently different in their way of handling stress (reviewed in Korte *et al.*, 2007). This was referred to as coping strategies, and applied ethologists soon realised that these traits are central if we want to understand how different individuals are affected by, for example, crowding in a fattening pig pen.

For many years, applied ethology was rather alone in taking interest in individual differences. The concept of coping strategies was extended to more general concepts of 'personality', and people started to use psychological analysis methods for assessing individual differences among pigs, chickens, cattle and dogs. My impression is that fundamental behavioural biologists tended to dislike the field, perhaps because the term 'personality' carried some anthropomorphic connotations (can a chicken really have 'personality'?). However, after some years, evolutionary biologists realised that many game-theory models developed in the 1970s actually predicted stable individual differences in populations to be the rule rather than the exception, and the research took hold also in the fundamental fields.

Terminology has changed back and forth, and some people prefer to talk about 'behavioural syndromes' and temperaments rather than personalities. Regardless of the words used, research on individual differences has become one of the most productive areas in contemporary evolutionary biology (Sih *et al.*, 2004), and still stands strong in applied ethology. It appears closely linked to, for example, feed efficiency, or the propensity to develop certain types of behaviour disorders. Including individual differences in such work may thus be critical for explaining variation and understanding the effects of different treatments, and may change our view on animal welfare rather dramatically – what is good for one individual in a group may actually be less so for another.

15.8 Cognitive science – what do they think about?

Ethology has a tradition of studying animal behaviour without considering the subjective experiences of animals. Going back to the pioneers, people such as Tinbergen and Lorenz used profoundly mechanistic terminology when describing behaviour and this has coloured much of the later science. Personally, I am convinced that neither Tinbergen nor Lorenz actually believed that animals lacked subjective, conscious experiences, but they chose to disregard them since they are inherently difficult to assess. This even led some ethologists to deny the existence of conscious thoughts and feelings in non-human animals.

However, this view started to change at the end of the 20th century. Interest grew in how animals actually perceive their worlds, and applied ethology was again in the forefront of this field. The question of animal feelings and subjective thoughts became central to animal welfare, and was therefore necessary to study in a scientific manner.

A pioneer in this respect was Marian Dawkins, who not only emphasised the importance of emotional experiences for animal welfare, but also envisaged methods for scientific and objective assessment of animal preferences and motivation. This has become known as the consumer-demand approach, and borrows its methods and theory from human micro-economy. Rather than only studying the effects on a chicken of having access or not to nest material (for example, levels of stress hormones, occurrence of stereotypies), Dawkins showed that asking the chickens to pay a price for access to a resource could help to determine how badly they wanted the material. By gradually raising the price for access (for example, by increasing the weight of a push door at the entrance to a nest box), the maximum effort can be determined and compared to, for example, what the chicken will pay for access to food.

The insight that subjective experiences can actually be assessed with quantitative methods has caused an increasing interest in finding new methods for doing this. Georgia Mason has taken the consumer-demand approach to very high standards and used it to show, for example, that mink are highly motivated to have access to swimming water. To do so, she introduced a number of quantitative measures for strength of motivation. Mike Mendl and co-workers used a different approach, and have explored the concept of cognitive bias as a measure of subjective feelings. When animals, just like humans, are depressed as a result of, for example, impoverished home environments, they tend to become pessimistic in their judgements of ambiguous stimuli. The degree of this cognitive bias can be used to quantify the experiences of animals, and the method has been used on various laboratory and farm animal species.

Yet another method was developed by Berry Spruijt and his students in the Netherlands. He realised that animals behave differently when they expect different things to occur, and formalised a method to quantify such anticipatory behaviour (Spruijt *et al.*, 2001). By assessing the behaviour when expecting a previously conditioned positive event, the anticipatory response can then be compared to other events. For example, male rats behave very similarly when they expect soon to meet a female in heat and when they expect access to environmental enrichment, indicating that enrichment is subjectively experienced as a very positive event. Similar methods have been used to study the emotional states of dogs, chickens and fish under different conditions. See Chapter 5 (Rault and Doyle, 2016) for further discussion of research on cognitive bias and anticipatory behaviour.

As in other fields, dogs have proven exceptionally interesting subjects for studying animal cognition. The fact that they are so willing to cooperate and communicate with humans makes them excellent study models in this respect. They are easily trained to perform different tasks and with clever experimental designs, detailed questions can be asked about their subjective experiences. Such studies have revealed that dogs possess complex emotions such as jealousy, dislike of unjust treatment, and shame, and are capable of relatively complex cognitive operations, such as numerical assessment, simple arithmetic, planning ahead of time, and differentiating different time courses (Miklosi, 2008).

15.9 Applications of applied ethology

A somewhat unfortunate development in the area of applied ethology is the extent to which it has come to be synonymous with animal welfare. Clearly, animal welfare is an extremely important field, and applied ethology can and should contribute extensively to it. However, applied ethology is much more than animal welfare (and animal welfare is much more than applied ethology). Knowledge of the mechanisms and functions of behaviour in farm animals is of course fundamental for making proper judgments about their welfare, but it also has many other potential applications.

For example, applied ethology could probably contribute much more than at present to the field of conservation biology. Breeding of threatened species in captivity for later reintroduction has proven to be a very difficult enterprise. Estimates have shown that more than 80% of all reintroduction projects fail, in the sense that virtually none of the released individuals survive to reproductive age. Applied ethologists have made important contributions to this field already. Inma Estevez and colleagues developed a novel breeding routine for captive bred North American Whooping Cranes (a threatened species), based on previous studies on the natural parental behaviour of the species. By modifying rearing practices in accordance with what was known about imprinting periods and other aspects of parent-offspring interactions, the survival rate of reintroduced birds was dramatically improved.

Another field which is poorly developed in applied ethology has to do with farming technology. Modern farms are indeed high-tech, and many contemporary dairy cows must learn to navigate a number of different technologies in their daily lives (for example, transponder feeders, milking robots, grooming brushes). Even getting water can be a tricky operation for many farm animals. Many farmers take for granted that a chicken or a pig should be able to easily learn how to drink from a nipple drinker, but research suggests that this can sometimes be a rather difficult task. Understanding how animals learn to interact with and operate various technical devices on the farms can help in designing better technology, increase production and probably also improve welfare.

This should not distract from the fact that animal welfare is the major field in which applied ethology research is used today. As shown in this chapter, this relationship has been highly fruitful for behaviour research at large. Without the concern for animal welfare, it is questionable whether there would have been such a large interest in areas such as animal personality and cognition. On the other hand, animal welfare has greatly benefitted from research triggered by a fundamental interest in behaviour, such as through genetics and domestication.

15.10 The synthesis

What will the future bring for applied ethology? I think it is safe to assume that it will be increasingly integrated into ethological research, and continue to develop by exchanging ideas and progress with fundamental science. I remember attending international ethology conferences 20 years ago, where small sessions on applied ethology were scheduled in remote rooms at obscure times of the day, leading to almost no attendance except the few people presenting. This has completely changed, and it is now uncommon to have separate sessions on applied ethology. Rather, research on farm animals is presented where it belongs scientifically, for example in sessions on cognition, behaviour genetics or personality. The somewhat patronising attitudes of some fundamental ethologists have definitely changed (such as those of my undergraduate teachers who strongly advised me against working on pigs, since 'real ethology cannot learn anything from studies of domesticated animals'), and the future will most likely see a fruitful integration of all aspects of behavioural biology.

Hence, what will happen to applied ethology is closely linked to the development of fundamental ethology and biology. It is almost impossible to predict future scientific development but some novel technologies are already starting to make their way into behaviour research, and these are fields that we can expect will grow over the coming years. GPS-technology is already used extensively in wildlife research, and as precision increases its usefulness under indoor farm conditions will grow. Soon, positions of animals may be assessed with a precision of centimetres or less, which will allow for continuous measurements of group interactions and dynamics and automated behaviour recording of animals in large groups. For additional discussion of GPS and other new technologies, see Chapter 5 (Rault and Doyle, 2016).

Continuous physiological measurements with remote recording are also developing quickly, and will add valuable data to behaviour recordings. It will soon be possible to monitor heart rate, body temperature, respiration and blood pressure continuously and the devices will be smaller and lighter, enabling measurements on small animals. Closely related to this is the increasing power of brain imaging techniques. So far, fMRI-cameras have severe limitations in that they are large and stationary, and animals normally need to be sedated to make brain scanning possible. However, recently dogs have been trained to remain motionless during fMRI scans, while presented with different stimuli. So far, this research has, for example, demonstrated that dogs have separate brain regions for processing stimuli from human voices, allowing them to distinguish owners from others, and that their brains contain a face recognition area in the same part of the temporal lobe as in humans (Dilks *et al.*, 2015).

A dream of many applied ethologists is of course to be able to monitor brain activity in animals as they move and behave freely. Here, the fNIRS-technology may turn out to be a solution. Some studies have already been published on dogs and sheep, but the method still has definite limitations in that it is only possible to monitor activity at the brain surface. However, the method is promising and even at the present stage it is likely to come into use in many experiments.

Another field which will probably play an important role in the future of applied ethology is epigenetics. For some decades, it has been known that the genome is more dynamic and responsive to environmental stimuli than previously thought. For example, stress can change the epigenetic state of different genes, thereby modifying patterns of gene expression. These modifications can be, for example, addition of methyl groups to cytosine-bases, or changes in the structure of histones (DNA-packaging proteins). Such changes may strongly affect the way in which the individual copes with future stress, and may also be transferred across generations, affecting the behaviour of the offspring.

Epigenetics will have to be taken into consideration when studying behaviour changes due to stress in different life phases, and also in behaviour genetics. Furthermore, it may turn out to be an important retrospective indicator of previous experiences. For example, in primates, early abuse causes life-long epigenetic modifications that can be spotted in various tissues, such as blood. It seems quite possible that a blood sample from a pig or a chicken at the slaughter house could reveal epigenetic marks of adverse life time events, which would be an important tool for applied ethology and for animal welfare.

Regardless of which of these prophecies come true, applied ethology seems destined to prosper more than ever. Many of us are looking forward to the future with great enthusiasm and expectation. When I started as a researcher in the 1970s, we did not have computers to record or analyse our data, and the sheer thought of quantitative assessment of subjective states of animals sounded like science fiction. Today, my group routinely uses automatic video tracking software and in the molecular lab my PhD students genotype animals and sequence DNA as part of their daily work. Whatever I speculate about the future, it seems most likely that reality will surprise us all. It is a great time to be an applied ethologist!

References

Agnvall, B., Ali, A., Olby, S. and Jensen, P., 2014. Red Junglefowl (*Gallus gallus*) selected for low fear of humans are larger, more dominant and produce larger offspring. Animal 8: 1498-1505.

Agnvall, B., Jöngren, M., Strandberg, E. and Jensen, P., 2012. Heritability and genetic correlations of fear-related behaviour in Red Junglefowl – possible implications for early domestication. PLoS ONE 7: e35162.

Algers, B. and Jensen, P., 1985. Communication during suckling in the domestic pig. Effects of continuous noise. Applied Animal Behaviour Science 14: 49-61.

Andersen, I.L., Nævdal, E. and Bøe, K.E., 2011. Maternal investment, sibling competition, and offspring survival with increasing litter size and parity in pigs (*Sus scrofa*). Behavioral Ecology and Sociobiology 65: 1159-1167.

Baxter, M.R., 1983. Ethology in environmental design for animal production. Applied Animal Ethology 9: 207-220.

Castrén, H., Algers, B., De Passillé, A.M., Rushen, J. and Uvnäs-Moberg, K., 1994. Nest building in sows in relation to hormone release. Applied Animal Behaviour Science 40: 74-75.

Duncan, I.J.H., Savory, C.J. and Wood-Gush, D.G.M., 1978. Observations on the reproductive behaviour of domestic fowl in the wild. Applied Animal Ethology 4: 29-42.

Dilks, D.D., Cook, P., Weiller, S.K., Berns, H.P., Spivak, M. and Berns, G.S., 2015. Awake fMRI reveals a specialized region in dog temporal cortex for face processing. PeerJ 3: e1115.

Fraser, D. and Thompson, B.K., 1991. Armed sibling rivalry among suckling piglets. Behavioral Ecology and Sociobiology 29: 9-15.

Gustafsson, M., Jensen, P., De Jonge, F.H., Illmann, G. and Špinka, M., 1999. Maternal behaviour of domestic sows and crosses between domestic sows and wild boar. Applied Animal Behaviour Science 65: 29-42.

Held, S.D.E. and Špinka, M., 2011. Animal play and animal welfare. Animal Behaviour 81: 891-899.

Hogan, J.A., 2014. A framework for the study of behavior. Behavioural Processes 117: 105-113.

Jensen, P., 1993. Nest building in domestic sows: the role of external stimuli. Animal Behaviour 45: 351-358.

Jensen, P., Gustafsson, M. and Augustsson, H., 1998. Teat massage after milk ingestion in domestic piglets: an example of honest begging? Animal Behaviour 55: 779-786.

Jensen, P., Vestergaard, K. and Algers, B., 1993. Nestbuilding in free-ranging domestic sows. Applied Animal Behaviour Science 38: 245-255.

Karlsson, A.-C., Mormede, P., Kerje, S. and Jensen, P., 2010. Genotype on the pigmentation regulating PMEL17 gene affects behavior in chickens raised without physical contact with conspecifics. Behaviour Genetics 41: 312-322.

Keeling, L., Andersson, L., Schütz, K.E. and Kerje, S., 2004. Chicken genomics: feather-pecking and victim pigmentation. Nature 431: 645-646.

Korte, S.M., Olivier, B. and Koolhaas, J.M., 2007. A new animal welfare concept based on allostasis. Physiology and Behavior 92: 422-428.

Miklosi, A., 2008. Dog behaviour, evolution, and cognition. Oxford University Press, Oxford, UK.

Persson, M.E., Roth, L.S.V., Johnsson, M., Wright, D. and Jensen, P., 2015. Human-directed social behaviour in dogs shows significant heritability. Genes, Brain and Behaviour 14: 337-344.

Rault, J.-L. and Doyle, R., 2016. Cognitive approaches and new technologies: changing methodologies in applied ethology. In: Brown, J.A., Seddon, Y.M. and Appleby, M.C. (eds.) Animals and us – 50 years and more of applied ethology. Wageningen Academic Publishers, Wageningen, the Netherlands, pp. 113-132.

Robinson, G.E., Fernald, R.D. and Clayton, D.F., 2008. Genes and social behavior. Science 322: 896-900.

Schrader, L. and Müller, B., 2009. Night-time roosting in the domestic fowl: the height matters. Applied Animal Behaviour Science 121: 179-183.

Scott, J.P. and Fuller, D.Q., 1965. Genetics and the social behavior of the dog. Chicago University Press, Chicago, IL, USA.

Sih, A., Bell, A. and Johnson, J.C., 2004. Behavioral syndromes: an ecological and evolutionary overview. Trends in Ecology and Evolution 19: 372-378.

Spruijt, B.M., Van den Bos, R. and Pijlman, F.T.A., 2001. A concept of welfare based on reward evaluating mechanisms in the brain: anticipatory behaviour as an indicator for the state of reward systems. Applied Animal Behaviour Science 72: 145-171.

Stolba, A. and Wood-Gush, D.G.M., 2010. The behaviour of pigs in a semi-natural environment. Animal Production 48: 419-425.

Toates, F., 2004. Cognition, motivation, emotion and action: a dynamic and vulnerable interdependence. Applied Animal Behaviour Science 86: 173-204.

Trut, L., Oskina, I. and Kharlamova, A., 2009. Animal evolution during domestication: the domesticated fox as a model. Bioessays 31: 349-360.

Vestergaard, K., 1982. Dust-bathing in the domestic fowl – diurnal rhythm and dust deprivation. Applied Animal Ethology 8: 487-495.

Vestergaard, K.S., Damm, B., Abbott, U. and Bildsøe, M., 1999. Regulation of dustbathing in feathered and featherless domestic chicks: the Lorenzian model revisited. Animal Behaviour 58: 1017-1025.

16. Future directions for applied ethology

J.N. Marchant-Forde[1] and T.B. Rodenburg[2]*
[1]USDA, Livestock Behaviour Research Unit, 125 South Russell St., West Lafayette, IN 47907, USA; jeremy.marchant-forde@ars.usda.gov
[2]Behavioural Ecology Group, Wageningen University, P.O. Box 338 6700 AH Wageningen, the Netherlands

Abstract

It is the expectation that the world population will grow to 9 billion people by 2050. There is no doubt that this dramatic increase will impact all of the animals with which we share this planet. A global population growing in size and affluence will greatly increase demand for food from animal sources, meaning global growth in livestock production. More livestock production means changes in land use, with more land being converted to grazing or animal feedstuff production, and more production of waste to be applied back to the land, together impacting natural habitats and wildlife species diversity. Greater affluence also means greater companion animal ownership, further increasing demand for food from animal sources, increasing feral populations with potential impacts on human health and wildlife diversity. In this chapter, we will attempt to gaze into the crystal ball and identify some future directions for applied ethology and for our Society. There is no doubt that applied ethologists can have exciting roles to play in shaping and safeguarding the welfare of domestic animals and wild animals in captivity, and also feral and wild animals living in their natural habitat, either impacting or being impacted by humans.

Keywords: animal behaviour, animal welfare, animal production, laboratory animals, companion animals, zoo animals

16.1 Introduction

It is the expectation that the world population will grow to 9 billion people by 2050. There is no doubt that this dramatic increase will impact all of the animals with which we share this planet. A global population growing in size and affluence will greatly increase demand for food from animal sources, meaning global growth in livestock production. More livestock production means changes in land use, with more land being converted to grazing or animal feedstuff production, and more production of waste to be applied back to the land, together impacting natural habitats and wildlife species diversity. Greater affluence also means greater companion animal ownership, further increasing demand for food from animal sources, increasing feral populations with potential impacts on human health and wildlife diversity. The emergence of new animal and zoonotic diseases, increased antibiotic resistance and changing distribution of disease vectors as climate changes will also fuel the use of laboratory animals.

DOI 10.3920/978-90-8686-828-5_16, © Wageningen Academic Publishers 2016

None of these scenarios are independent of each other and although we applied ethologists can and should continue to address both fundamental and applied questions within our research, it is also essential that we bear in mind the broader applicability of our work and the way in which it can mesh with other disciplines so as to produce a sustainable future for humans and animals alike (Marchant-Forde, 2015). In this chapter, we will attempt to gaze into the crystal ball and identify some future directions for applied ethology and for our Society. There is no doubt that applied ethologists can have exciting roles to play in shaping and safeguarding the welfare of domestic animals and wild animals in captivity, and also feral and wild animals living in their natural habitat, either impacting or being impacted by humans.

16.2 Farm animals

To feed the expected 9 billion people, a considerable increase or intensification of food production is needed. Even if people in the Western world significantly decrease their meat consumption, the large population increase in Africa and Asia is expected to lead to a strong increase in animal production. One of the major challenges we see in the near future is realising that increase in production in a sustainable way.

The question that should be addressed by applied ethologists is how this increase in production affects animal behaviour and welfare. We need to understand better how farm size and flock or herd size affects animals. One part of that could be comparing animal welfare in extensive and intensive systems and identifying strengths and weaknesses of each system and using this information to improve systems. We do not think that, by definition, larger or more intensive farms have poorer welfare[. Much depends on the system and management in place and factors like group size and stocking density. All the same, we feel ISAE and its members have an obligation to keep on informing the industry, decision-makers and stakeholders at large about the importance of keeping animals in an environment that ensures good feeding, good housing, good health and appropriate behaviour, defined as key welfare principles by the EU project Welfare Quality (Blokhuis *et al.*, 2010) based on the Five Freedoms (Brambell, 1965). A risk we see is that the intensification of animal production will mainly take place in regions of the world where animal welfare is not yet high on the agenda. This could lead to a large investment in traditional intensive systems, such as the conventional cage for laying hens or individual gestation stalls for sows.

To prevent this, applied ethologists worldwide need to be active in informing the various stakeholders that there are alternatives that can yield similar production results with better animal welfare, scoring better for the societal aspects of sustainability. In the past decades, Europe has been very focused on increasing the welfare level of farm animals kept within the European Union. We can see that Europe runs the risk to isolate itself on the world market and has only limited impact on animal production systems worldwide. In that scenario, Europe would have high quality animal production with high welfare standards for the EU market, but not play a role of importance in more intensive animal production in other parts of the world. Applied ethologists worldwide could play an important role in putting animal welfare on the agenda and aid in identifying systems that provide a good balance between productivity and animal welfare. Applied ethologists can help to take the emotion out of the debate and to focus on facts rather than feelings: if we observe animals housed in the different systems, where do we measure differences? How could systems be further improved?

The optimal system both regarding productivity and welfare could well be a hybrid of current intensive and extensive systems. A nice example of that is the Roundel system that was developed as a sustainable system for laying hens in the Netherlands, with attention to the hens' behavioural needs, societal concerns and farmer income. This resulted in a design where hens have no actual free range access, but they do have a large indoor foraging area with sand, grains and roughage (Figure 16.1). Hens also have more space than in a conventional aviary system (1,425 cm^2 compared with 1,111 cm^2 per bird) and flock size is limited to 6,000 birds. In this system, hens are kept without beak trimming and no major problems with feather pecking or cannibalism have occurred so far. The eggs are sold under the freedom foods scheme of the Dutch Organisation for Animal Protection (Dierenbescherming) and were awarded the maximum three stars in this welfare label, the same number of stars as organic eggs.

This example shows that there are possibilities to design systems in such a way that intensive production can be combined with good animal welfare, for instance by taking the qualities provided by an outdoor run and taking them indoors. In many countries we see transitions from more traditional intensive systems, such as battery cages for hens and gestation stalls for sows, to modern group housing systems. In these group housing systems, social behaviour plays a much more important role than in the traditional systems. Of course, these systems also allow a greater accommodation of normal behaviour.

ISAE members have a clear role here in guiding the industry regarding which aspects and indicators they should pay attention to in order to avoid behaviour and welfare problems. If no guidance is provided, behaviours such as tail biting and aggression in pigs and

Figure 16.1. Laying hens in the Dutch Roundel system (Photo: courtesy of Dr. Janice Siegford).

feather pecking in laying hens can cause major welfare problems and lead to increased mortality rates and poor performance. ISAE members play, and should continue to play, an important role in several major research areas:

- breeding and genetics of animals suited to modern production systems;
- epigenetics and the link between animal and environment (G×E interaction);
- improvement of housing systems by observing animal behaviour and performance;
- welfare assessment in various production systems;
- development of novel behavioural and welfare indicators;
- contribution to Precision Livestock Farming: using sensor data to monitor animal behaviour, welfare and performance.

This list illustrates that the field of applied animal behaviour and welfare links to all areas and aspects of animal production.

In more complex animal production systems, where animals are kept in large groups or flocks, behavioural measures are becoming increasingly important and valuable for farm management. In these large-group housing systems, an outbreak of damaging behaviour such as tail biting or feather pecking can have devastating effects. This is especially true in countries where procedures like tail docking and castration in pigs or beak trimming in laying hens are prohibited, like Norway and Sweden. These measures, while used by the respective industries as a form of damage control, are painful procedures that are not well accepted by society. As a result they have been banned in several European countries and are under discussion in others. So how do you keep groups of intact boars together, or how do you successfully manage a flock of laying hens with intact beaks? We foresee that the areas described above will remain core areas of research for many of our members in the coming years, and the knowledge they develop will be needed by the industry.

So, is being active in applied ethology research on farm animals only about solving welfare problems and avoiding negative affective states or is there also a positive side to this coin? We would definitely say there is. In recent years there has been much development in cognitive bias research, driven by Mike Mendl and Liz Paul from the University of Bristol (Paul *et al.*, 2005). Their work has sparked a renewed interest in tests where we aim to 'ask the animal' how it evaluates a given situation or environment. These methods are also discussed in Chapters 5 (Rault and Doyle, 2016) and 15 (Jensen, 2016). Work on this topic has been done on a wide range of species, from rodents to sparrows, chickens, dogs and pigs. The success of these studies in measuring cognitive bias has been varying, as developing an appropriate test is quite challenging. What we see as an important gain from the surge in studies on cognitive bias is the increased attention for measuring animal feelings, building on the foundations laid by Ian Duncan (Duncan, 1998). For instance, an increasing number of studies is measuring positive social interactions, mother-young interactions, emotional contagion and empathy. Many ISAE members are currently active in this field and we predict that this field of work will remain important in the coming years, as it provides invaluable information on how farm animals experience their social and physical environments. In turn, this knowledge can help us to develop new systems that better fit the animals' own preferences, for instance by keeping farrowing sows in groups with several litters rather than in individual farrowing crates, allowing piglets to learn from their mothers and resulting in a more natural social system. An example of recent research is the development of a multi-suckling system with five sows and their litters, shown in Figure 16.2. The system has five farrowing pens which are connected to a communal area, divided into spaces for lying, defecating and eating. Piglets are stimulated to explore solid feed at an early age, by the possibility to eat together with the sows in the communal eating area.

Such systems come closer to the pigs in the Edinburgh and Swedish pig parks, as studied by Dr. David Wood-Gush (Stolba and Wood-Gush, 1989) and Dr. Per Jensen (Jensen, 1986), where domestic pigs were set free in a large natural habitat. Like that early work, studies of animals in more complex social and physical environments can yield a wealth of scientific information on social behaviour and behavioural development that can be used to further improve modern housing systems.

Another area that is of major importance in applied animal behaviour and welfare research is that of human animal interactions. Work from Dr. Paul Hemsworth's group has shown that positive contact between animals and stockpersons not only leads to reduced stress and fearfulness in farm animals, but also to increased productivity (see Hemsworth *et al.*'s discussion in Chapter 9). If we think back to the observation that overall, farm size continues to increase, a note of caution may be in order to ensure that workers on larger farms continue to pay attention to their animals. For larger farms, novel technologies may offer perspective, such as employing robots with a human appearance for flock management in poultry, or using sensor technology to identify locations where problems arise and take appropriate measures in that specific pen or compartment. Automation of standard procedures may also free the workers to pay more or closer attention to the animals themselves. We feel that the area of human-animal interaction and good stockmanship will remain an important area for ISAE members in the years to come. This also applies to procedures around transport and slaughter. Here, Temple Grandin's work has had an enormous impact on the design of slaughterhouse facilities and of procedures around lairage and slaughter. Temple Grandin's unique perspective as

Figure 16.2. The multi-suckling system for five sows and their litters at the Pig Innovation Centre in Sterksel, the Netherlands. Photos courtesy of Sophie van Nieuwamerongen.

an autistic person has allowed her to experience slaughterhouse conditions more from the animal's perspective. Through her work Temple has inspired many scientists, and continues to have a major impact on changes within industry. The area of transport and slaughter may not always be the most attractive area to work in as a researcher in animal behaviour and welfare, but it most certainly is an area where one can have a major impact on animal welfare, and there remains much work to be done in this area.

So, what makes the work of ISAE members on farm animals unique? We think some of the major strengths of our research are that it covers multiple disciplines, that it bridges both fundamental and applied studies, and that it engages many different groups of stakeholders, ranging from farmers to industry, to policymakers and finally, society. Firstly, the multidisciplinarity: in many of our studies we not only focus on the effects of a given treatment on behaviour, but also on the physiological or immunological traits and on animal performance in general. By taking this approach, we feel that differences between treatments can be interpreted in a much more meaningful way. This also allows us to collaborate with many different disciplines, as evidenced by multiple examples throughout this book, and is definitely a strength we need to keep. Secondly, as emphasised in Chapters 6 (Lawrence, 2016) and 15 (Jensen, 2016), the combination of fundamental and more applied research is complementary. Fundamental research is crucial for developing new ideas and new research directions. This approach allows us to begin from scratch and come up with innovations or with completely novel ways of looking at existing problems. In turn, fundamental research will eventually come to a stage where it can be brought into practice, allowing the industry to interact with researchers and work towards application. In addition, on-farm studies should remain at the centre of our activities: development of welfare assessment methods, such as those developed in the EU project, Welfare Quality®, but also smaller projects where researchers work with innovative farmers to develop or test innovations. Fundamental and applied studies can thus benefit from each other and we would strongly advocate that researchers continue working along these lines. Thirdly, the many different stakeholders that are involved in our research make it unique. This aspect shows that our work is important to many different groups of stakeholders and that we are able to adapt our scientific questions to areas where knowledge gaps exist. In European research programmes, it is becoming increasingly important to engage different groups of stakeholders and to demonstrate what the relevance for and impact on society will be. This probably also applies to other areas of the world. This is positive, for such stakeholder involvement can support application of the research results in due course.

16.3 Companion animals

What constitutes a 'companion animal' can vary depending on the context. Animal species that are perceived as companion animals in some countries may be agricultural or working animals in others – for example, equids. For the purpose of this section, we are using companion animals to describe what could also be termed pets. Most commonly, these are cats and dogs, but can also include small mammals, birds, ornamental fish, and some reptiles. They may be wholly indoor pets, they may be a combination of indoor and outdoor, or they may be wholly outdoor, barn or pond animals. We also include feral populations of these species within this section.

Companion/pet animal ownership can be beneficial for our health and well-being (Rowan and Beck, 1994) and pet ownership is increasing, but globally, the exact numbers are very difficult to quantify. Data from the European Pet Food Industry Federation estimates that there are 75 million households across Europe owning at least one pet animal (FEDIAF, 2014) and a total of about 280 million animals. In terms of numbers of animals by type, the breakdown is given in Figure 16.3, though the numbers for fish are given as numbers of aquaria, rather than numbers of individual fish. A WSPA report (Batson, 2008) indicated total numbers of 282 million cats and 315 million dogs from 81 and 93 countries respectively. Given that these are the two most popular pet species, at around 60% of reported pets, we can perhaps extrapolate that the total number of pets within the 90+ countries reported (out of 196 total countries) will be in the region of 1 billion. This compares to around 27 billion livestock animals at any one time.

Across the board, these numbers are increasing with the rate of increase being greatest in those countries whose economies are also developing quickly. The rate of pet population increase in Western Europe was about 7-8% from 2002 to 2012, but increased by as much as 30% over the same time period in Russia (USDA FAS, 2013). This compared with developing economies, such as India, where although the total dog population may only be around 11 million, there has been a 95% increase in dog population between 2002 and 2012. Similarly, the Philippines have seen a 60% growth to 12 million dogs and Brazil has seen a 35% growth to 36 million dogs over this time period. These increases in

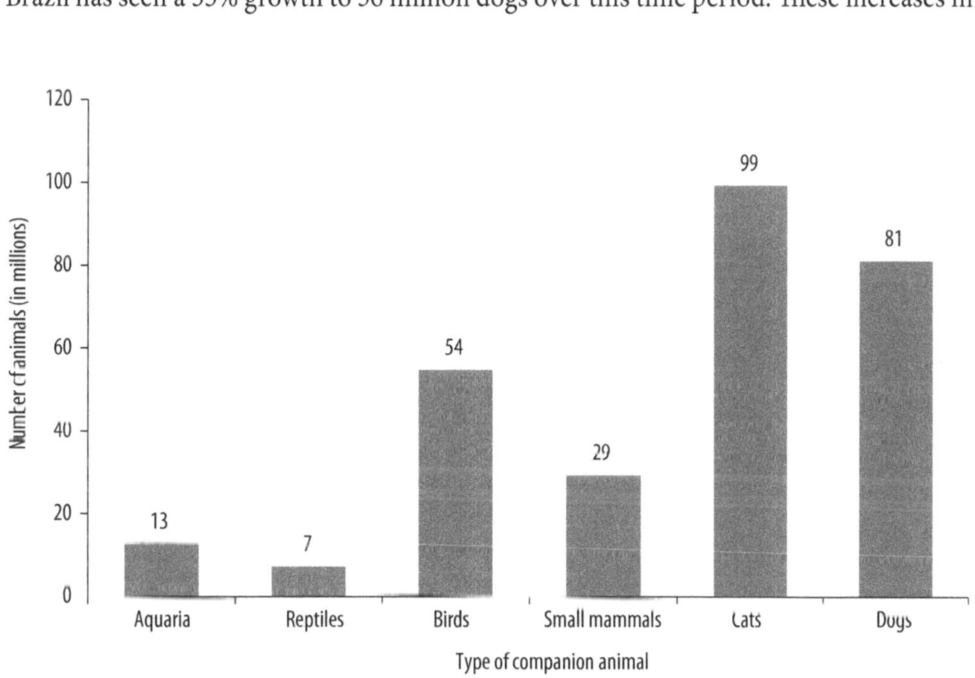

Figure 16.3. Estimated numbers of companion animals within Europe (FEDIAF, 2014).

pet ownership may be related to the increasing urbanisation of the world's population, a greater reliance on pets for social support as families change from traditional structures and become geographically spread, and a resulting self-reinforcing change in attitudes to animals away from the rural, utilitarian perspectives, and towards a more anthropomorphic attitude that alters our perception of animals away from 'things' and towards 'persons' (Serpell, 2013; Serpell and Paul, 2011). Thus, the pets we have in the home become 'animal ambassadors' (Serpell, 2000) serving to act as a moral link between us and all animals, and increasing our awareness of, and engagement in, the debates surrounding all animal welfare issues, be they concerning farm, laboratory, zoo or wildlife animals. Also, as companion animal populations grow, so do the stray/feral populations of dogs and cats, and these have become an increasing ecological, social and human health problem in themselves.

So, given the increasing population of companion animals, there is a large and diverse number of areas in which applied ethologists and welfare scientists should become more engaged in the future. Historically, the field of companion animal behaviour and welfare has been poorly funded; reliant on charitable funding or funding from pet-related industry while governmental investment in this field of research has been minimal. The main reasons for this are that: (1) the animals themselves do not provide a 'product' as farm or laboratory animals do – although there may be considerable financial advantages in terms of human health benefits of pet ownership for governments that provide a nationalised healthcare system (Clower and Neaves, 2015); (2) the overall numbers of animals are relatively low, and (3) they also live in varied, small-scale holdings (homes) that would make any type of standardisation or enforcement of legislation extremely difficult. Where government funding has been apparent, it has been associated with potential impacts on human health and well-being. Given the increasing numbers of companion animals, it is important for applied ethologists and welfare scientists to continue to emphasise that companion animal welfare is deserving of funding from broader sources, and that companion animal populations impact society in numerous ways that deserve continued study.

16.4 The welfare of companion animals – housing and husbandry

Many companion animals are kept in housing conditions that would be unacceptable were they EU-legislated farm animals. The main issues may include size of enclosure (stocking density), lack of species-specific environmental enrichment, and inappropriate thermal and social environments. In terms of other aspects of husbandry, there are deficiencies in species-specific knowledge prior to acquisition, lack of preventative veterinary care, poor nutrition, and insufficient meeting of behavioural needs. A recent report from the People's Dispensary for Sick Animals (PDSA) – a UK animal welfare charity – highlights some of these issues (PDSA, 2015). For example, only 4% of pet owners sought advice from veterinarians prior to acquiring a pet, though advice was sought via the internet (e.g. 29% of dog and rabbit owners), family and friends (e.g. 18% of dog owners), pet shelter staff (e.g. 18% of cat owners) or pet shop staff (e.g. 26% of rabbit owners). Eighteen percent of dog owners, 25% of cat owners and 12% of rabbit owners did no research at all before bringing their pet home. Regarding vaccination, 17% of dogs, 26% of cats and 50% of rabbits have never been vaccinated and 10, 16 and 35%, respectively, are not registered with veterinarians.

A major factor that could affect an owner's ability and knowledge to look after their pet appropriately is an overwhelming underestimation of the actual cost of pet ownership. PDSA estimates that the lifetime cost of a dog is between £16,000 and 31,000, of a cat is £17,000 and a rabbit is £9,000 – which includes purchase cost, nutrition, vaccinations, worming and flea treatment, micro-chipping, toys, cat litter and monthly pet insurance cost. Only 5% of dog owners, 2% of cat owners and 0% of rabbit owners estimated these costs correctly (PDSA, 2015). The rest all underestimated, many by a very large amount. This could be either because the PDSA costs included spending on items which they do not consider including, such as monthly insurance, or because there is ignorance about appropriate animal care and its associated cost. Regardless, there is sufficient evidence to indicate that many companion animals within home environments may experience a degree of poor welfare and there is a real role for applied ethologists and welfare scientists to determine the impact of husbandry deficiencies on companion animal welfare and engage with governmental and NGO bodies to educate pet owners to improve companion animal quality of life – i.e. a concept that includes not only meeting an animal's physical and physiological needs, but also incorporates psychological aspects of wellbeing (Taylor and Mills, 2007a). Dogs and cats have had the majority of research attention, given their popularity, but there are certainly welfare concerns about the husbandry of other common pets such as small mammals (Mullan and Main, 2006) and fish (Oldfield, 2011), which have been largely ignored and which deserve greater attention.

16.5 The welfare of companion animals – stray populations, relinquishment and shelters

Another impact of the increase in companion animals in homes is a corresponding increase in companion animals being abandoned or relinquished or straying. Thus, there are large populations of stray/feral companion animals in urban and rural communities and large populations in animal shelters, where these facilities exist. The stray and feral populations present both a risk to human health (e.g. dog bites, rabies; Dalla Villa *et al.*, 2010) and a risk to ecosystems (e.g. feral cats in Australia; Doherty *et al.*, 2015). However, views and methods, and indeed legislation about how to control feral populations can vary widely and culturally (e.g. dog control in the UK compared with India; Srinivasan, 2013), with control options such as euthanasia, trap & relocate, trap neuter & return with or without community 'adoption', and rehoming as a household pet. Certainly more research is needed to quantify the impact of stray/feral populations and develop sustainable strategies of control that are situationally specific, given the species, size of population, culture and habitat.

The number of companion animals entering or cycling through animal shelters is also very difficult to determine but represents a large economic burden which is perhaps unappreciated by governments. For example, a survey of UK charities involved in animal shelters showed that the sheltering, rehoming or euthanasia of 250,000 cats and dogs was carried out with a total expenditure of £340 million or £1,360 per animal (Stavisky *et al.*, 2012). The American Society for the Prevention of Cruelty to Animals (ASPCA) has estimated numbers for the United States of around 7.6 million companion animals entering shelters per year, of which 3.9 million are dogs and 3.4 million are cats (ASPCA, 2015). These populations include relinquished animals (2.5 million) and stray animals (5 million), which may be temporarily separated from owners, abandoned or feral. This number represents roughly 5% of the total owned companion animal populations for these species. If we extrapolate our initial population estimates of 1 billion companion animals within the 90 countries reported earlier, we might expect 50 million animals to cycle through shelters in a given year. For the animals themselves, the shelter experience may result in exposure to a combination of positive and negative events and for a significant proportion, it will also result in euthanasia. The ASPCA estimate that 36% or 2.7 million animals entering shelters in the US are euthanised annually (ASPCA, 2015) with 650,000 animals returned to their owners of which only 100,000 are cats.

The reasons for relinquishment are multiple and can include owner circumstances (moving house, new baby, financial hardship, etc.), allergies, pet health and behavioural reasons. A study of initial relinquishment and return of rehomed cats found that, excluding stray or abandoned animals, behavioural problems accounted for nearly 11% of relinquishments (Casey *et al.*, 2009). For animals returned to shelters, this figure jumped to over 41% with aggression towards other cats in the household and nervousness/anxiety issues being the primary behavioural reasons given. Studies on dogs also show behavioural problems as a primary factor in relinquishment, ranging from 10.8% (Marston *et al.*, 2004) to 34.3% (Diesel *et al.*, 2010), with aggression, destructivity, disobedience and fearfulness/anxiety listed among reasons (Coe *et al.*, 2014).

Within the shelters themselves, companion animals are often housed in conditions that are far removed from a household environment. Historically, shelter housing has been characterised as being fairly 'barren', individual housing with limited contact between conspecifics. The space may be quite limited, with little complexity and little opportunity to interact with humans or other animals or to exercise. There is a perceived need for rigorous hygiene to prevent spread of disease, which is a real risk (Pedersen *et al.*, 2004), and also segregation to prevent aggression (Taylor and Mills, 2007b). More recent developments have highlighted advantages of modified housing, in terms of welfare and adoptability, for cats (Rochlitz, 2007) and dogs (Taylor and Mills, 2007b) and the increased use of improved behavioural assessment tools to screen animals for behavioural problems prior to adoption, especially in dogs (Duffy *et al.*, 2014). This type of screening should be extended to cats (Moore and Bain, 2013), given the high incidence of behavioural reasons for cats being returned after adoption. As well, improving education programmes for those adopting pets would help towards reducing behavioural problems and relinquishment.

Clearly there is a role for applied animal behaviour within companion animal care and welfare and it is important that scientifically-trained applied ethologists take the lead. The companion animal industry is worth billions of dollars a year, yet investment in behavioural research and welfare science has been generally lacking. The bulk of investment has been carried out by charities and some industry-leading companies, with governments slow to provide funding, even though the ownership of pets has implications for human health and the environment. Pet behaviour is a field awash with amateur ethologists and it is important that ISAE continues its role in the certification of clinical animal behaviourists (CCAB) within the UK, but should also be spear-heading a move towards more global standardisation of certification in other countries.

16.6 Zoo animals and wildlife

Zoos are complex animal holding facilities, in that they usually contain small numbers of many different species (for example, the Smithsonian National Zoological Park in the US states a population of about 1,800 animals from about 300 species), and they serve multiple functions in terms of entertainment, education and conservation. An estimate of the global total number of animals kept in zoos is not readily accessible, but it is far less than farm, companion and laboratory animal numbers, with perhaps between 1-2 million (Hill and Broom, 2009). The American Zoo and Aquarium Association (AZA) estimates that there are 10,000 zoos worldwide of which about 2,400 'animal exhibits' are licenced in the US. Of these, only just over 200 are under the oversight/accreditation of the AZA. Within these types of institutions, and in corresponding zoos across the world (e.g. those 300 member zoos of the World Association of Zoos and Aquariums), there should be some adherence to policies on animal welfare and some true efforts in conservation. However, there are a great many zoos across the world that are outside accredited bodies and thus a great many zoo animals existing without good welfare safeguards, and with little purpose other than to be seen by visitors and generate profit for owners.

There has been growing public antipathy towards animals being used purely as entertainment – most animals are now gone from circuses, tourists are being lobbied to avoid animal rides, aquariums are being questioned about mammal shows, etc. If the entertainment role of zoos becomes more and more questioned, then the focus in all zoos must shift towards education and conservation for zoos to remain relevant, functioning and supported. Certainly the Association-member-type zoos have adapted over the last two decades, and there has been a real shift towards safeguarding or improving the welfare of the individual animals and towards improving education and conservation. Zoos have moved from cages to dioramas to immersion exhibits, and WAZA's objectives include to 'promote and use the highest standards of animal welfare and husbandry', helped by their animal welfare strategy (WAZA, 2015).

There are however, further challenges and changes ahead. At the individual animal level, there is still a need for species-specific study to improve welfare in captive environments. Whereas farm and laboratory animal welfare studies can incorporate relatively large numbers of animals, with the application of control and sometimes numerous experimental treatments, the study of the welfare of zoo animals is more problematic in a number of identified ways (Swaisgood, 2007):

- a relative lack of knowledge about the biological needs of a wild animal not selected for captive environments and what might constitute appropriate behavioural and physiological baselines to ensure good welfare;
- small, non-independent numbers of animals often housed within the same enclosure;
- a prevalence of studies applying multiple treatments simultaneously, making identification of key positive elements difficult.

There remain many opportunities for applied ethologists to contribute to zoo animal welfare. There has been heavy focus on large mammals such as bears, primates, big cats and elephants. These represent only a handful of the species kept in zoos and there remain big holes in our knowledge. Filling these holes does of course take time, money and labour, but for those of us in universities with many interested undergraduate students looking to engage with animal behaviour and welfare research, there are great benefits to partnering with local zoos – many of which are keen to establish such relationships – and generating both immediate, local data and also data which could be used to generate broader applicability in terms of databases and indexes (e.g. Koene, 2013). Working across institutions is another approach to overcoming the small, non-independent numbers issue. Examples of this are the Elephant Welfare Project carried out in the USA which collected data from 291 elephants kept in 72 out of the 73 total AZA member zoos that had elephants (Carlstead *et al.*, 2013) and an okapi behaviour project which collected data from 45 okapis (more than half the N. American population) kept in 16 different zoos (Bennett *et al.*, 2015). These projects have yielded excellent information, but at the cost of between US$3,000 to 6,000 per animal studied. Given the number of different species kept in zoos across the world, this major grant-funded approach is likely only to be feasible for high-profile species, and again opportunities exist to partner with educational institutions to carry out much lower-cost and less extensive epidemiological-style research that could still yield useful information on factors affecting welfare of less 'visible' species.

As mentioned in our opening paragraph, human growth continues to impact habitats and the wildlife that live in them. Accredited zoos have increased the scale and scope of their conservation-related efforts over the last few decades and many are now very actively involved in both *ex situ* and *in situ* conservation. The need for such efforts is without question. The 2000 International Union for the Conservation of Nature (IUCN) Red List contained 599 Critically Endangered, 1,021 Endangered and 2,184 Vulnerable species of mammals, birds, reptiles, amphibia and fish. The corresponding numbers in the 2008 IUCN Red List are 1,228, 1,967 and 2,771 species. Applied ethologists can play an important role both in the zoo and in the field. For *ex situ* conservation, investigating the effects of environment on reproductive behaviour and parental care is an area that applied ethologists working with farm and laboratory species has been working on for many years. We have also been working on ways to equip animals with behavioural skills needed to thrive in changing environments. Thus helping to understand the challenges for, and impacts of, reintroduction of captive-bred animals into natural habitats is another area in which we should be working more. For *in situ* conservation, the human-animal interface is important, and we have developed skills to assess the impacts of human behaviour on animals and ways to modify both human attitudes and behaviour towards animals through education. Animals in zoos should be acting as ambassadors for their species and we can help both in educating the zoo-going public (estimated at 700 million people visiting just the WAZA-listed institutions) to foster concern about protection of natural habitats and also in the field, at the 'coal face', minimising direct effects of human impact on animals in the wild.

16.7 Laboratory animals

As with companion animals, the numbers of animals classified and used as laboratory animals over any given year are difficult to quantify. The most used estimate is that of Taylor *et al.* (2008), who collated 2005 animal use data from 37 countries and developed and applied a statistical model based on publication rates for a further 142 countries. Their estimate gave 115.3 million animals, including animals killed to provide tissue, animals used to maintain genetically modified (GM) populations and surplus animals bred for laboratory use but euthanised. An analysis of 2012 data using similar methods yielded an increase to 118.4 million (Anonymous, 2014). Reporting of animal use varies greatly between countries, making accurate quantification difficult. Taylor *et al.* (2008) conclude that their figure is likely to be an underestimate.

In terms of species, data from the EU (EU, 2013) would suggest that mice, rats and rabbits account for about 80% of laboratory animals (Figure 16.4) and that that number

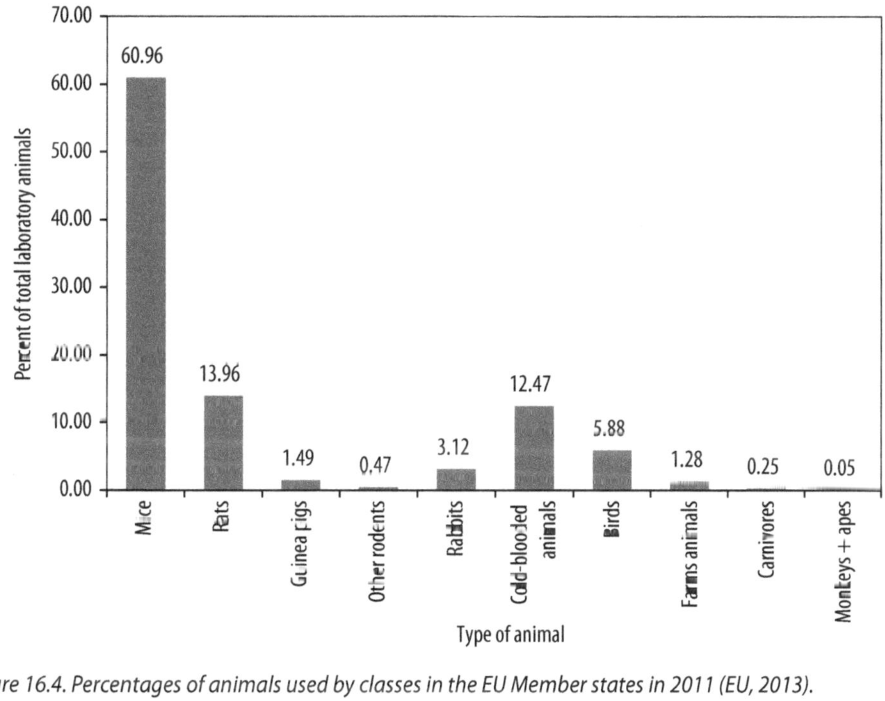

Figure 16.4. Percentages of animals used by classes in the EU Member states in 2011 (EU, 2013).

has remained fairly constant over the last 15 years. The Third Report based on 1999 data showed the same species make up about 83% of all lab animals (EU, 2003). Other trends may include an increase in the use of cold-blooded animals – mostly fish and amphibians – from 6% to 12%, and a decrease in the use of carnivores – mostly cats and dogs – from 0.36% to 0.25% and monkeys/apes from 0.09% to 0.05%. Certainly the use of cats, dogs, monkeys and apes is likely to continue to decline as public antipathy towards use of these species is particularly strong. However, overall animal use remains fairly constant or indeed slightly increasing and this would, on the surface, appear to be in opposition to the fact that there has been widespread commitment to replacement, reduction and refinement – the Three Rs – proposed by Russell and Burch (1959) over half a century ago. What would appear to be a discrepancy may be more related to the expansion of scientific knowledge and research into new areas, thereby increasing absolute numbers, even though efficiency may have greatly improved thereby decreasing relative numbers in long-standing research areas (Olsson *et al.*, 2012).

The types of studies using animals within the EU in 2011 are illustrated in Figure 16.5. Nearly two-thirds of all animal use is for studies of fundamental biology, a lot of which is related to human and animal health, and the development of treatments for human and animal disease. The use of animals as models for human disease and the clinical testing of drugs are subject to scrutiny based on inefficiencies and failure rates of translational research. For example, in drug development, it is estimated that only 1 in 10,000 potential drug molecules or new chemical entities (NCEs) make it to market, and that the R&D cost may be as much as $2.16 billion, though probably nearer a median $60 million per marketable drug (Light and Warburton, 2011). The vast majority of the drop-out is in the pre-clinical phase, which includes animal testing (Figure 16.6). However, even among compounds that reach clinical testing, only 10-20% reach market (Kola and Landis,

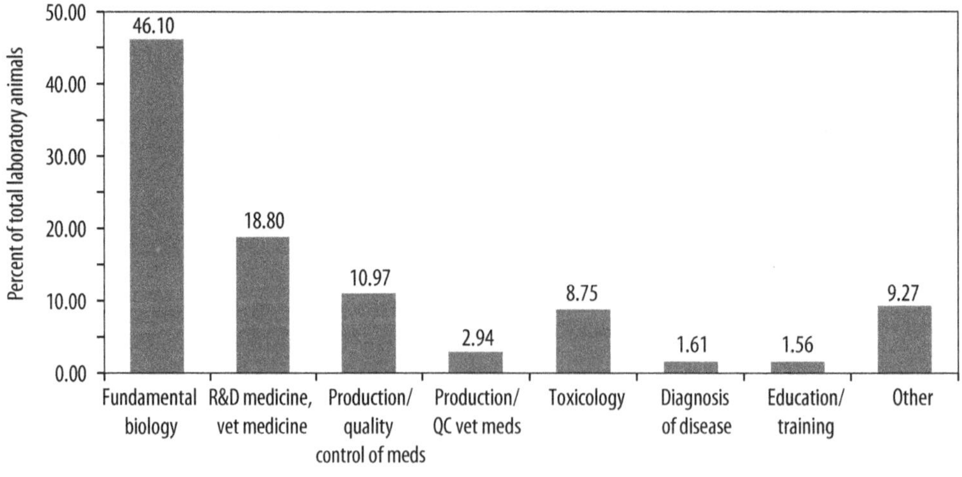

Figure 16.5. Percentages of animals used by experimental purpose in the EU Member states in 2011 (EU, 2013).

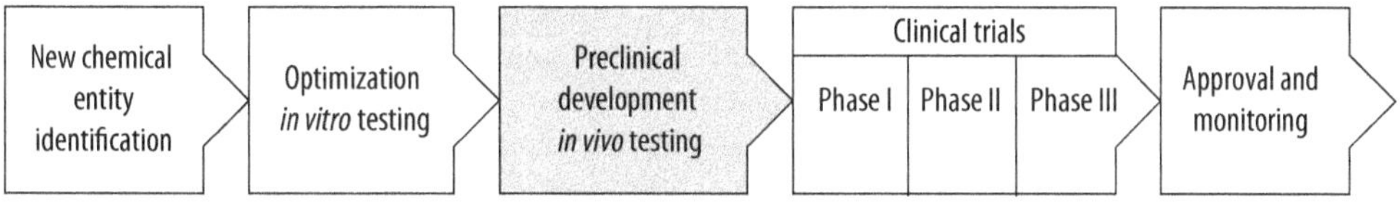

Figure 16.6. Representation of the stages of drug development, highlighting the stage involving testing on laboratory animals.

2004), thereby meaning that the translation from 'successful' animal studies to human application is relatively poor (Van der Worp *et al.*, 2010).

Can we make further progress on the Three Rs? Absolutely, and we could argue that those involved in laboratory animal experimentation need to interact with the applied ethologists and animal welfare scientists much more than they currently do. As Medina and Hawkins (2014, p. 45) wisely state, 'the validity of animal models could also be improved by better communication and sharing of ideas between those operating in different disciplines, such as animal behaviour scientists and researchers in specific biomedical fields'. Within drug development, reasons for failure in the process may be multiple, but there does need to be greater scrutiny of the animal testing phase. The animal model needs to be applicable and relevant (Medina and Hawkins, 2014) and there needs to be greater emphasis on both internal and external validity (Van der Worp *et al.*, 2010). The animal species used needs to be closely examined (Swindle *et al.*, 2012), and the housing conditions that the animal is kept in need to be examined in terms of thermal comfort (Gaskill *et al.*, 2009), environmental enrichment, and social environment (DeBoer *et al.*, 2015). Rather than aiming for uniformity and standardisation, the implementation of systematic variation may improve scientific validity rather than reducing it (Richter *et al.*, 2010).

As with companion animal funding, there is relatively little money directly available for welfare science in relation to laboratory animals. There are charitable sources and some of the major laboratory animal suppliers/contract research companies are funding studies or offering support in terms of providing animals or facilities, but there is again little direct governmental support specifically for lab animal welfare and little support from the major users of laboratory animals. Within ISAE, laboratory animal welfare would appear to be a minor area of research interest (20 out of 543 members listing 'lab or laboratory animals' as a keyword on their profiles). There is a need for more members either to get involved in lab animal research, or at the very least to become staunch advocates of this field of study and its importance, not only for the welfare of the laboratory animals themselves, but also for the ramifications that good laboratory animal science has on biomedical research and translational science. Lab animal welfare scientists should be looking for mutually beneficial ways to engage with biomedical scientists, as this could open up new funding streams. Also, those of us working with non-traditional laboratory species should keep in mind opportunities such as the 'Dual purpose with dual benefit: research in biomedicine and agriculture using agriculturally important domestic animal species' scheme run jointly by the National Institutes of Health (NIH) and United

States Department of Agriculture (USDA) within the USA, which aims to highlight the development of better animal models for human disease (NIH, 2013).

16.8 Leadership: ISAE and the promotion of applied ethology and animal welfare science globally

Although the foundation of ISAE was based on applied ethology, the Society has also become intertwined with the field of animal welfare science. Whereas animal behaviour remains a core discipline within our membership, our members also have listed expertise in other disciplines that contribute to animal welfare science, such as immunology, stress physiology, health, endocrinology, etc. We are culturally diverse and have members active in research, teaching and outreach in 34 countries across the world, although membership numbers are variable with relatively little representation in Asia, Latin America and Africa (Chapter 1; Petherick and Duncan, 2016).

Our collective expertise has long been acknowledged within Europe, where the ISAE has held advisory relationships with the Council of Europe and the European Union, offering advice through individuals who are designated as ISAE experts. The ISAE provides experts to the Board of Trustees of the Association for Assessment and Accreditation of Laboratory Animal Care International (AAALAC International), to the Advisory Board of the Animal Welfare Judging and Assessment Contest, and to the Association for the Study of Animal Behaviour (ASAB) Certification scheme for Animal Behaviourists. As individuals, ISAE members sit on committees or boards of various other national and international bodies, or act as advisors, and thus the value of the knowledge base that we represent is appreciated and respected.

However, in certain parts of the world, there is still something of a barrier between animal welfare science and veterinary science. Even within veterinary science, there is reluctance in some countries to acknowledge animal welfare science as a veterinary specialty, let alone an independent scientific specialty. Animal welfare science is still young, only surfacing in veterinary school curricula within the UK during the 1980s and spreading first through Europe and then to other continents. Its rise as a stand-alone degree subject has occurred subsequently, and although there is perhaps a 30-year history in Europe, animal welfare science is much younger elsewhere. Within the veterinary profession, the Royal College of Veterinary Surgeons within the UK was the first to offer specialised post-graduate qualifications and a professional association, the Animal Welfare Science, Ethics and Law Veterinary Association (AWSELVA) was formed in 1995. Other countries have similar recognised veterinary specialties, such as the formation of the Animal Welfare Chapter of the Australian and New Zealand College of Veterinary Scientists (ANZCVS) in 2000, the provisional recognition of the European College of Animal Welfare and Behavioural Medicine in 2002 (full recognition 2013) and the establishment of the American College of Animal Welfare with recognition from the American Veterinary Medical Association in 2012, following initial rejection of establishment in 2011. However, even though national veterinary bodies may be unwilling to recognise animal welfare science as a distinct veterinary specialty, they are usually keen to keep welfare issues within their domain. Where animal welfare legislation is in place, the majority of enforcement or oversight is placed under veterinary inspection and in many cases this means that when a welfare issue surfaces into public discourse, the 'default' response of governmental organisations or industry representative bodies is to obtain veterinary opinion and guidance, rather than engage with non-veterinarian experts in animal welfare science. Within multinational organisations, such as the World Organisation for Animal Health (OIE), there is also a strong veterinary influence that as a body, ISAE has not yet convinced of its expertise and worth. Although health and welfare are intrinsically linked, animal welfare must encompass broader expertise than those trained in clinical veterinary medicine. ISAE, both collectively and as individual members, must continue to engage when asked, must show our capabilities and must keep lobbying for greater inclusion.

How do we increase the profile of our Society and the work we do, and become effective leaders of applied behaviour and animal welfare science across the world? Firstly, we have to ensure the quality of the science we do, and then we have to become more effective at self-promotion and fully engage with the teaching and outreach aspects of the science. We have excellent established networks, but we need greater presence in under-represented geographic regions and countries. As we reach our 50th anniversary, it is a perfect time for the Society's self-reflection and to identify the key areas for future expansion, both of our science and its influence. In this electronic age, there is an overwhelming abundance of 'animal welfare' information out there, and it is essential that ISAE evolves to fill knowledge voids with sound science and impartial opinion.

Although we remain a member-run organisation and reliant on members filling Council positions and taking on their associated workloads, there is certainly a need for us not to remain static but to transform as we see fit. We should be looking to expand our education and outreach roles – perhaps through website content in partnership with current collaborative projects or other providers, and/or moving to Senior/Junior positions for our Education Officer and Communications Officer. We should be looking to boost membership in under-represented areas of the globe – perhaps by appointing a Development Officer on Council. We should be more proactive in our interactions with those organisations such as OIE and FAO, who have true global coverage and try to shift our 'European/North American-centric' outlook towards one that is truly global.

Finally, we would pose this question. Should we move towards 'certification'? David Fraser has recently asked the question 'Could animal production become a profession?' (Fraser, 2014), with the idea that, in part, public trust is increased with regulated certification. Certainly within animal science, you can become a PAS (Professional Animal Scientist) certified by the American Registry of Professional Animal Scientists in the US or a R.Anim.Sci. (Registered Animal Scientist) certified by the British Society for Animal Science and the Royal Society of Biology in the UK. And indeed, ISAE is already involved in the certification of clinical animal behaviourists (CCAB) by the Association for the Study of Animal Behaviour in the UK. Would ISAE's influence and scope increase if we were to establish standards for a Certified Applied Ethologist or Certified Animal Welfare Scientist?

16.9 Concluding remarks

The Danish philosopher, Søren Kierkegaard, once said that 'Life can only be understood backwards; but it must be lived forwards'. In other words, in order to make future progress, we must reflect on past performance and take the lessons learned and apply them to what we strive to achieve. Most of us, especially those of us working in universities and research institutes, have perhaps become more acutely aware in recent years of such processes as performance reviews and strategic planning, and although we may see them as rather burdensome at the time their deadlines approach, their usefulness in powering our future accomplishments must also be acknowledged! The 50th Anniversary of the International Society for Applied Ethology has given us the ideal opportunity as a Society, to engage in reflection, bask in the glories of past achievements and plot a course for the future. As can be seen throughout this volume, the academic study of applied ethology has impacted and improved the lives of managed animals. There is more to be achieved and some further fields of study and clear approaches identified. There is especially a need for us to focus on the 'International' part of our title. Importantly, we remain a friendly, inclusive society, encouraging and nurturing young scientists and we need to keep supporting risk-taking and novelty in our research endeavours. Here's to the next 50 years and beyond.

References

Anonymous, 2014. A global view of animal experiments 2014. Available at: http://tinyurl.com/h6ytnmb.

American Society for the Prevention of Cruelty to Animals (ASPCA), 2015. Shelter intake and surrender; pet statistics. Available at: http://tinyurl.com/o9ns6dz.

Batson, A., 2008. Global companion animal ownership and trade: project summary. June 2008. WSPA, London, UK.

Bennett, C., Torgerson-White, L, Fripp, D., Watters, J. and Petric, A. 2015. A multi-institutional assessment of factors influencing locomotion and pacing in captive okapis (Okapia johnstoni). Journal of Applied Animal Welfare Science 18: S43-S61.

Blokhuis, H.J., Veissier, I., Miele, M. and Jones, B., 2010. The Welfare Quality (R) project and beyond: safeguarding farm animal well-being. Acta Agriculturae Scandinavica A 60: 129-140.

Brambell, F.W.R., 1965. Report of the technical committee to require into the welfare of animals kept under intensive husbandry systems. HMSO, London, UK.

Carlstead, K., Mench, J.A., Meehan, C. and Brown, J.L. 2013. An epidemiological approach to welfare research in zoos: the elephant welfare project. Journal of Applied Animal Welfare Science 16: 319-337.

Casey, R.A., Vandenbussche, S., Bradshaw, J.W.S. and Roberts, M.A., 2009. Reasons for relinquishment and retrun of domestic cats (*Felis silvestris catus*) to rescue shelters in the UK. Anthrozoös 22: 347-358.

Clower, T.L. and Neaves, T.T., 2015. The healthcare cost savings of pet ownership. Human Animal Bond Research Initiative (HABRI), Washington, DC, USA, pp. 1-12.

Coe, J.B., Young, I., Lambert, K., Dysart, L., Nogueira-Borden, L. and Rajic, A., 2014. A scoping review of published research on the relinquishment of companion animals. Journal of Applied Animal Welfare Science 17: 253-273.

Dalla Villa, P., Kahn, S., Stuardo, L., Ianetti, L., Di Nardo, A. and Serpell, J.A., 2010. Free-roaming dog control among OIE-member countries. Preventive Veterinary Medicine 97: 58-63.
DeBoer, S.P., Garner, J.P., McCain, R.R., Lay Jr., D.C., Eicher, S.D. and Marchant-Forde, J.N., 2015. An initial investigation into the effects of social isolation and enrichment on the welfare of laboratory pigs housed in the PigTurn System assessed using tear staining, behaviour, physiology and haematology. Animal Welfare 24: 15-27.
Diesel, G., Brodbelt, D. and Pfeiffer, D.U., 2010. Characteristics of relinquished dogs and their owners at 14 rehoming centers in the United Kingdom. Journal of Applied Animal Welfare Science 13: 15-30.
Doherty, T.S., Davis, R.A., Van Etten, E.J.B., Algar, D., Collier, N., Dickman, C.R., Edwards, G., Masters, P., Palmer, R. and Robinson, S., 2015. A continental-scale analysis of feral cat diet in Australia. Journal of Biogeography 42: 964-975.
Duffy, D.L., Kruger, K.A. and Serpell, J.A., 2014. Evaluation of a behavioral assessment tool for dogs relinquished to shelters. Preventive Veterinary Medicine 117: 601-609.
Duncan, I.J.H., 1998. 'Pleasures', 'pains' and animal welfare: toward a natural history of affect. Animal Welfare 7: 383-396.
European Union (EU), 2003. Third report on the statistics on the number of animals used for experimental and other scientific purposes in the member states of the European Union. Available at: http://tinyurl.com/gosdrt4.
European Union (EU), 2013. Seventh report on the statistics on the number of animals used for experimental and other scientific purposes in the member states of the European Union. Available at: http://tinyurl.com/p5k3gyk.
European Pet Food Industry Federation (FEDIAF), 2014. Facts and figures 2014. Available at: http://www.fediaf.org/facts-figures.
Fraser, D., 2014. Could animal production become a profession? Livestock Science 169: 155-162.
Gaskill, B.N., Rohr, S., Pajor, E.A., Lucas, J.R. and Garner, J.P., 2009. Some like it hot: mouse temperature preferences in laboratory housing. Applied Animal Behaviour Science 116: 279-285.
Hemsworth, P.H., Mellor, D.J., Coleman, G.J., Beausoleil, N.J., Fisher, A.D. and Stafford, K.J., 2016. Australian and New Zealand perspectives. In: Brown, J.A., Seddon, Y.M. and Appleby, M.C. (eds.) Animals and us – 50 years and more of applied ethology. Wageningen Academic Publishers, Wageningen, the Netherlands, pp. 193-209.
Hill, S.P. and Broom, D.M., 2009. Measuring zoo animal welfare: theory and practice. Zoo Biology 28: 531-544.
Jensen, P., 1986. Observations on the maternal behaviour of free-ranging domestic pigs. Applied Animal Behaviour Science 16: 131-142.
Jensen, P., 2016. New answers – old questions; new questions – old answers: how applied ethology is cross-fertilised by other disciplines. In: Brown, J.A., Seddon, Y.M. and Appleby, M.C. (eds.) Animals and us – 50 years and more of applied ethology. Wageningen Academic Publishers, Wageningen, the Netherlands, pp. 279-295.
Koene, P., 2013. Behavioral ecology of captive species: using behavioral adaptations to assess and enhance welfare of nonhuman zoo animals. Journal of Applied Animal Welfare Science 16: 360-380.
Kola, I. and Landis, J., 2004. Can the pharmaceutical industry reduce attrition rates? Nature Reviews Drug Discovery 3: 711-716.

Lawrence, A.B., 2016. Applied animal behaviour science and animal welfare: seeking the best balance between our science and its application. In: Brown, J.A., Seddon, Y.M. and Appleby, M.C. (eds.) Animals and us – 50 years and more of applied ethology. Wageningen Academic Publishers, Wageningen, the Netherlands, pp. 133-151.

Light, D.W. and Warburton, R., 2011. Demythologizing the high costs of pharmaceutical research. BioSocieties 6: 34-50.

Marchant-Forde, J.N., 2015. The science of animal behavior and welfare: challenges, opportunities and global perspective. Frontiers in Veterinary Science 2: 16-21.

Marston, L.C., Bennett, P.C. and Coleman, G.J., 2004. What happens to shelter dogs? An analysis of data for 1 year from three Australian shelters. Journal of Applied Animal Welfare Science 7: 27-47.

Medina, L.V. and Hawkins, P., 2014. Contemporary issues in laboratory animal welfare. In: Bayne, K. and Turner, P.V. (eds.) Laboratory animal welfare. Elsevier Inc., Philadelphia, PA, pp. 39-53.

Moore, A.M. and Bain, M.J., 2013. Evaluation of the addition of in-cage hiding structures and toys and timing of administration of behavioral assessments with newly relinquished shelter cats. Journal Veterinary Behaviour 8: 450-457.

Mullan, S.M. and Main, D.C.J., 2006. Survey of the husbandry, health and welfare of 102 pet rabbits. Veterinary Record 159: 103-109.

National Institutes of Health (NIH), 2013. Dual purpose with dual benefit: research in biomedicine and agriculture using agriculturally important domestic animal species (R01) overview. Available at: http://tinyurl.com/z2rg743.

Oldfield, R.G., 2011. Aggression and welfare in a common aquarium fish, the Midas cichlid. Journal of Applied Animal Welfare Science 14: 340-360.

Olsson, A., Franco, N., Weary, D. and Sandøe, P., 2012. 'The 3Rs principle – mind the ethical gap!' ALTEX. Proceedings of WC8. Available at: http://tinyurl.com/zpyyavn.

Paul, E.S., Harding, E.J. and Mendl, M., 2005. Measuring emotional processes in animals: the utility of a cognitive approach. Neuroscience and Biobehavioral Reviews 29: 469-491.

The People's Dispensary for Sick Animals (PDSA), 2015. Animal wellbeing report. Available at: http://tinyurl.com/pa8gf96.

Pedersen, N.C., Sato, R., Foley, J.E. and Poland, A.M., 2004. Common virus infections in cats, before and after being placed in shelters, with emphasis on feline enteric coronavirus. Journal of Feline Medicine and Surgery 6: 83-88.

Petherick, J.C. and Duncan, I.J.H., 2016. The International Society for Applied Ethology: going strong 50 years on. In: Brown, J.A., Seddon, Y.M. and Appleby, M.C. (eds.) Animals and us – 50 years and more of applied ethology. Wageningen Academic Publishers, Wageningen, the Netherlands, pp. 27-50.

Rault, J.-L. and Doyle, R., 2016. Cognitive approaches and new technologies: changing methodologies in applied ethology. In: Brown, J.A., Seddon, Y.M. and Appleby, M.C. (eds.) Animals and us – 50 years and more of applied ethology. Wageningen Academic Publishers, Wageningen, the Netherlands, pp. 113-132.

Richter, S.H., Garner, J.P., Auer, C., Kunert, J. and Wurbel, H., 2010. Systematic variation improves reproducibility of animal experiments. Nature Methods 7: 167-168.

Rochlitz, I., 2007. Housing and welfare. In. Rochlitz, I. (ed.) The welfare of cats. Springer, Berlin, Germany, pp. 177-203.

Rowan, A.N. and Beck, A.M., 1994. The health benefits of human-animal interactions. Anthrozoös 7: 85-89.

Russell, W.M.S. and Burch, R.L., 1959. The principles of humane experimental technique. Methuen, London, UK.
Serpell, J.A., 2013. How social trends influence pet ownership and vice versa. Available at: http://tinyurl.com/zkbzbr4.
Serpell, J.A. and Paul, E.S., 2011. Pets in the family: an evolutionary perspective. In: Salmon, C. and Shackelford, T. (eds.) The oxford handbook of evolutionary family psychology. Oxford University Press, Oxford, UK, pp. 297-309.
Serpell, J.A., 2000. Creatures of the unconscious: companion animals as mediators. In: Podberscek, A.L., Paul, E.S. and Serpell, J.A. (eds.) Companion animals and us; exploring the relationships between people and pets. Cambridge University Press, Cambridge, pp. 108-124.
Srinivasan, K., 2013. The biopolitics of animal being and welfare: dog control and care in the UK and India. Transactions of the Institute of British Geographers 38: 106-119.
Stavisky, J., Brennan, M.L., Downes, M. and Dean, R., 2012. Demographics and economic burden of un-owned cats and dogs in the UK: results of a 2010 census. BMC Veterinary Research 8: 163.
Stolba, A. and Wood-Gush, D.G.M., 1989. The behaviour of pigs in a semi-natural environment. Animal Production 48: 419-425.
Swaisgood, R.R., 2007. Current status and future directions of applied behavioral research for animal welfare and conservation. Applied Animal Behaviour Science 102: 139-162.
Swindle, M.M., Makin A., Herron, A.J. Clubb Jr, F.J. and Frazier, K.S., 2012. Swine as models in biomedical research and toxicology testing. Veterinary Pathology 49: 344-356.
Taylor, K., Gordon, N., Langley, G. and Higgins, W., 2008. Estimates for worldwide laboratory animal use in 2005. ATLA 36: 327-342.
Taylor, K.D. and Mills, D.S., 2007a. Is quality of life a useful concept for companion animals? Animal Welfare 16, Suppl. 1: 55-65.
Taylor, K.D. and Mills, D.S., 2007b. The effect of the kennel environment on canine welfare: a critical review of experimental studies. Animal Welfare 16: 435-447.
United States Department of Agriculture – Foreign Agricultural Service (USDA-FAS), 2013. Processed product spotlight: pet food. Available at: http://tinyurl.com/hu8yrec.
Van der Worp, H.B., Howells, D.W., Sena, E.S., Porritt, M.J., Rewell, S., O'Collins, V. and Macleod, M.R., 2010. Can animal models of disease reliably inform human studies? PLoS Medicine 7: e1000245.
World Association of Zoos and Aquariums (WAZA), 2015. Caring for wildlife: the world zoo and aquarium animal welfare strategy. WAZA Executive Office, Gland, Switzerland, 87 pp.

Editors

Michael C. Appleby

Mike Appleby (BSc Zoology, PhD Animal Behaviour) is Chief Scientific Adviser with World Animal Protection, based in London, UK. At the Poultry Research Centre and the University of Edinburgh, UK, he carried out research for 20 years on behaviour and welfare of farm animals, before a period with The Humane Society of the United States in Washington, DC. His most recent book is '*Dilemmas in Animal Welfare*' (co-editor, 2014). Mike is a member of the Farm Animal Welfare Committee, a Visiting Professor at the University of Plymouth and the Scottish Rural College and an Honorary Fellow at the University of Edinburgh. He was a member of SVE from 1982, on the Council from 1987 and Secretary of ISAE from 1991-1995. He has been to about 23 Congresses, and is Anniversary Coordinator for ISAE 2016.

Jennifer A. Brown

Jennifer showed an early interest in animals, caring for the family dogs during childhood and bringing home a variety of reptiles, amphibians and small mammals. During her undergraduate work she studied the behavioural ecology of bats in eastern Canada, and was the first to identify the Northern Long-Eared Bat (*Myotis septentrionalis*) in Prince Edward Island. Before entering the field of ethology, Jennifer obtained BSc and MSc degrees at the University of Prince Edward Island, and worked in clinical chemistry and toxicology. She completed a PhD in Applied Ethology at the University of Guelph in 2009 under the supervision of Dr Tina Widowski. Her thesis studied the effects of temperament and handling experience on the stress response in pigs, and the relationship with pork quality. Currently she is a Research Scientist in Ethology at the Prairie Swine Centre, and an Adjunct Professor at the University of Saskatchewan. She and her husband Peter live on a 320 acre farm in rural Saskatchewan surrounded by an endlessly entertaining assortment of livestock and wildlife.

Yolande M. Seddon

Yolande M. Seddon is Assistant Professor of swine behaviour and welfare at the Western College of Veterinary Medicine, University of Saskatchewan, Canada. Ever curious about animals, she trained in the UK, studying animal behaviour and welfare during her BSc and MSc, with a postgraduate exchange in the Netherlands. Specialising in pigs, she received her PhD in 2011 from the University of Newcastle, UK, researching management strategies to promote health in finisher pigs. Yolande moved to Canada in January 2012 and spent four years as a research associate in swine ethology at the Prairie Swine Centre, Saskatchewan. She conducts applied swine behaviour and welfare research covering all aspects of swine production. She is the section editor (farmed pigs category), for the scientific journal, Animal Welfare. Yolande loves attending ISAE conferences and in 2012 she helped organise the North American Regional meeting.

Authors

Bo Algers

Bo Algers graduated from the Veterinary College in Stockholm in 1974. He was employed at the Veterinary College, later the Swedish University of Agricultural Sciences, Faculty of Veterinary Medicine, from 1975 until his retirement in 2014. During the first ten years of his university career he also worked as a district veterinarian and at a small abattoir in meat inspection. Algers received his PhD on 'Vocal and tactile communication during suckling in pigs' in 1989 and achieved a chair as professor in animal hygiene and applied ethology in 1994. His research focused on the effects of housing, animal transport and slaughter on farm animal health, behaviour and welfare, and later also on the development of an Animal Welfare Risk Assessment methodology. Algers was a member of EFSA's expert panel on Animal Health and Animal Welfare from 2004 to 2009.

Jun Bao

Dr Jun Bao is a professor of farm animal behaviour and welfare in the College of Animal Science and Technology, Northeast Agricultural University (NEAU), in Harbin, Heilongjiang Province, PR China. He has been President of NEAU since 2013. He graduated from NEAU in 1984 with a BSc in Animal Sciences. Then he entered the PhD programme at University College Cork in Ireland in 1989, and achieved his Doctoral degree on Grazing Behaviour of Dairy Cows in 1993. Returning to NEAU in 1994, he started his teaching career in farm animal ethology and conducted research on behaviour and welfare of pigs, poultry and dairy cattle. He has been awarded many research grants both from the national and provincial governments and industries. Dr Bao has published over 100 papers, in both English and Chinese, and three textbooks as a chief editor in animal behaviour and animal protection. He also authored the chapter 'Developing Country – China' in the Encyclopedia of Dairy Science (2nd edition, 2011). Dr Bao is currently Chairman of the China Association of Environment and Hygiene in Farm Animals, and President of the Heilongjiang Association of Animal Science and Veterinary Medicine.

Ngaio Beausoleil

After completing her PhD in 2006, Ngaio Beausoleil was appointed lecturer in animal physiology at Massey University, New Zealand. She is currently senior lecturer and Deputy Director of Massey's Animal Welfare Science and Bioethics Centre, an OIE collaborating centre. Ngaio's research focuses on scientific assessment of animal welfare and employs behavioural and physiological methods to investigate various aspects of welfare in farm, companion and wild animal species. Major research themes include: systematic scientific evaluation of animal welfare; breathlessness in animals; welfare impacts of various killing methods; development, recognition, modulation and alleviation of pain in farm animals; humane vertebrate pest control; and welfare impacts associated with wildlife conservation activities. Ngaio is currently the animal welfare expert member of the New Zealand Veterinary Journal Editorial Board and the Wellington Zoo Trust Animal Welfare Committee. She is also an independent scientific member of the New Zealand Animal Behaviour and Welfare Consultative Committee, and the Massey University liaison for the Universities Federation for Animal Welfare. She teaches BVSc and BSc undergraduates and supervises postgraduate research students.

Donald Broom

Donald M. Broom is Emeritus Professor of Animal Welfare, Cambridge University, Department of Veterinary Medicine. Broom has developed concepts and methods of scientific assessment of animal welfare and studied: cognitive abilities of animals, the welfare of animals in relation to housing and transport, behaviour problems, attitudes to animals, sustainable livestock production and ethics of animal usage. He has published over 300 refereed papers, lectured on animal welfare in 45 countries and served on UK (Farm Animal Welfare Committee, Animal Procedures Committee) and Council of Europe committees. He was Chairman or Vice Chairman of EU Scientific Committees on Animal Welfare from 1990 to 2009 and a member of the European Food Safety Authority Scientific Panel on Animal Health and Welfare until June 2012. He chaired the OIE working group on Welfare of Animals during Land Transport and is a former President and Honorary Fellow of ISAE. Amongst his eight books are '*Sentience and Animal Welfare*' (CABI 2014), '*Stress and Animal Welfare*' (Kluwer), '*The Evolution of Morality and Religion*' (Cambridge University Press), and '*Domestic Animal Behaviour and Welfare*' (CABI 5th edition 2015).

Rebecca Doyle

Dr Rebecca Doyle is a research fellow with the Animal Welfare Science Centre at the University of Melbourne. Rebecca initially studied animal science at Sydney University, graduating with honours following a project in crocodile endocrinology. After graduating, Rebecca started a PhD with the CSIRO and The University of New England in Armidale, Australia, where she conducted the first studies of cognitive bias in livestock. After completing her PhD, Rebecca was employed as a lecturer at Charles Sturt University where she taught physiology and animal welfare, and continued her research in applied ethology and animal welfare. Since joining the AWSC, Rebecca has continued to teach animal behaviour and welfare to undergraduates and graduates. Rebecca's research spans both fundamental areas, including cognitive measures of welfare, and applied research, including some work in developing and emerging countries. She is a recipient of the Australian Government's Animal Welfare Science and Innovation award. Rebecca has been a proud member of the ISAE since 2008 and is the Australasian/African regional secretary. Rebecca loves working in the area of sheep cognition, but also has a soft spot for Galapagos tortoises, sloths and other slow moving creatures that occupy slightly obscure ecological niches.

Ian Duncan

Ian Duncan is Professor Emeritus and Emeritus Chair in Animal Welfare at the University of Guelph in Canada. He has been carrying out research into the welfare of animals since 1965 and was thus one of the first people to bring a scientific approach to solving animal welfare problems. His approach has been to develop methods of 'asking' animals what they feel about the conditions in which they are kept and the procedures to which they are subjected. He has published more than 160 scientific papers most of which are connected to animal welfare. Ian has had a long-standing connection with SVE/ISAE having served as Honorary Secretary, Honorary Librarian, Junior President, President, Senior President and Ethics Officer.

Roger Ewbank

Roger Ewbank qualified as a Veterinary Surgeon in 1957. After three years in veterinary practice he was appointed as Lecturer (later Senior Lecturer) in Animal Husbandry in the Veterinary Faculty of the University of Liverpool. His first published paper, 'The Behaviour of Cattle in Crushes', appeared in the Veterinary Record in 1961, and received a William Hunting Award from the British Veterinary Association. He went on to publish research on topics including nursing and suckling behaviour in sheep, social behaviour of pigs in relation to tail biting, dog-sitting in cattle and sheep, lying behaviour in cattle and its relation to mastitis distribution in the udder, and the formal teaching of ethology in the veterinary undergraduate course. Ewbank was the Founder Vice-President of the SVE in 1966 (Andrew Fraser was Founder President), becoming SVE President in 1967 and again in 1981, and was made an Honorary Fellow of the ISAE in 1999. He was appointed as Director of the Universities Federation for Animal Welfare (UFAW) in 1979, and in 1992 established UFAW's journal '*Animal Welfare*' as Founder/Editor. He also served as a member of the UK Government's Farm Animal Welfare Council from 1970 to 1987, and on the Animal Procedures Committee from 1987 to 1995. Ewbank was named an Officer of the Order of the British Empire (OBE) in 1990. He retired in 1996, and over the years has produced some 150 publications, of which about 40 are original scientific papers in refereed journals.

Andrew Fisher

Andrew Fisher graduated as a veterinarian and worked in practice in Australia and in northern England. After a PhD in beef cattle health in Dublin, he worked for five years with AgResearch in New Zealand, conducting research aimed at improving dairy cow management. In 2002, Andrew joined the CSIRO in Australia, and researched practices for pain management in farm animals, livestock transportation and improved measurement of animal welfare. At the start of 2009, Andrew took up a position at The University of Melbourne, and in 2011 he was appointed Chair of Cattle and Sheep Production Medicine. He conducts research through the Animal Welfare Science Centre, with a focus on animal health and welfare in the livestock industries.

Francisco Galindo

Francisco Galindo obtained a degree in Veterinary Medicine from the National Autonomous University of Mexico (UNAM) and later a PhD in Animal Behaviour and Welfare (Cambridge, UK, 1996). In 1995 he was appointed as Head of the Department of Ethology at UNAM and subsequently started teaching Animal Behaviour and Welfare. He has supervised several MSc and PhD theses on areas related to Applied Ethology and Welfare of Domestic animals and Wildlife, and is co-editor of *Etología Aplicada,* one of the first publications on the topic in Spanish. He is coordinator in Mexico of the OIE Collaborating Centre on Animal Welfare and Livestock Production Systems for the Americas. Through this work he has contributed to the elaboration of Animal Welfare Legislation in several Latin American countries. He has a strong interest in the integration of animal welfare and sustainability.

Paul Hemsworth

Paul Hemsworth is a professor at the University of Melbourne and is Director of the Animal Welfare Science Centre, a joint centre of the University of Melbourne, the University of Adelaide, the Department of Economic Development, Jobs, Transport and Resources (Victoria), the South Australian Research and Development Institute and The Ohio State University. He is also an Adjunct Professor in the Department of Animal Science, The Ohio State University. Paul is an agricultural graduate with a PhD in animal behaviour and physiology. He is most recognised internationally for his research on the role of human-animal interactions in farm and zoo animal welfare. He has also collaborated with numerous colleagues studying the influence of a wide range of husbandry and housing systems on farm and companion animal welfare. Paul has more than 200 peer-reviewed full publications and he teaches undergraduate and postgraduate subjects on animal behaviour and animal welfare at the University of Melbourne.

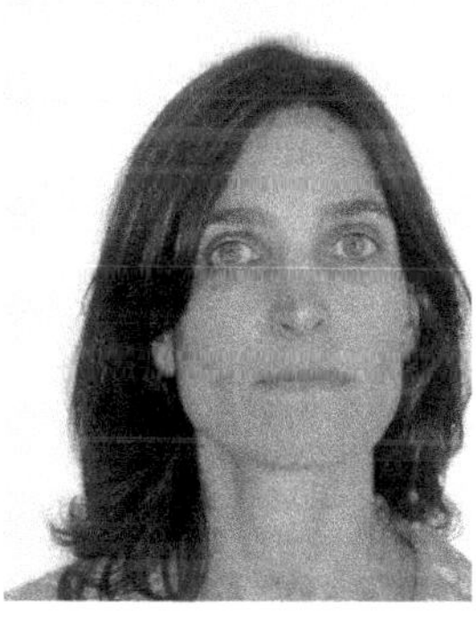

Maria Hötzel

María José Hötzel is a veterinarian with a Master's degree in veterinary medicine and PhD degree in animal sciences. She joined the Laboratory of Applied Ethology at the Federal University of Santa Catarina, Brazil, in 1997, and has been an Assistant Professor of the Department of Animal Science and Rural Development at the same university since 2006. Her research interests are agroecology and farm animal applied ethology and welfare.

Per Jensen

Per Jensen is professor of Ethology at Linköping University, Sweden. He conducted his PhD-training (until 1983) under the supervision of Ingvar Ekesbo in Skara, during which he studied social and maternal behaviour in pigs. He also spent a period in Edinburgh to study free-range pigs under the supervision of David Wood-Gush, and the experiences gathered there led him to start a similar, large-scale project in Sweden. During the 1980s and 1990s, he published a series of papers on the natural behaviour of pigs under free-range conditions. Following this, his research interest has been on domestication effects on behaviour, and the main species of interest has been the chicken. He has for the last 20 years mainly worked on behaviour genetics and epigenetics in relation to domestication and animal welfare, using chickens, and lately dogs, as main study subjects. Per Jensen is editor-in-chief of Applied Animal Behaviour Science since 1997 and has served for more than 10 years in scientific committees under the EU Commission in Brussels. He has published more than 100 scientific papers and edited two text-books in English, in addition to writing a number of popular science books in Swedish (photo by Örjan Karlsson).

Alistair Lawrence

Alistair was born into a farming family in Perthshire, Scotland and studied Zoology at St Andrews University. As a postgraduate he moved to the University of Edinburgh to study for a PhD under Professor David Wood-Gush, one of the pioneers of farm animal behaviour and welfare research. Following this he became responsible over a number of years for behaviour and welfare research at the Scottish Agricultural College (now Scotland's Rural College (SRUC)) with involvement in a wide range of topic areas (see http://www.sruc.ac.uk/alawrence). Later he managed SRUC's Animal & Veterinary Science Group. However, recently Alistair stepped away from these management roles, and currently holds a joint chair with SRUC and the University of Edinburgh in Animal Welfare. With colleagues he is pursuing research into positive animal welfare (with a focus on animal play), and young people's attitudes to animal welfare. He also contributes to teaching of under- and post-graduate students including the animal welfare masters courses at the University of Edinburgh and SRUC (http://tinyurl.com/j3xdksa).

Yuzhi Li

Dr Yuzhi Li is an associate professor of swine behaviour and alternative swine production at the University of Minnesota's West Central Research and Outreach Center, Morris, Minnesota, USA. Her current research projects include management of group-housed gestating sows, including determining their minimal space requirement, identifying behavioural indicators of slow growing pigs, understanding tail biting in pigs, and automated monitoring of swine behaviours. Her extension programmes focus on swine management in different housing systems and improving performance and welfare of swine in alternative housing systems. Li has published more than four dozen publications in peer-reviewed journals relating to swine behaviour, welfare, and management. She was born and raised in Shenyang, Liaoning Province of China where she studied Animal Science and graduated from Shenyang Agricultural University. She earned her Master's degree from the Northeast Agricultural University, studying the effects of thermal environment on energy metabolism of poultry before moving to Japan to pursue her PhD in animal bioenergetics at Hiroshima University. She then worked at the University of Alberta and the Prairie Swine Center, Saskatoon, Canada, for 9 years before joining the University of Minnesota in 2005.

Jeremey Marchant-Forde

Dr Jeremy Marchant-Forde is a Research Animal Scientist with the United States Department of Agriculture – Agricultural Research Service. He is based in West Lafayette, Indiana at the Livestock Behaviour Research Unit. Jeremy has researched welfare and behaviour of pigs and dairy cattle for 25 years, following his PhD on gestating sow welfare at the University of Cambridge. He has worked mainly in the United Kingdom (11 years) and the United States (14 years), with research visits to Denmark and Canada. His major research interests include social and maternal behaviour of sows, the manipulation of pig behaviour by diet, and pig and systems factors influencing piglet mortality. He also has a major interest in the development of novel indicators of animal welfare, and actively engages in inter-disciplinary studies allied to the concept of One Health, collaborating with kinesiology and pharmacology researchers. He is the author/editor of 3 books and 190 refereed articles and abstracts on farm animal welfare and his research has attracted funding of over $7.5 million. He is Specialty Section Editor for Frontiers in Veterinary Science and is also the current Senior Vice President of the International Society for Applied Ethology.

David Mellor

Professor David Mellor's research and scholarly interests include: foetal and neonatal physiology, stress physiology, pain assessment and management, livestock slaughter, conceptual frameworks in animal welfare science, and bioethics as it applies to animal welfare. He has more than 515 publications in these areas, at least 285 of which are significant works of scholarship, including 6 books. He has published on foetal and neonatal physiology for 46 years and has contributed to animal welfare science and bioethical thinking throughout the last 28 years. He was lead author with E. Patterson-Kane and K.J. Stafford of '*The Sciences of Animal Welfare*', published by Wiley-Blackwell in 2009, and chief editor of the OIE (World Organisation for Animal Health) volume '*Animal Welfare: Focusing on the Future*' (OIE Scientific and Technical Review 2014, 33, 1-358). He also authored or co-authored eight of nine reviews in the New Zealand Veterinary Journal (2015, vol. 63/1) on the theme of animal welfare. Currently he is Professor of Applied Physiology and Bioethics, Professor of Animal Welfare Science, and Foundation Director of the Animal Welfare Science and Bioethics Centre at Massey University in New Zealand where he has occupied various professorial positions since 1988.

Joy Mench

Joy Mench received her PhD in Ethology from the University of Sussex, UK, in 1983. In 1985, after conducting postdoctoral research at Cornell, she was hired as the first poultry welfare faculty member in the USA, at the University of Maryland. She moved to University of California-Davis in 1995, where she is Professor and Director of the Center for Animal Welfare. Dr Mench conducts research on animal behaviour and welfare, focusing on poultry, companion birds, and laboratory animals, and has published more than 150 papers, book chapters, and books on these and related topics. She teaches courses on animal ethics and welfare. She has served on numerous committees and boards, including for United Egg Producers, the National Chicken Council, McDonald's, ALDI, Sysco, Certified Humane, American Humane Certified, the Food Marketing Institute, OIE, FAO and the National Academy of Sciences. She was a member of the AAALAC Council on Accreditation, and President of ISAE from 1999-2001. Mench is the recipient of the Poultry Science Association (PSA) Poultry Welfare Research Award and the UC Davis Distinguished Scholarly Public Service Award, and has recently been named a Fellow of the PSA.

Giuliana Miguel-Pacheco

Giuliana is a Peruvian veterinarian with MSc and PhD degrees in Animal Behaviour and Welfare from the University of Edinburgh and University of Notttingham, respectively. During her MSc, she studied calf play behaviour during weaning under the supervision of Anne Marie de Passillé and Jeff Rushen at the Agriculture and Agri-Food Research Centre in Canada. She then joined the Dairy Herd Health group of the University of Nottingham in October of 2011 as a PhD student, completing her studies in 2014. Since then, she has worked as a research assistant and postdoc looking at the effects of maternal nutrition during pregnancy on cow and calf behaviour. Currently, Giuliana collaborates with the OIE Animal Welfare Collaborating Centre at the National Autonomous University of Mexico. Her research interest focuses on understanding how animal behaviour is affected by farm management and veterinary interventions.

Daniel Mills

Daniel Mills was Europe's first professor of veterinary behavioural medicine and as such he has pioneered both the research and practice of companion animal problem behaviour management internationally. Daniel is recognised specialist (RCVS, Europe and Association for Study of Animal Behaviour) in clinical animal behaviour, who represents ISAE on the ASAB accreditation panel for the certification of clinical animal behaviourists. He has been developing and exploring new interventions for behaviour problems at Lincoln for over 20 years. He has a strong research interest in the comparative psychology underpinning behaviour and behavioural interventions, with a particular interest in what makes an individual different and how this arises from their interaction with the environment. This links his applied and fundamental research, for example by examining how we and non-human animals recognise and respond to the emotional state of another. More recently he has had opportunities to explore scientifically his interests in the potential value of our relationships with animals. His research in this area focuses on the benefits from pet keeping using a multidisciplinary approach, for example through collaborations with biologists, health care professionals, psychologists, lawyers and economists. Daniel's success owes a lot to the quality of collaborators he has worked with over the years, but he is also conscious of his father's observation relating to academic success: 'If you find a field no-one knows anything about, you can soon become the world's expert!'

Ruth Newberry

Ruth C. Newberry is a Professor of Ethology in the Department of Animal and Aquacultural Sciences at the Norwegian University of Life Sciences. From origins in Ghana, she survived long-distance prenatal emigration to hatch in the shocking cold of a Canadian winter. Her postnatal development took place with shoulder-perching hens and fence-jumping cows and horses on a farm near Ottawa. As a zoology student at the University of Edinburgh, she was drawn into the study of animal behaviour by the inspirational lectures of Professor Aubrey Manning, and pre-internet life in the library lapping up Jane Goodall's 'In the Shadow of Man'. A persuasive David Wood-Gush steered her away from following in the shadow of Jane to the far more practical study of who grunts after whom in the Edinburgh Pig Park. Riding on David's fowl fame, she landed a position in Agassiz, British Columbia, where she pretended to know something about chickens (and eventually did). In 1996, she canoed south to Palouse country, home of cowboys, Washington State University Cougars and captive grizzly bears. Having finally habituated to Stetson hats in her classroom, the northern lights beckoned once more. Consequently, since 2013 she has been tickled by pickled fish and the sight of Nordic skiers swooshing past her window.

Carol Petherick

Carol obtained a BSc (Hons) in Psychology/Zoology, specialising in ethology, from The University of Reading, England, under the guidance of Don Broom. She obtained an MSc from The University of Aberdeen, Scotland, through research on the spatial requirements of livestock, using the pig as a model. She continued her research on pig housing for her PhD, which she obtained from The University of Queensland, Australia. She returned to Scotland to work with Ian Duncan at what was then the Poultry Research Centre in Roslin near Edinburgh, investigating the cognitive abilities, motivation and welfare of laying hens, with a focus on dustbathing behaviour. In 1993 Carol migrated to Australia to take up a position with the Queensland government, leading research and development in beef cattle behaviour and welfare pertaining to the extensive, rangeland production systems found in northern Australia. In 2010 Carol moved to The University of Queensland through an alliance between the university and Queensland government. In the last 10 years Carol's research has concentrated on the stressful and painful husbandry procedures performed on the cattle. In 2009 she was part of the team that received the prestigious Eureka Prize for science that contributes to the protection of animals, for research on alternatives to/more humane methods for dehorning beef cattle. Carol has been a member of ISAE since 1982 and has held several positions on Council.

Jean-Loup Rault

Jean-Loup Rault is a Senior Research Fellow at the Animal Welfare Science Centre, University of Melbourne, Australia, after training in Europe and America under fellow ISAE colleagues. He specialises in pig and poultry behaviour and welfare, with broad research interests that lead him beyond these species to investigate novel methods of behaviour and welfare assessment. He has a keen research interest in the hormone oxytocin for its involvement in positive social behaviour such as social support. He currently officiates as the Secretary of ISAE.

Bas Rodenburg

Dr Bas Rodenburg is Associate Professor in Behavioural Ecology at Wageningen University in the Netherlands. Within the Behavioural Ecology Group, he coordinates research and teaching on farm animal behaviour and welfare. His research focuses on poultry behaviour and welfare, with a focus on laying hen research. He obtained his PhD on feather pecking in laying hens in Wageningen in 2003 and continues to work on this topic. His current research focuses on the effects of the early-life environment on behavioural development of laying hens. He is also developing methodology for automatic measurement of social behaviour. He is Junior Vice President of the International Society for Applied Ethology (ISAE) and member of the Editorial Boards of the journals Applied Animal Behaviour Science, Animal Welfare, and Poultry Science. Bas is vice-chair elect of the EU COST Action GroupHouseNet (2016-2020), focusing on solutions for tail biting in pigs and feather pecking in laying hens. He is also involved in the EU COST Action KeelBoneDamage (2016-2020), aiming to find solutions for the problem of keel bone fractures in laying hens. He is co-organiser of the WAFL congress in Wageningen in 2017.

Vicki Sandilands

Victoria Sandilands did a PhD on laying hen behaviour at the University of Glasgow while based at the Scottish Agricultural College (now SRUC). She was supervised by John Savory, a poultry ethologist who had previously worked for many years at the Poultry Research Centre (now Roslin Institute). A post-doc position was followed by another and another. Fifteen years later (and still at SRUC) she is still researching poultry behaviour and welfare issues. Her work has looked at preening behaviour, welfare of broiler breeders, housing systems for laying hens and whether or not they can be housed without beak trimming, assessing lameness in broiler chickens, welfare of layers at depopulation and methods of culling poultry. Her work involves not only research but also advisory work to government and the farming community, and teaching at both undergraduate and postgraduate level.

Shusuke Sato

Dr Shusuke Sato is an emeritus professor in Tohoku University and a professor of applied animal behaviour science in Teikyo University of Science in Japan. He has studied the behaviour of feral cattle and free-ranging cattle, pigs, and chickens, investigating topics such as: the normal behaviour of domestic animals; allo-grooming behaviour, its relationship with oxytocin in cattle and effects on group stability; temperament and its relationship with growth hormone in cattle; improving human-animal interactions; and alternative rearing systems for cattle, pigs, and broilers. He has published over 100 peer-reviewed papers, over 100 popular papers and 35 books. He was a founding president of the Japanese Society for Applied Animal Behaviour Science, established in 2002. Now he puts his heart into unifying '*aigo*' (the Japanese ethics of respect for animals) with animal welfare. Sato was born in 1949 and raised in Sendai in northern Japan where he studied Animal Science and received his PhD from the School of Agricultural Science in Tohoku University. During his career he worked at Miyazaki University in southern Japan, the National Institute of Livestock and Grassland Science in central Japan, and then at Tohoku University in northern Japan for 18 years. He studied the development of behaviour in calves with Professor David Wood-Gush at Edinburgh University in 1985-1986. Currently, Sato teaches veterinary nursing and animal welfare at Teikyo University of Science in the capital city of Tokyo.

Kevin Stafford

Kevin Stafford is Professor of Veterinary Ethology at Massey University, Palmerston North, New Zealand. Kevin is a veterinary graduate from University College Dublin. He has an MSc in Tropical Animal Health and Production (University of Edinburgh) and a PhD (National University of Ireland). He spent 10 years working in developing countries: Yemen 1978-1981; Belize 1981-1984; Zambia 1988-1990. In 1990 Kevin moved to Massey University, teaching sheep and beef medicine. In 2004 Kevin was made a Fellow of the Royal College of Veterinary Surgeons for his work on pain in cattle and in 2010 a Fellow of the Australian College of Veterinary Scientists in Animal Welfare Science, Ethics and Law. He has carried out research in veterinary science, livestock health, welfare of ruminants, pain in animals, lamb survival, ruminant physiology, nutrition and behaviour, wild horse biology, wallaby behaviour, companion animal behaviour and human-animal relations. He has authored more than 250 refereed papers. He has authored two textbooks, '*The welfare of dogs*' (2006) and '*Animal welfare in New Zealand*', and co-authored '*The sciences of animal welfare*' (2009). In his spare time Kevin farms, fishes and gardens.

Janice Swanson

Janice Swanson received her PhD (applied ethology) from the University of Maryland in 1988, and a masters and baccalaureate degree from the University of Connecticut. Her career includes five years as a technical information specialist in the US Department of Agriculture's Animal Welfare Information Center and 15 years in the Department of Animal Science and Industry at Kansas State University (1992-2007), including as interim department head (2005-2007). In 2007 Swanson assumed the position of professor and Director of Animal Welfare at Michigan State. She coordinates outreach, teaching and research in the area of animal welfare with a focus on social responsibility in the food system. She is a member of the MSU Animal Behaviour and Welfare Group and has been serving as the chair of the Department of Animal Science since 2010. Swanson also provides scientific service to government, industry and scientific animal welfare advisory committees including: United Egg Producers, McDonalds, Food Marketing Institute and Tysons. She was co-chair of the scientific revision committee for the third edition of the FASS Guide for the Care and Use of Agricultural Animals in Research and Teaching, chairs the taskforce for annual review and revision of the Michigan Generally Accepted Agricultural and Management Practices for Farmed Animals, and recently served as the scientific co-director of the Coalition for Sustainable Egg Supply Project.

Tamara Tadich

Tamara Tadich was born and educated in the city of Valdivia in the south of Chile. In 2004 she graduated as veterinary surgeon at the Universidad Austral de Chile, and from 2005 to 2006 studied at the University of Edinburgh for a Master of Science in Equine Science, for which the welfare of working horses was investigated. Once back in Chile she pursued a Doctor in Veterinary Science degree at the Universidad Austral de Chile, researching the behaviour and welfare of Chilean rodeo horses, obtaining the degree in 2012. Nowadays she works as Assistant Professor at the Veterinary Faculty of the Universidad de Chile in Santiago. Her main lines of research include behavioural measures of animal welfare and behavioural disorders in equids. Teaching areas include applied behaviour and animal welfare at undergraduate and postgraduate levels. Collaborative work is also done with the OIE animal welfare and sustainable animal production collaborating centre shared between Chile, México and Uruguay.

Rodolfo Ungerfeld

Rodolfo Ungerfeld is the Head of the Departamento de Fisiología, Facultad de Veterinaria, Universidad de la República, Uruguay. He studied Biology and Veterinary Science in the Universidad de la República, and later worked for an MSc in Physiology (PEDECIBA, Uruguay) and PhD (Swedish University of Agricultural Sciences, Uppsala, Sweden). He is listed in the highest category (level III) of researchers in the National Researcher System. Up to 2013 he had published more than 130 peer-reviewed articles and 3 books. Rodolfo is a member of the Editorial Board of several international journals (Animal, Animal Production Science, Acta Veterinaria Scandinavica), and has reviewed articles for more than 40 different journals. He has advised postgraduate students in Uruguay, Argentina, Brazil and Mexico.

Tina Widowski

Tina Widowski earned MS and PhD degrees in the Department of Animal Sciences at University of Illinois-Urbana where she trained with Stan Curtis in the late 1980s. She then spent three years at the University of Wisconsin-Madison managing two colonies of endangered primates and conducting post-doctoral research on their socio-endocrinology. In 1990 she moved to the University of Guelph, Canada, where she conducted post-doctoral work with Ian Duncan. She was appointed to a faculty position in 1998. Widowski has studied diverse topics such as the endocrinology of nest building in sows, the behavioural responses of hens to different lighting systems, the ontogeny of feeding, drinking and abnormal behaviour in piglets and motivation for dust bathing and nesting in laying hens. Her research group has tackled some difficult issues including handling and transport of market pigs and methods for euthanasia of piglets and poultry. She served as ISAE Editor from 2001-2004 and on the editorial board for Applied Animal Behaviour Science from 2000-2008. Widowski was appointed Director of the Campbell Centre for the Study of Animal Welfare at Guelph in 2007, University Chair in Animal Welfare in 2009 and Egg Farmers of Canada Research Chair in Poultry Welfare in 2011. She serves on numerous scientific and advisory committees for setting policy, codes and standards of care for farm animals.

Acknowledgements

The editors wish to thank the many sponsors who contributed funds in support of this project. Your donations have supported the publication of the book, just in time to celebrate the 50th anniversary of the International Society for Applied Ethology, held in Edinburgh, Scotland in July 2016. Like the book, the congress will be a great celebration of the science of applied ethology. We also thank our contributing authors for their time and reflections on the history, application and future of this science, and for their dedication and enthusiasm for the project. We hope you will find your efforts are rewarded. Finally we would like to thank Mike Jacobs and Wageningen Academic Publishers for supporting the project from the outset and guiding it to completion.

www.ingramcontent.com/pod-product-compliance
Lightning Source LLC
Chambersburg PA
CBHW081125180926
51611CB00004B/52
* 9 7 8 9 0 8 6 8 6 2 8 2 5 *